Biogenic Nanomaterial for Health and Environment

The book titled *Biogenic Nanomaterial for Health and Environment* covers the synthesis, characterization, and applications of nanomaterials synthesized using living organisms, like bacteria, algae, fungi, plants, and biomolecules derived from them. This book is intended to meet the needs of undergraduate and graduate-level students of chemistry, biology, nanotechnology, and chemical engineering disciplines. The book will also serve as a useful reference work for researchers working in the fields of nanochemistry, material science, biology, and industrial chemistry.

FEATURES

- A systematic overview of the biogenic synthesis nanomaterial.
- Recent research results and pointers to the advancement in the field.
- Discussion of putative applications of biogenic nanomaterials in health and the environment, with a main emphasis on biocidal activity, disease diagnosis, drug delivery, and sensing and remediation of pollutants from the environment.

This book is compiled in such a way that it aids in understanding the underpinning concepts of the biogenic synthesis of nanoparticles. The biogenic synthesis of nanoparticles delivers more important and effective prospects for nanotechnology researchers. The biomass derived from various organisms acts as a template for the synthesis of nanoparticles with desired structural and featural aspects. The rewards of employing biomass and molecules derived from organism of choice in the synthesis process of nanoparticles are able to enhance the electrochemical consistency, control particle size, reduce toxicity, and escalate reactivity in an eco-friendly way. This book will provide the latest insights into the synthesis of nanomaterials employing biomass, cell extract, or as a whole, of various organisms and their roles in the health and remediation of the environment.

Biogenic Nanomaterial for Health and Environment

Edited by
Rahul Badru, Yadvinder Singh,
Narinder Singh, and Deepak Dubal

CRC Press
Taylor & Francis Group
Boca Raton London New York

CRC Press is an imprint of the
Taylor & Francis Group, an **informa** business

First edition published 2024
by CRC Press
2385 NW Executive Center Drive, Suite 320, Boca Raton FL 33431

and by CRC Press
4 Park Square, Milton Park, Abingdon, Oxon, OX14 4RN

CRC Press is an imprint of Taylor & Francis Group, LLC

Library of Congress Cataloging-in-Publication Data
Names: Badru, Rahul, editor. | Singh, Yadvinder, editor. | Singh, Narinder (Chemistry professor), editor. | Dubal, Deepak P., editor.
Title: Biogenic nanomaterial for health and environment / edited by Rahul Badru, Yadvinder Singh, Narinder Singh and Deepak Dubal.
Description: Boca Raton, FL : CRC Press, 2024. | Includes bibliographical references and index.
Identifiers: LCCN 2023036698 (print) | LCCN 2023036699 (ebook) | ISBN 9781032553160 (hardback) | ISBN 9781032553184 (paperback) | ISBN 9781003430087 (ebook)
Subjects: LCSH: Nanostructured materials—Industrial applications. | Biosynthesis.
Classification: LCC TA418.9.N35 B558 2024 (print) | LCC TA418.9.N35 (ebook) | DDC 620.1/15—dc23/eng/20231024
LC record available at https://lccn.loc.gov/2023036698
LC ebook record available at https://lccn.loc.gov/2023036699

ISBN: 978-1-032-55316-0 (hbk)
ISBN: 978-1-032-55318-4 (pbk)
ISBN: 978-1-003-43008-7 (ebk)

DOI: 10.1201/9781003430087

Typeset in Times
by Apex CoVantage, LLC

Contents

Preface

The book titled *Biogenic Nanomaterial for Health and Environment* covers the description of various modes of synthesis of nanoparticles using biota or biomaterials and the role played by them in drug delivery, precise medicine, biocidal activity, health, medical fields, and environmental applications. This book is intended to meet the needs of undergraduate and graduate students in chemistry, biology, nanotechnology, and chemical engineering. The book will also serve as a useful reference work for analytical and biology researchers as well as industrial chemists. It is created in such a way that it aids in the understanding of the concepts underpinning the biogenic synthesis of nanoparticles and their applications in the divergent fields of health and environment sciences. The biogenic synthesis of nanoparticles delivers more important and effective prospects for nanotechnology researchers. The natural plant assets in the form of biomass act as templates for the synthesis of nanoparticles with desired structural and featural aspects. The rewards of employing biomass in the synthesis of nanoparticles is to enhance the electrochemical consistency, control particle size, increase conductivity, reduce toxicity, and escalate reactivity in an eco-friendly way. Nanomaterials synthesized in this manner can be implemented toward wastewater remediation, biochemical sensors to indicate the quality of water, medical diagnostics, and the treatment of many metabolic disorders or ailments. This book will provide current information about the synthesis of nanoparticles by employing biomass of various organisms and the use of biogenic metal/metal oxide nanomaterials in health and the environment. The book then explores how biogenic nanomaterials are being used for remediation of contaminants from the environment, in cancer diagnosis, precision medicine, and targeted drug delivery, and as biocidal agents in integrated pest management.

Editor Biographies

Dr. Rahul Badru is working as an Assistant Professor, Department of Chemistry, Faculty of Basic and Applied Sciences, SGGSW University, India (www.sggswu.edu.in/dept-faculty.php?id=20). He is a multidisciplinary researcher in his career. He started his research journey in 2009 and earned his PhD degree in synthetic organic chemistry for a study titled "Synthesis of Heterocyclic Compounds through Cycloadditions of Nitrones and Azomethines with Active Dipolarophiles" from Punjabi University, Patiala. His PhD research focused on the synthesis of heterocyclic compounds of pharmacological importance by dipolar cycloadditions. Later, he gained experience in the fields of nanocomposites of inorganic–inorganic and inorganic–organic hybrid nature, electrochemical sensors, adsorptive and degradative wastewater remediation, ionic liquids, and metal–organic frameworks. His current research focuses on nanoparticle synthesis, and he has published a number of research articles in SCI-indexed journals from Elsevier, RSC, ACS, Wiley, and Springer in the past five years. He is a lifetime member of several scientific societies and a reviewer of various peer-reviewed journals.

Dr. Yadvinder Singh is working as an Assistant Professor (Botany) in the Department of Botany and Environmental Science of Sri Guru Granth Sahib World University, Fatehgarh Sahib, Punjab (India) (www.sggswu.edu.in/dept-faculty.php?id=19). He completed his PhD in botany at Punjabi University, Patiala. Dr. Singh is the recipient of some national and international awards and fellowships. In his 14 years of research career, he has published extensively on various aspects of extremophilic cyanobacteria and algae, cyanobacteria for value-added products, freshwater limnology and ecology, bioremediation potential of cyanobacteria, cyanobacterial systematics, and phylogenetics. He has published a number of research papers in journals of international repute and several book chapters, having more than 475 citations with 12 h-index and 13 i10-index. Dr. Singh is also awarded the Early Career Research Award and is a lifetime member of various scientific societies as well as a reviewer of various peer-reviewed journals.

Dr. Narinder Singh received his PhD from Guru Nanak Dev University, Amritsar. He joined Osaka City University Japan as a postdoctoral fellow in 2005 and, afterward, worked as a postdoc fellow at Yonsei University, Korea, and Robert Gordon University, Scotland, UK. In 2009, he returned to India to join the Indian Institute of Technology–Ropar (IIT–Ropar), and currently, he is a professor at the Chemistry Department as well as in the Department of Biomedical Engineering of IIT–Ropar. He served the institute as the chairperson of the Chemistry Department from October 2011 to March 2015; chairman, GATE Exam; and currently serving as a member, Senate, IIT–Ropar. He has completed various national and international projects and is currently handling projects (including consultancy) worth several crores. He has published 315 research papers in international peer-reviewed journals. Most of them are reported in the Royal Chemical Society and American Chemical Society. He has guided 14 PhD students, 5 postdoc fellows, and several MSc/MTech students. His research work is concerned with the organic material synthesis of supramolecular assemblies and evaluation of photo-physical properties for sustainable development. In December 2016, he was awarded with the Prof. SS Sandhu Medal from the Indian Chemical Society for his excellent contribution in analytical and environmental chemistry. In 2018, he was awarded with the Faculty Research Innovation Award from IIT–Ropar.

Dr. Deepak Dubal is an Associate Professor (Australian Future Fellow) at Queensland University of Technology (QUT) and has a strong research background in the field of nanomaterials for clean energy conversion and storage systems. Briefly, Deepak's research interest is focused on designing and engineering of functional materials, such as new oxides/nitrides, polyoxometalates (POMs), conducting polymers and their hybrids for energy storage applications, with special emphasis on supercapacitors, Li-ion batteries, beyond Li-ion batteries, etc. At QUT, his research is to develop an integrated system as self-charging power source for wearable electronics and implantable medical devices. Before joining QUT, Deepak worked as the vice-chancellor fellow at the School of Chemical Engineering, University of Adelaide. He is also a former recipient of two of the most prestigious fellowships: Marie-Curie at ICN2, Spain (2014), and Alexander von Humboldt at TUC, Germany (2012). So far, Deepak has coauthored more than 140 research articles, including in *Chemical Society Reviews*, *Advanced Materials*, *Advanced Functional Materials*, *Nano Energy*, *NPG Asia Materials*, etc., and achieved over 9,767 citations, "h" index 53 and "i10" index 122 (Google Scholar). He has edited one book, contributed several book chapters, and filed several patents, including world, US, European, and Spanish. He has also participated as a founding partner of the Earthdas graphene mobility spin-off company, Spain, with Prof. Gomez-Romero.

Contributors

Muhammad A. Abuelmagd
Department of Botany, Faculty of Science
Mansoura University
Egypt

Md Arifuzzaman
Institute of Tissue Banking and Biomaterial Research
Atomic Energy Research Establishment
Savar, Dhaka 1349, Bangladesh

Rahul Badru
Department of Chemistry
Sri Guru Granth Sahib World University
Fatehgarh Sahib, Punjab 140406, India

Karina Bano
Department of Chemistry
Sri Guru Granth Sahib World University
Fatehgarh Sahib, Punjab, India

Dimbeswar Das
Department of Botany
Rajiv Gandhi Central University
Rono Hills 79112, Doimukh
Arunachal Pradesh, India

Badr E. El Bialy
Department of Forensic Medicine and Toxicology
Faculty of Veterinary Medicine
University of Sadat City
Sadat City 32897, Egypt

Mayyada M. H. El-Sayed
Chemistry Department
American University in Cairo, AUC Avenue
New Cairo 11835, Egypt

Marwa M. Eltarahony
Environmental Biotechnology Department
Genetic Engineering and Biotechnology Research Institute (GEBRI)
City of Scientific Research and Technological Applications (SRTA–City)
Alexandria (Egypt)

Hebatullah H. Farghal
Chemistry Department
American University in Cairo, AUC Avenue
New Cairo 11835, Egypt

Balasubramanian Mythili Gnanamangai
Department of Biotechnology
K. S. Rangasamy College of Technology
Tiruchengode 637215, Tamil Nadu, India

Hania A. Guirguis
Chemistry Department
American University in Cairo, AUC Avenue
New Cairo 11835, Egypt

Ragaa A. Hamouda
Department of Biology
College of Sciences and Arts, Khulais
University of Jeddah
Jeddah, Saudi Arabia
Department of Microbial Biotechnology
Genetic Engineering and Biotechnology Research Institute (GEBRI)
University of Sadat City
Sadat City, 32897, Egypt

Md Monir Hossain
Department of Biotechnology and Genetic Engineering
Jahangirnagar University
Savar, Dhaka 1342, Bangladesh

John Joseph
Department of Biotechnology
Karpagam Academy of Higher Education
Coimbatore, Tamil Nadu (India)

Ayman Kamal
Environmental Biotechnology Department
Genetic Engineering and Biotechnology Research Institute (GEBRI)
City of Scientific Research and Technological Applications (SRTA–City)
Alexandria (Egypt)

Kamalpreet Kaur
Department of Chemistry
Indian Institute of Technology Ropar
Punjab 140001, India

Navneet Kaur
Department of Chemistry
Panjab University
Chandigarh 160014, India

Sandeep Kaushal
Department of Chemistry
Sri Guru Granth Sahib World University
Fatehgarh Sahib, Punjab, India

Güllü Kaymak
Kütahya Health Sciences University
Simav Vocational School of Health Services
Kütahya, Turkey

Meliha Koldemir-Gündüz
Department of Basic Sciences of Engineering, Faculty of Engineering and Natural Sciences
Kutahya Health Sciences University
Kutahya, Turkey

Komal
Department of Botany and Environmental Science
Sri Guru Granth Sahib World University
Fatehgarh Sahib Punjab 140406, India

Ranvijay Kumar
University Centre for Research and Development
Chandigarh University, Gharuan
Mohali, Punjab, India

Muthuchamy Maruthupandy
Lab of Toxicology
Department of Health Sciences
the Graduate School of Dong-A University
37, Nakdong-Dearo 550 Beon-Gil, Saha-Gu
Busan 49315, South Korea

Rajamanickam Mohanraj
Department of Biotechnology
K. S. Rangasamy College of Technology
Tiruchengode 637215, Tamil Nadu, India

Sunita Patil
Department of Biotechnology
Karpagam Academy of Higher Education
Coimbatore, Tamil Nadu (India)

Shakil Ahmed Polash
Ian Potter NanoBiosensing Facility
NanoBiotechnology Research Laboratory (NBRL)
School of Science, RMIT University
Melbourne, Victoria 3001, Australia

Franck Quero
Nanocellulose and Biomaterials Laboratory
Department of Chemical Engineering, Biotechnology, and Materials
Faculty of Physical Sciences and Mathematics
Universidad of Chile
Avenida Beauchef 851, Santiago, Chile

Govindan Nadar Rajivgandhi
Nanocellulose and Biomaterials Laboratory
Department of Chemical Engineering, Biotechnology, and Materials
Faculty of Physical Sciences and Mathematics, Universidad of Chile
Avenida Beauchef 851, Santiago, Chile

Chandrasekaran Rajkuberan
Department of Biotechnology
Karpagam Academy of Higher Education
Coimbatore, Tamil Nadu (India)

Satya Ranjan Sarker
Department of Biotechnology and Genetic Engineering
Jahangirnagar University
Savar Dhaka 1342, Bangladesh

Arun Sharma
Department of Chemistry
Indian Institute of Technology
Ropar, Punjab 140001, India

Ashutosh Sharma
Department of Chemistry
Sri Guru Granth Sahib World University
Fatehgarh Sahib, Punjab 140406, India

Tamer Shoeib
Chemistry Department
American University in Cairo
AUC Avenue New Cairo 11835, Egypt

Gagandeep Singh
Department of Biomedical Engineering
Indian Institute of Technology Ropar
Punjab 140001, India

Narinder Singh
Department of Chemistry
Indian Institute of Technology Ropar
Punjab 140001, India

Yadvinder Singh
Department of Botany and Environmental Science
Sri Guru Granth Sahib World University
Fatehgarh Sahib, Punjab 140406, India

Sumpam Tangjang
Department of Chemistry
Sri Guru Granth Sahib World University
Fatehgarh Sahib 140407 Punjab, India

Khushboo Verma
Department of Botany
Eastern Karbi Anglong College
Sarihajan 782480, Karbi Anglong, Assam, India

Hu Yang
State Key Laboratory of Pollution Control and Resource Reuse
School of the Environment
Nanjing University
Nanjing 210023, P. R. China

1 A Comprehensive Perspective of Conventional and Biogenic Nanoparticles

Dimbeswar Das, Khushboo Verma, and Sumpam Tangjang

1.1 INTRODUCTION

The field of nanotechnology, which involves the creation, manipulation, and synthesis of materials on a scale smaller than 1 μm, has seen significant growth due to its diverse applications in various sectors of science and technology. The term *nano* originates from the Greek word for *dwarf* or *extremely small*. Nanotechnology is a multidisciplinary field that firmly refers to the atomic-, molecular-, and macromolecular-scale research and development that enable us to control the manipulation and study of minute structures and devices. The nanomaterials (NMs) possess various unique electrical, optical, and magnetic features due to their minute size. In the 21st century, nanotechnology has made significant strides in various disciplines, like medicine, genomics, robotics, and communications [1–2]. The explosive advancement of nanotechnology in the recent years has developed a new cost-effective and effortless technique for treatment of wastewater [3–4]. Nanotechnology is used to clean wastewater in a variety of ways. The size-dependent properties, like high specific surface area, high reactivity, high absorption, and fast dissolution, are used for the detection and purification of various contaminants, including heavy metal contamination [5].

As the field of nanotechnology continues to advance, a large number of NMs have been produced. They can be classified according to their characteristics, that is, their structures, shapes, sizes, and methods of chemical synthesis.

1.1.1 Carbon Nanomaterials

Carbon nanomaterials, such as fullerenes (0D), carbon nanotubes (CNTs) (1D), and graphene (2D), have attracted a strong study interest for almost three decades. This is mainly because of their unique optical, electrical, thermal, and chemical capabilities, which present important chances for advancing basic and applied science as well as creating novel technologies and applications. Because of this, these nanomaterials have been in the vanguard of groundbreaking discoveries and have the potential to spur revolutionary developments across a range of industries. Controlling the experimental environment provides stable operation of the size, shape, and diameter of carbon products at the nanoscale, enabling the production of a wide variety of carbon nanomaterials. Graphite and metal-graphite rods were evaporated in an electric arc to create fullerenes and multiwalled CNTs. Numerous shapes of carbon nanofibers and nanotubes were created during the catalytic thermal cracking of methane and ethylene. Other graphene forms were generated by chemically reducing graphite oxide [6]. CNTs and nanofibers (CNF) have remarkable tensile strength, electrical and thermal conductivity, elasticity, and low thermal expansion coefficient, which make them highly important in a variety of sectors. The use of these carbon nanomaterials (CNMs) to mimic biological processes, such as protein adsorption, DNA and therapeutic molecule binding, and red blood cell stabilization, is also very beneficial to the pharmaceutical and healthcare industries, particularly in the area of controlled drug administration [7].

DOI: 10.1201/9781003430087-1

1.1.2 Metal and Metal Oxide Nanomaterials

Metal oxides and metals are used in the fabrication of NMs, formally known as inorganic nanomaterials. These NMs comprises of zinc oxide (ZnO), titanium dioxide (TiO_2), gold (Au), silver (Ag), and other metal oxide–based NMs. The extraordinary physical and chemical characteristics of metal- and metal oxide–based nanozymes, alongside their minimal cost, great resilience, and ease of archiving, have resulted in their exponential advancement in the last decade [8]. The fabrication of nanoparticles with a wide range of compositions, size, and morphologies is made possible by the use of a variety of synthesis techniques, including chemical vapor deposition, electrochemical deposition, chemical reduction, and hydrothermal methods. A simple, green, and environmentally acceptable method may be provided by the biogenic production of nanoparticles by sea algae and other marine plants. This eco-friendly bio-based methodology has the ability to offer an alternative to traditional production techniques, which frequently employ hazardous substances and solvents [9–10].

1.2 ORGANIC NANOMATERIALS

Various kinds of nanostructures, such as liposomes, dendrimers, micelles, and polymers, can be easily synthesized by transforming organic compounds. Multiple studies examined the formation, characterization, bioactivity, and biotoxicity of organic nanomaterials. The advent of nanotechnology and a deeper comprehension of various ONMs results in a wide range of viable substrates for biomedical applications [11]. Polyethylenimine (PEI) nanoparticles and chitosan nanoparticles are organic nanocompounds used for antimicrobial purposes. In order to add antibacterial and antifungal qualities, PEI nanoparticles with quaternary ammonium groups have been employed in root canal sealers, glass ionomer cements, and dent nanocompound-filling resin composites. Chitosan nanoparticles, which are biocompatible and biodegradable, have broad antibacterial activity and have been used as organic nanofillers in nanocomposites to improve mechanical properties and inhibit biofilm formation. Conductive polymer-based nanoparticles, such as polyaniline, polypyrrole, and polythiophene, can also be used for antimicrobial activities. Conductive polymer-based nanoparticles nanocompounds are synthesized by chemical or electrochemical polymerization methods and can be shaped into various nanostructures [12].

1.2.1 Nanocomposites

Nanocomposites are solid materials made up of several phases, at least one of which is nanoscale in size. They consist of nanoparticles, nanofibers, and nanoclays and have improved properties compared to other composites because of remarkably high surface-area-to-volume ratio and interactions between phase interfaces. Nanocomposites can be categorized into three groups on the basis of matrix material: ceramic matrix nanocomposites (CMNC), polymer matrix nanocomposites (PMNC), and metal matrix nanocomposites (MMNC). The main advantages of nanocomposites are better mechanical properties, improved optical properties, and a higher surface/volume ratio. However, they also have disadvantages, such as toughness and impact performance, limited understanding of formulation/property/structure relationship, and cost-effectiveness [13].

1.3 CLASSIFICATION OF NANOMATERIALS

Classifying nanomaterials is crucial to determine their appropriate applications. Nanomaterials are mostly composed of solid particles, and their dimensions can be measured through various methods.

1.3.1 Zero-Dimensional Nanomaterials

Zero-dimensional (0D) nanomaterials are typically spherical in shape, with various diameters ranging from 1 to 50 nm. Some cube and polygon shapes have been discovered to be 0D nanomaterials which retain all their dimensions in the nanoscale; 0D nanoparticles have more functional interface points for each unit mass as a result of the fundamental structural properties of 0D, including extremely small diameters and high specific surface area ratios. Additionally, 0D NMs possess enhanced values for photoluminescence (PL), chemiluminescence, and quantum efficiency [14–15], in comparison to other materials.

1.3.2 1D Nanomaterials

The 1D (one-dimension) nanomaterials have at least one dimension in the nanoscale (1–100 nm), and the other dimensions are in the macroscale. These include nanowires, nanofibers, nanorods, and nanotubes. Other materials that can form 1D nanostructures include metals such, as gold, silver, and silicon; metal oxides, such as zinc oxide, titanium dioxide, and cerium oxide; as well as quantum dots and others [18]. Zinc oxide (ZnO) nanostructures in the form of nanowires and nanorods are extensively sought after for a range of applications due to their advantageous features. The production of these 1D ZnO nanostructures is relatively straightforward and results in materials

TABLE 1.1
Zero-Dimensional (0D) Nanomaterials and Their Uses [16–17]

0D Nanoparticle	Description	Potential Uses
Graphene quantum dots (GQDs)	A novel class of 0D graphene nanomaterials called graphene quantum dots (GQDs) is distinguished by atom-sized planes.	Bioimaging, energy storage and conversion, catalysis, drug delivery
Carbon quantum dots (CQDs)	Carbon quantum dots (CQDs), formerly abbreviated carbon dots (CDs), are fluorescent nanoparticles with diameters less than 10 nm.	Bioimaging, cancer diagnosis
Fullerenes	0D nanoparticles composed of carbon atoms arranged in a closed cage-like structure. A molecular allotrope of carbon called fullerene was found in 1985 by Kroto et al. The most prevalent fullerene, C60, is a truncated icosahedron composed of five to six sp^2 hybrid carbon rings.	Bioimaging, cancer diagnosis, electronics
Inorganic quantum dots (QDs)	Inorganic quantum dots (QDs) are fluorescent semiconductor NMs often created using groups II–VI or III–V elements.	Fluorescence, chemiluminescence, and bioluminescence sensors; pathogen detection
Magnetic nanoparticles (MNPs)	0D nanoparticles with magnetic properties, high-saturation magnetization.	Medical imaging, magnetic hyperthermia, magnetic separation and purification, drug delivery
Noble metal nanoparticles	0D nanoparticles made of noble metals such as gold or silver.	Bioimaging, catalysis, sensing, plasmonics
Up-conversion nanoparticles (UCNPs)	NIR light can be transformed into multicolored, high-energy ultraviolet/visible light by UCNPs, a specific class of lanthanide-doped nanoparticles.	Bioimaging, photodynamic therapy
Polymer dots (Pdots)	0D nanoparticles made of polymers.	Bioimaging, biomolecular detection, clinical analysis (miRNA)

with excellent charge transport properties and a high degree of crystalline quality. These properties make ZnO nanowires/nanorods particularly useful for applications such as gas sensors, biosensors, solar cells, and photocatalysts [19]. Si nanowires are attractive for gas sensing; they are label-free, offer real-time detection capabilities for fast results, and have high sensitivity to gases, allowing for the detection of very low concentrations. Applicable for environmental monitoring, industrial process control, and medical diagnosis [20].

1.3.3 2D Nanomaterials

This class of nanomaterials has two dimensions at the nanoscale and one at the macroscale. Nanosheets, nanowalls, thin-film multilayers, or two-dimensional nano-thin films (2D) bulk supplies nanomaterials. A 2D nanomaterial can contain many square micrometers of surface area while always having a thickness in the nanoscale range [22]. Moreover, 2D NMs have unique electronic, physical, chemical, and optical properties due to their one-dimensional spatial confinement. These materials have strong in-plane bonds and a high degree of flexibility, making them desirable for use in various devices. The frictional characteristics of 2D materials are predominantly governed by their feeble interlayer interaction, flexibility, and periodic atomic arrangement. Before reaching a stable state, the static frictional force displayed by these materials steadily increases during a few early atomic periods. Thereafter, it tends to decrease as the number of 2D layers rises. The frictional properties of 2D materials can also be impacted by factors such as the low bending stiffness of the thinnest layers (known as the puckering effect), electrical or electron–phonon coupling, and the creation of tribofilm at the point of contact between sliding surfaces [23].

1.3.4 3D Nanomaterials

Three-dimension nanomaterials (3DNMs) don't have any dimension in the nanoscale. Bulk materials are three-dimensional nanomaterials that consist of separate units that can range from nanometer scale (1–100 nm) to larger sizes [14]. Theranostics applications frequently employ 3D nanostructured materials like nanocubes and nanostars. Due to their special qualities, these materials can be used for both diagnostics and treatment. Using the idea of surface plasmon resonance, the photothermal applications of silver- and gold-based NMs have been explored. The nanoparticle's outermost electrons oscillate when a laser beam strikes them, absorbing near-infrared light and creating heat powerful enough to destroy malignant cells. Cobalt–iron oxide NMs were also evaluated for their theranostic applications due to their high coercivity and magnetization property, which is used for magnetic hyperthermia. Lanthanum hexabromide nanocubes (LaB6-NCs) were used as a

TABLE 1.2
Respective Fields and Their Applications of One-Dimensional (1D) Nanomaterials

	Field	Applications
Emerging use of nanofiber technology [21]	Building materials and construction	Reinforced polymers, self-healing materials
	Generation, transmission, and energy conservation	Supercapacitors, fuel cells, photovoltaic cells, and batteries
	Environmental conservation and enhancement	High-efficiency water filtering, treatment of wastewater, and air filtering
	Biomedical applications	Blood cleansing, synthetic biology, biopharmaceuticals, and tissue regeneration
	Additional applications	Protection and national defense

theranostic agents in the CT- and T2-weighted MRI techniques, which helped overcome the challenge of deep tissue attenuation in cancer therapy [24].

1.4 CONVENTIONAL NANOTECHNOLOGY, OR CONVENTIONAL NANOPARTICLE SYNTHESIS

The synthesis technique used to synthesize nanomaterials has a major impact on the final characteristics of nanomaterials. The two major approaches used for the preparation of nanomaterials are bottom-up and top-down approaches. In the former, bulk materials get first mechanically machined, then divided into fine nanostructured particles. Bottom-up approaches involve the use of co-precipitation methods to assemble fine particles into NMs [25]. Top-down methods involve the breaking of van der Waals attraction forces between the stacking of bulk components in order to achieve a thin layer of crystals. Top-down methods require a large amount of energy and have a significant environmental impact. Top-down methods involve mechanical milling, ion sputtering, laser ablation, ball milling, cutting, grinding, and so on. Even though the top-down approach is easier to carry out, it cannot achieve extremely small-sized particles. The main drawback affiliated with the top-down approach of nanoparticle synthesis is in the changes in physicochemical properties of nanoparticles. Bottom-up methods rely on molecular recognition, that is, self-assembly. In the self-assembly, precursors are put together in random orientations and positions and provide energy to attain sample configuration space. Chemical vapor deposition, hydrothermal method, laser pyrolysis, physical vapor deposition, solgel method, and other methods are examples of bottom-up approaches. The attainment of extreme pressure and temperatures, and the use of chemicals in physical and chemical techniques of nanoparticles formation, results in harmful effects on the environment. Most of the chemical procedures utilize harmful chemicals, like reducing agents, such as sodium borohydride, ethylene glycol, hydrazine, and sodium citrate, resulting in poor-sized particles and poor regulation of size distribution; thus, it is necessary to use capping agents, such as ascorbic acid, dendrimers, surfactants, polymers, phospholipids, Gemini surfactants, and polyvinyl pyrrolidone, to prevent nanoparticles agglomeration. The high cost and the toxic quality of the synthesis approach limit its use on a large scale. Moreover, the products developed have poor efficiency.

The synthetic methods employed in NM synthesis regulate their structure, shape, and size. Mechanical milling, hydrothermal method, solution combustion, and solgel method are examples of conventional synthetic modes of NMs [26]. Among these the hydrothermal method, the solgel method and wet chemical synthesis are quite common in use. These techniques are widely used since they are rapid, easy to manufacture, economical, and reproducible. Solgel is considered as the least-expensive method of producing nanostructures and can regulate the stoichiometry of the nanomaterial proficiently. Because it has high control over the growth rate of the nanoparticles across the gel, the nanomaterials synthesized in this approach exhibited tiny particle sizes and a restricted size distribution. A hydrothermal approach is another prominent method in nanomaterial fabrication. This is a flexible and straightforward approach for synthesizing nanomaterials under high pressure and temperature-confined conditions. Hybrid nanomaterials can also be easily made

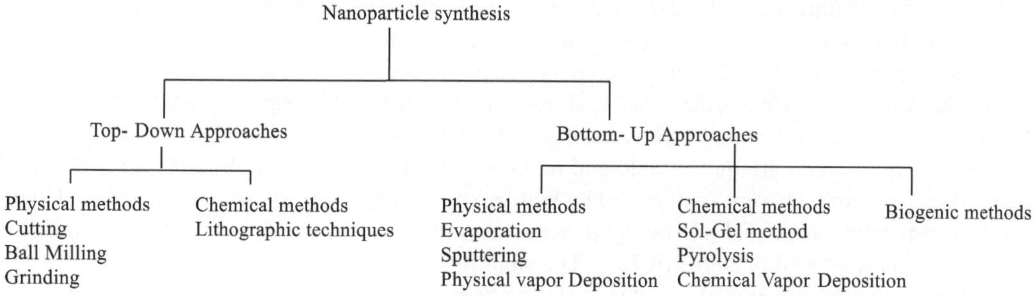

using hydrothermal processes. The hydrothermal method's adaptability makes it possible to create hybrid nanomaterials made of both metal and non-metallic elements. Wet chemical synthesis has aroused the interest of researchers owing to the simple procedure with a significant control on the penultimate physical properties of the nanostructures formed [27].

1.4.1 Advantages of Conventional Techniques

The ball milling method has many benefits, including producing fine powder and being suitable for grinding harmful and abrasive substances [28]. Physical vapor deposition is considered as a more eco-friendly approach than others. This technique involves the use of inorganic and organic materials and thus improves the properties of the final nanostructured substances comparable to the starting material. Highest-purity NMs are produced using the solgel method. Moreover, it synthesizes nanostructures with ultra-high uniformity even at low temperatures. The chemical vapor deposition process produces abrupt junctions and thin films of ultra-high purity [24].

1.4.2 Drawbacks of Conventional Techniques

The top-down approaches suffer from a major disadvantage – that it can cause surface imperfection of nanostructured materials. Henceforth, top-down approaches are limited to some applications in comparison to bottom-up approaches. Yan et al. [29], in their recent study, have described some shortcomings in the conventional methods for the preparation of lignin nanomaterials. For example, to control the shape and size of particles, a neutralization–sedimentation method is stringent to adjust pH values. Mechanical size reduction is not so much reliable for tuning of particle shapes. Solgel and electrospinning methods are uneconomical and consume a large amount of chemicals [30]. Also, the solgel method and the ball milling method produce a high level of impurities by the formation of by-products and wear–tear in the collisions of balls [31]. Also, the ball milling method is time-consuming and creates machine noise. The key downside of the physical vapor deposition technique, complex procedures, is low output, and it requires extra costs [32]. The chemical vapor deposition process also faces few difficulties, such as the properties of nanomaterials being dependent on precursors and producing poor uniformity [33]. The lithography technique is also inconstant to the production of free radical and since it causes DNA damage in photo-initiation processes [34].

1.4.3 Advantages of Biogenic Methods over Conventional Methods

One of the plant parts that can be used extensively for the large-scale scaling-up of the synthesis of nanostructured materials are leaves. Leaves are plant materials that are always available in all weather conditions. Furthermore, the consumption of leaves doesn't show any effect on plants, but the use of other resources may be harmful. The use of root, flower, seed, fruit, and latex must not affect the quantity of materials and output of the crop [25]. Additionally, plant-based NMs are reusable, stable, and safe for the environment. The size of plant-mediated nanoparticles will depend on a number of factors, including pH of the solution, concentration of extract and precursors, duration, and others. The optimization of these factors will finally result in the production of nanostructures with the desired shape and size, because plant ingredients vary between species and genera [25]. After the synthesis, stability of synthesized nanostructures becomes a significant concern. For their use in the long term, nanoparticle colloidal stability is crucial. The capping agents control the stability of NPs created through biological means, in contrast to the nanoparticles created through chemical means, which are quite stable and long-lasting. In a recent study by Mohammad et al., silver (Ag) nanoparticles were synthesized by both biogenic and chemical modes. The value of zeta potential was found to be 17.8 mV for Ag nanoparticles created chemically, while it is 15.2 mV for Ag nanoparticles created biologically [35]. The stability of the nanoparticle solution is completely dependent on the stabilization of metal particles by biomolecule involved in their biosynthesis.

1.5 FACTORS AFFECTING SYNTHESIS OF NANOPARTICLES BY BIOGENIC METHODS

The pH, capping agents, and functionalization procedures used during biogenic synthesis can all have an impact on the size, shape, and stability of the nanoparticles. The impact of pH on the synthesis of gold NPs was covered in a study by Armendari et al. [36]. At different pH values, the nanoparticles tend to exhibit different shapes. For instance, at pH values of 2, 3, and 4, gold nanoparticles exist in tetrahedral, hexagonal, decahedral, icosahedral, irregular, and rod shapes. Therefore, the development of crystal structures in NPs is initiated at the stage of nucleation, which was thermodynamically favored with particle sizes in the desirable range [37]. Another study by Chandran et al. has discussed the effect of the concentration of precursors and reducing agents on the characteristics of synthesized silver NPs. The effect on shape and size of NPs was prominent due to the excessive use of reducing agents, which get adhered to the nuclei surface. Secondarily, it magnified the reduction process of Ag^+ ions on the pre-formed nuclei and led to the formation of NPs of bigger sizes. An excess of reducing agents also enhances the bridging effect in the formation of nanoparticles, which results in the aggregation of nanostructures [38]. Thus, controlling the concentration of both the precursor and reducing agent is a primary check in order to get NPs of desired size and should be necessarily optimized.

By regulating the rate of nucleation, temperature is just as crucial to the synthesis of nanoparticles as the concentration of reducing agents. In general, nucleation for large-sized NPs is favorable at higher temperatures [39]. In biogenic methods, the formation of nanoparticles is greatly characterized by the incubation time of the reaction media. During long incubation periods, an aggregation or shrinkage of NPs can occur, which in turn can affect the shelf life and potential of synthesized NPs [40].

1.6 BIOGENIC NANOMATERIALS

Biogenic nanoparticles (BNPs) have attracted a lot of study interest in the last few decades, ever since the concept of microbial interaction with metals was developed [41]. BNPs are primarily employed in biomedicals, electronics, and water and wastewater treatment. The use of microbial interaction to engineer BNPs is now widely documented [42]. The use of hazardous chemical components and volatile compounds in nanoparticle synthesis and subsequent procedures poses the threat of serious environmental problems [43]. Substituting volatile organic substances and a vast array of dangerous industrially produced compounds for the manufacture and encapsulation of nanoparticles has sparked the scientific and academic communities' interest. As a result, the idea of environment management and the concern for ecologically suitable production has energized scientists to progress more toward environmentally friendly and ecologically sustainable pathways to synthesize super-efficient nanomaterials for the intervention and separation of a range of different hazardous contaminants that pollute water resources [44–45]. The chemical production of nanoparticles has been a focus for the past few decades [46]. Synthetic compounds are employed as reducing, capping, and stabilizing agents in the chemical process. Hydrazine hydrate, sodium borohydride amines, and thiols are among these chemicals that are widely used and recognized for their severe ill effects on people's health and the environment [47–48]. Microbes can be used as a reductant to make BNPs in a safe setting. The biogenic technique is environmentally and economically feasible since it replaces toxic substances with natural components for the production of NPS [49–50]. Plants, algae, fungi, bacteria, yeast, actinomycetes, and other biological entities are used to make these natural NM products. For the manufacture of metallic nanoparticles, natural polyphenols, amino acids, vitamins, and biopolymers like carbohydrates have become increasingly popular [51–53]. The bioreduction by these compounds acquired from living creatures is fundamental to the biogenic creation of nanomaterials in living organisms [54, 45]. These chemicals, in addition to reduction, also operate as capping and stabilizing agents [55]. Microorganisms such as bacteria, fungi, and actinomycetes potentially make zero-valent or magnetic nanomaterials by secreting compounds with such a high capability to reduce or oxidize metal ions [56–57]. Plant extracts have already been effectively used to produce Fe [58], Zn [59], Au [60], Ag [61], Pd [52], and Cu [62] NPs in an array of configurations, sizes, and forms.

1.7 SYNTHESIS OF BIOGENIC NANOMATERIALS

1.7.1 Plant-Based Biosynthesis

Because of its feasibility and efficiency, plant-mediated nanomaterial synthesis has become a well-known and approved approach [63–64]. More significantly, the synthesis process is resource-efficient because it does not involve strong compression or heating [65]. Salem [65] found that the contents of plant extract comprise diverse secondary metabolites, proteins, enzymes, polysaccharides, and reductants

TABLE 1.3
Plant Origins and Morphology of Biogenic Nanoparticles

BNPs	Biological Materials	Biological Agents	Characteristics	References
TiO_2 NPs	Leaf extract	*Jatropha curcas L.*	10–20 nm; spherical;	[70]
	Leaf extract	*Catharanthus roseus L.*	phase, anatase	[71]
	Gel	*Aloe vera L.*	25–110 nm; uneven and	[72]
	extract	*Hibiscus rosasenansis L.*	clustered	[73]
	Flower aqueous extract		80–90 nm; almost spherical	
			7 nm; monodispersed and spherical	
AuNPs	Plant leaf extract	*Lagerstroemia speciosa*	41–91 nm; hexagonal	[74]
AgNPs	Leaf extract	*Ficus benjamina*	60–105 nm	[75]
	Petal extract	*Rosa "Andeli"*	4–29 nm; spherical	[76]
	Stem extract	*Breynia rhamnoides*	Ag, ~64 nm; spherical	[77]
	Leaf extract	*Zanthoxylum armatum*	15–50 nm; spherical	[78]
Fe NPs	Leaf extract	*Azadirachta indica,*		[79]
Iron oxide NPs	Leaf extract	*Magnolia champaca,*	96–150nm,	[80]
	Leaf extract and juices	*Mangifera indica,* and	spherical–50–60 nm;	[81]
	Leaf extract	*Murraya koenigii*	spherical	[82]
	Peel extract	*Camellia sinensis and* pomegranate, *Punica granatum*	20–80 nm; spheroidal ~50 nm; spherical	[83]
		Extracts of *Camellia sinensis, Syzygium aromaticum, Mentha spicata), Punica granatum* juice, and red wine		
		Eucalyptus sp.		
		Tangerine		
ZnO NPs	Lemon juice	Lemon fruits	21.5 nm; spherical	[84]
Potassium Zinc Hexacyanoferrate nanocubes	Leaf extract	*Plectranthus amboinicus*	88 nm; rod shape–	[85]
	Natural surfactant	*Sapindus mukorossi*		
Cu NPs	Peel extract	*Citrus grandis*	22–27 nm; spherical	[86]
Reduced graphene oxide (RGO)/Fe3O4 nanocomposites	Leaf extract	*Solanum trilobatum*	18 nm; spherical	[87]

which aid in the production and stabilization of NMs. Many physicochemical parameters, including ionic strength, temperature, and others, have an impact on the production of nanoparticles during phytoreduction by plant extracts [66]. The nucleation rate of metal ions can be reduced at low pH values, leading to the agglomeration of metal nanoparticles. This phenomenon can result in the formation of large-sized nanoparticles due to the low pH value. Conversely, a higher pH value can promote the synthesis of small-sized nanoparticles [68]. Plant components such as leaves, flower blooms, seeds, pods, fruit waste, stems, and roots were used to generate nanomaterials (NMs) [68]. Multiple species of plants, both terrestrial and aquatic, have already been effectively used to synthesize nanomaterials [69].

1.7.2 Bacteria-Mediated Synthesis

A vast array of bioprocesses, such as bioleaching, biomineralization, and bioaccumulation, are used by bacterial cells to solubilize and accumulate metal ions [88]. Various types of microorganisms and

TABLE 1.4
Biogenic Nanoparticle Synthesis: Bacterial Origin, Precursors, Size, and Shape of BNPs

BNPs	Precursor	Microbes	Condition	Size/Shape	References
Pd	Na_2PdCl_4	*Shewanella oneidensis MR-1*	Anaerobic	28 nm–4.78 nm; short	[91]
Pd	Na_2PdCl_4	*butyricum (LMG 1217),*	Anaerobic	rod	[92]
Pd/Au	Pd/Au	*Citrobacter braakii (ATCC*	Aerobic	50 nm–	[93]
Pd	Na_2PdCl_4	*6750), Enterococcus faecium*	Aerobic		[94]
Pd	Na_2PdCl_4	*(PhIP-M1- a), Klebsiella*	Anaerobic		[95]
Pd-BnM/H_2	Fe(III)$Cl_3.6H_2O$	*pneumoniae (DSM 2026), E. coli (ATCC 25922),* and *Bacteroides vulgatus (LMG 177672(2)* *Shewanella oneidensis MR-1* Microscopic granules from aerobic microorganism Gram⁻ *D. desulfuricans NCIMB 8307, E. coli MC4100, Serratia* sp. *NCIMB 40259, Shewanella oneidensis MR1 NCIMB 146063* and *Cupriavidis metallidurans NCIMB 14178.* Gram⁺*Arthrobacter oxydans NCIMB 10504* and *Micrococcus luteus NCIMB 9278* *Geobacter sulfurreducens*	Anaerobic		[96]
MnOx	Bio-Ag MnB_6	*Pseudomonas putida*	Aerobic	–	[97]
MnOx	Manganese chloride $MnCl_2$	*Pseudomonas putida MnB6*	Anaerobic		[94]
ZnS	$Na_2S_2O_3$	*Shewanella oneidensis MR-1*	Anaerobic	5 nm, spherical	[98]
	$ZnCl_2$	*P. aeruginosa BS01*	Aerobic	10–15 nm, spherical	[92]
Se	Na_2-SeO_3	*Bacillus* sp.	Aerobic	50–200 nm, spherical	[90]
	Sodium selenite	Anaerobic granules *(UASB)*	Anaerobic	160–180 nm, spherical	[89]
CuS	$CuSO_4$	*Shewanella oneidensis MR-1*	Aerobic	D: 17.4 nm, L: 80.8 nm; nanorods	[88]
Hydroxides biogenic Fe(II) and magnetite	$FeCl_3.6H_2O$ and $FeCl_2.4H_2O$	*Shewanella oneidensis* MR-1	Aerobic	–	[91]

metal ions solutions were used in the manufacturing of BNPs; the form and properties of BNPs differ, depending on the microorganisms, precursors, and parameters under which they are used. The species of microbe, the mixing concentration of electron source and acceptor, pH, incubation time, and incubation temperature are all factors that can influence BNP synthesis [89–90].

1.7.3 Biosynthesis of BNMs (Biogenic Nanomaterials) by Algae

Microalgal nanomaterials are a new, innovative field of research for improving wastewater treatment efficiency. Algae are extremely varied and may be found practically anywhere on the planet. These photoautotrophic species include a large source of biologically active compounds with hydroxyl, carboxyl, and amine functional groups, serving as a reducing and capping tool to create metallic NMs [99–100]. Biosynthesis of nanoparticles is concentration-dependent and also is associated to algal species. Metal ions are reduced by a variety of biocompounds, including pigments, peptides, and polysaccharides. In aqueous medium, sulfated polysaccharides and proteins (amino group or cysteine residues) are employed to bind and cap metallic nanoparticles. The manufacture of algal nanomaterials requires a fraction of the time it takes to synthesize conventional biomaterials. Microalgae may indeed be considered a strong nanofactory, capable of manufacturing nanoparticles containing not just Ag ions but also Au, Cd, and Pt [101].

1.7.4 Fungi-Mediated Synthesis of BNMs

Fungal use for biosynthesis is a cost-effective process; fungi can generate stable BNMs with size control [115–116]. Due to wide variety and high metal tolerance of fungi, the mycosynthesis of nanosized materials is receiving a strong interest [117]. To create mycogenic BNMs, fungal species use biomineralization and biotransformation modes [118]. Several publications on the effective

TABLE 1.5
The Algal Method of Production and Morphology of Biogenic Nanoparticles

Nanoparticle	Algal Types	Species	Synthetic Routes	Size/Shape	References
Au	Chlorophyta	*Chlorella vulgaris*	*In vivo*	75 nm, mostly spherical	[102]
Au	Chlorophyta	*Tetraselmis kochinensis*	*In vivo*	35 nm, mostly spherical	[103]
Au	Chlorophyta	*Rhizoclonium fontinale*		16 nm, spherical	[104]
Au	Rhodophyta	*Kappaphycus alvarezii*	*In vitro*	10–40 nm, spherical	[105]
Au	Phaeophyta	*Sargassum wightii*	*In vitro*	8–12 nm, mostly planar	[106]
Ag	Chlorophyta	*Scencedesmus sp.*	*In vitro*	15–20 nm	[107]
Ag	Chlorophyta	*Chlorella pyrenoidosa*	*In vitro*	5–20 nm	[108]
Ag	Chlorophyta	*Caulerpa taxifolia*	*In vitro*	10–100 nm, spherical	[109]
Ag	Phaeophyta	*Sargassum myriocystum*	*In vitro*	20±2.2 nm, hexagonal	[110]
Ag	Phaeophyta	*Padina sp.*	*in vitro*	25–60 nm	[111]
Pt	Cyanophyceae	*Plectonema boryanum UTEX 485*	*In vivo*	30–300 nm, spherical	[112]
Pd	Cyanophyceae	*Plectonema boryanum UTEX 485*	*In vivo* and *in vitro*	≤30	[112]
Pd	Phaeophyta	*Sargassum bovinum*	*In vitro*	5–10 nm, octahedral	[113]
Pd	Chlorophyta	*Chlorella vulgaris*	*In vitro*	5–20 nm, spherical	[114,109]

TABLE 1.6
Fungal Nanoparticle Morphology and the Pollutant Removed

BNPs	Species	Size/Shape	Pollutant Removed	References
AgNPs	*F. Oxysporum*	–	Textile effluent	[119]
AgNPs	*Peniciilium Citreonigum* Dierck	–	Microbial contaminant	[120]
AgNPs	*Scopulaniopsos brumptii salvanet* Duval	–	Microbial contaminant	[120]
FeNPs	*Itajahia* sp.	Rod shape	Pesticides lindane	[121]
AuNPs	*C. oxysporum*	Quasi-spherical, 50–90 nm	Rhodamine B	[122]
AuNPs	*A. japonicus*	Composite	4 nitrophenol and hexacyanoferrate	[122]

fungal production of metallic nanomaterials for the pre-concentration of common ecological contaminants are firmly available.

1.8 TECHNICAL CHALLENGES FACED IN THE SYNTHESIS OF NANOPARTICLES

After the existence of NMs for more than 50 years, researchers still have not fully gained knowledge of how they are perceived. The green chemistry concept supports the synthesis of nanoparticles as an environmentally benign replacement for current NP synthesis techniques. Additionally, a greener approach to the synthesis of NPs is truly based on the environment-friendly procedure that uses renewable resources and eliminates the use of harmful substances and should be non-toxic and economical. Plants are commonly used for the synthesis of nanoparticles due to their ethnobotanical significance, active chemical components, ease of access, effortless procedure, and being economical. Irrespective of the facts, there are other significant concerns and technological obstacles that researchers must solve in order to effectively synthesize green nanoparticles. Conventional synthesis of nanoparticles, including physical and chemical-based procedures, achieve uniformity, homogeneity, and mono NPs, although this is still uncertain in biogenic-based processes.

1.9 FUTURE ASPECTS AND KEY ISSUES ABOUT BIOGENIC NANOPARTICLE SYNTHESIS AND DEVELOPMENT

There is an insufficient knowledge to produce green nanoparticles that use plant components. To synthesize green NPs with definite size, shape, and uniformity, a reasonable technique should be followed. The biogenically synthesized NPs have shown a wider range of size, shape, and structure modifications. The fundamental problem that has arrived in the case of biogenically produced nanoparticles is the conversion of salt to its ionic form [25]. This conversion of salt into its ionic form should be achieved in an optimum quantity during biogenic synthesis. The specific role of natural components of plants in the synthesis of nanostructured materials should be determined, that is, whether they act as surfactants, reducing agents, or stabilizing agents. The homogeneity of the NPs must be controlled while using biological resources for the purpose of synthesis. The transfer of biogenic methods from research laboratories to industry should focus on easy and reproducible synthetic procedures, maximum usage of resources, and formation of NPs of uniform and monodispersed nature. The synthetic procedures should be able to tackle waste management while completely avoiding the use of toxic materials. Additionally, the biogenically synthesized metallic NPs may be harmful to humans or other creatures or the environment. For environmental applications, it is equally crucial to learn the ecotoxicological aspect of metallic NPs. It is also important to conduct researches on non-target creatures, the aquatic ecosystem, and various habitats for niche populations [25]. So research should be encouraged to synthesize NPs by biogenic means while considering all the aforementioned factors, which is fundamental and a must for the sustainable growth of the society.

1.10 CONCLUSION

The research, development, and usage of NMs and BNMs is a fast-expanding field with much room for additional investigation, particularly in terms of making synthesizing techniques more environment-friendly and for economical nanomaterials. Green nanoparticles have a number of benefits in respect of biocompatibility and responsiveness, which might contribute to their use in biomedicine, consumer products, and environmental remediation. ZnO NPs, TiO_2 NPs, silica NPs, and iron NPs are significantly employed in wastewater treatment, textile and food processing, and other sectors. Future research aimed at closing various gaps will be critical in commercializing NMs and biogenic NM production and reaping the complete value of it. Furthermore, improved nanomaterials planning and management are critical in reducing the negative effects of NMs on people's health and the environment.

REFERENCES

1. Thrall, J. H. **2004**. Nanotechnology and medicine. *Radiology* 230: 315–318.
2. Sanchez, F.; Sobolev, K. **2010**. Nanotechnology in concrete – A review. *Construction and Building Materials* 24: 2060–2071.
3. Bora, T.; Dutta, J. **2014**. Applications of nanotechnology in wastewater treatment – A review. *Journal of Nanoscience and Nanotechnology* 14: 613–626.
4. Rajesh Prasanna, P.; Selvamani, P.; Gomathi, E. **2013**. Waste water treatment through dendrimer – Conjugated magnetic nanoparticles. *International Journal of Chemtech Research* 5: 1239–1245.
5. Qu, X.; Alvarez, P. J. J.; Li, Q. **2013**. Applications of nanotechnology in water and wastewater treatment. *Water Research* 47: 3931–3946.
6. Tarasov, B. P.; Muradyan, V. E.; Volodin, A. A. **2011**. Synthesis, properties, and examples of the use of carbon nanomaterials. *Russian Chemical Bulletin* 60: 1261–1273.
7. Coville, N. J.; Mhlanga, S. D.; Nxumalo, E. N.; Shaikjee, A. **2011**. A review of shaped carbon nanomaterials. *South African Journal of Science* 107: 1–15.
8. Liu, Q.; Zhang, A.; Wang, R.; Zhang, Q.; Cui, D. **2021**. A review on metal- and metal oxide-based nanozymes: Properties, mechanisms, and applications. *Nano-Micro Letters* 13: 154.
9. Fawcett, D.; Verduin, J. J.; Shah, M.; Sharma, S. B.; Poinern, G. E. J. **2017**. A review of current research into the biogenic synthesis of metal and metal oxide nanoparticles via marine algae and seagrasses. *Journal of Nanoscience* 2017: 1–15.
10. Maduraiveeran, G.; Sasidharan, M.; Jin, W. **2019**. Earth-abundant transition metal and metal oxide nanomaterials: Synthesis and electrochemical applications. *Progress in Materials Science* 106: 100574.
11. Zhang, S.; Pei, X.; Gao, H.; Chen, S.; Wang, J. **2020**. Metal-organic framework-based nanomaterials for biomedical applications. *Chinese Chemical Letters* 31: 1060–1070.
12. Wang, C.; Makvandi, P.; Zare, E. N.; Tay, F. R.; Niu, L. **2020**. Advances in antimicrobial organic and inorganic nanocompounds in biomedicine. *Advanced Therapeutics* 3: 2000024.
13. Omanović-Mikličanin, E.; Badnjević, A.; Kazlagić, A.; Hajlovac, M. **2020**. Nanocomposites: A brief review. *Health and Technology (Berlin)* 10: 51–59.
14. Malhotra, B. D.; Ali, M. A. **2018**. Nanomaterials in biosensors. *Nanomaterials for Biosensors* 2018: 1–74.
15. Wang, Z.; Hu, T.; Liang, R.; Wei, M. **2020**. Application of zero-dimensional nanomaterials in biosensing. *Frontiers in Chemistry* 8: PMID: 32373593.
16. Chuah, S.; Pan, Z.; Sanjayan, J. G.; Wang, C. M.; Duan, W. H. **2014**. Nano reinforced cement and concrete composites and new perspective from graphene oxide. *Construction and Building Materials* 73:113–124.
17. Sinha, S.; Kim, H.; Robertson, A. W. **2021**. Preparation and application of 0D-2D nanomaterial hybrid heterostructures for energy applications. *Materials Today Advances* 12. p.100169.
18. Chopra, N.; Gavalas, V. G.; Hinds, B. J.; Bachas, L. G. **2007**. Functional one-dimensional nanomaterials: Applications in nanoscale biosensors. *Analytical Letters* 40: 2067–2096.

19. Galdámez-Martinez, A.; Santana, G.; Güell, F.; Martínez-Alanis, P. R.; Dutt. A. **2020**. Photoluminescence of ZnO nanowires: A review. *Nanomaterials* 10: 857.
20. Akbari-Saatlu, M. **2020**. Silicon nanowires for gas sensing: A review. *Nanomaterials* 10: 1–57.
21. Barhoum, A.; Pal, K.; Rahier, H.; Uludag, H.; Kim, S.; Bechelany M. **2019**. Nanofibers as new-generation materials: From spinning and nano-spinning fabrication techniques to emerging applications. *Applied Materials Today* 17: 1–35.
22. Pokropivny, V. V.; Skorokhod, V. V. **2017**. Classification of nanostructures by dimensionality and concept of surface forms engineering in nanomaterial science. *Materials Science and Engineering C* 27: 990–993.
23. Uzoma, P. C.; Hu, H.; Khadem, M.; Penkov. O. V. **2020**. Tribology of 2D nanomaterials: A review. *Coatings* 10: 897.
24. Paramasivam, G.; Palem, V. V.; Sundaram, T.; Sundaram, V.; Kishore, S. C.; Bellucci, S. **2021**. Nanomaterials: Synthesis and applications in theranostics. *Nanomaterials* 11: 3228.
25. Patil, S.; Chandrasekaran, R. **2020**. Biogenic nanoparticles: A comprehensive perspective in synthesis, characterization, application and its challenges. *Journal of Genetic Engineering and Biotechnology* 18: 67. https://doi.org/10.1186/s43141-020-00081.
26. Darroudi, M.; Sabouri, Z.; Oskuee, K. R.; Zak, A. K.; Kargar, H.; Abid Hamid, M. H. N. **2013**. Sol–gel synthesis, characterization, and neurotoxicity effect of zinc oxide nanoparticles using gum tragacanth. *Ceramics International* 39: 9195–9199. https://doi.org/10.1016/j.ceramint.2013.05.021.
27. Pottathara, B. Y.; Grohens, Y.; Kokol, V.; Kalarikkal, N.; Thomas, S. **2019**. Synthesis and processing of emerging two-dimensional nanomaterials. *Micro and Nano Technologies, Nanomaterials Synthesis*: 1–25. ISBN 9780128157510; https://doi.org/10.1016/B978-0-12-815751-0.00001-8.
28. Sun, J.; Wang, M.; Zhao, Y.; Li, X.; Liang, B. **2009**. Synthesis of titanium nitride powders by reactive ball milling of titanium and urea. *Journal of Alloys and Compounds* 482: 29–31. https://doi.org/10.1016/j.jallcom.2009.04.043.
29. Yan, Z.; et al. **2021**. Bringing material concepts into conventional biorefineries: Considerations of sources, preparations, and applications of lignin nanomaterials. *ACS Sustainable Chemistry & Engineering* 9: 10403–10423. https://doi.org/10.1021/acssuschemeng.1c02954.
30. Tillotson, T.; Gash, A.; Simpson, R.; Hrubesh, L.; Satcher, J. Jr.; Poco, J. **2001**. Nanostructured energetic materials using sol–gel methodologies. *Journal of Non-Crystalline Solids* 285: 338–345. https://doi.org/10.1016/S0022-3093(01)00477-X.
31. Delogu, F.; Gorrasi, G.; Sorrentino, A. **2017**. Fabrication of polymer nanocomposites via ball milling: Present status and future perspectives. *Progress in Materials Science* 86: 75–126. https://doi.org/10.1016/j.pmatsci.2017.01.003.
32. Mattox, D. M. **1999**. Physical vapor deposition (PVD) processes. *Metal Finishing* 97: 417–430. https://doi.org/10.1016/S0026-0576(99)80043-9.
33. Sun, L.; et al. **2021**. Chemical vapor deposition. *Nature Reviews Methods Primers* 1: 1–20. https://doi.org/10.1038/s43586-020-00005-y.
34. Sha, D.; Hsieh, L.; Chen, K. **2001**. Wafer rework strategies at the photolithography stage. *The International Journal of Industrial Engineering: Theory, Applications and Practice* 8: 122–130.
35. Khattat, M. M.; Keyhanfar, M.; Razmjou, A. **2018**. A comparative study of stability, antioxidant, DNA cleavage and antibacterial activities of green and chemically synthesized silver nanoparticles. *Artificial Cells, Nanomedicine, and Biotechnology* 46: 1022–1031.
36. Armendariz, V.; Herrera, I.; Peralta-videa, J. R.; Jose-Yacaman, M. **2004**. Size controlled gold nanoparticle formation by Avena sativa biomass: Use of plants in nanobiotechnology. *Journal of Nanoparticle Research* 6: 377–382.
37. Das, J.; Velusamy, P. **2013**. Biogenic synthesis of antifungal silver nanoparticles using aqueous stem extract of banana. *Nano Biomedicine and Engineering* 5: 34–39. https://doi.org/10.5101/nbe.v5i1.p34-38.1.
38. Chandran, S. P.; Chaudhary, M.; Pasricha, R.; Ahmad, A.; Sastry, M. **2006**. Synthesis of gold nanotriangles and silver nanoparticles using aloe vera plant extract. *Biotechnology Progress* 22: 577–583.
39. Yang, N.; Li, W. **2013**. Mango peel extract mediated novel route for synthesis of silver nanoparticles and antibacterial application of silver nanoparticles loaded onto non-woven fabrics. *Industrial Crops and Products* 48: 81–88. https://doi.org/10.1016/j.indcrop.2013.04.001.
40. Patra, J. K.; Baek, K. **2014**. Green nanobiotechnology: Factors affecting synthesis and characterization techniques, *Journal of Nanomaterials Synthesis* 41: 7305.
41. Rai, M.; Gade, A.; Yadav, A. **2011**. Metal nanoparticles in microbiology. *Germany Springer Berlin Heidelberg* 201: 1–14.

42. Holmes, A. B.; Gu, F. X. **2016**. Emerging nanomaterials for the application of selenium removal for wastewater treatment. *Environmental Science: Nano* 3: 982–996.
43. Ju-Nam, Y.; Lead, J. R. **2008**. Manufactured nanoparticles: An overview of their chemistry, interactions and potential environmental implications. *Science of the Total Environment* 400: 396–414.
44. Rezxici, I. **2011**. Determination of engineered nanoparticles on textiles and in textile wastewaters. *TrAC – Trends in Analytical Chemistry* 30: 1159–1167.
45. Gautam, P. K.; Singh, A.; Misra, K.; Sahoo, A. K.; Samanta, S. K. **2019**. Synthesis and applications of biogenic nanomaterials in drinking and wastewater treatment. *Journal of Environmental Management* 231: 734–748.
46. Salavati-Niasari, M.; Soofivand, F.; Sobhani-Nasab, A.; Shakouri-Arani, M.; Yeganeh Faal, A.; Bagheri, S. **2016**. Synthesis, characterization, and morphological control of $ZnTiO_3$ nanoparticles through sol-gel processes and its photocatalyst application. *Advanced Powder Technology* 27: 2066–2075.
47. Hannah, W.; Thompson, P. B. **2008**. Nanotechnology, risk and the environment: A review. *Journal of Environmental Monitoring* 10: 291–300.
48. Jortner, J.; Rao, C. N. R. **2002**. Nanostructured advanced materials: Perspectives and directions. *Pure and Applied Chemistry* 74: 1491–1506.
49. Vaseghi, Z.; Nematollahzadeh, A.; Tavakoli, O. **2018**. Green methods for the synthesis of metal nanoparticles using biogenic reducing agents: A review. *Reviews in Chemical Engineering* 34: 529–559.
50. Gade, A.; Gaikwad, S.; Duran, N.; Rai, M. **2014**. Green synthesis of silver nanoparticles by Phoma glomerate. *Micron* 59: 52–59.
51. Mohanpuria, P.; Rana, N. K.; Yadav, S. K. **2008**. Biosynthesis of nanoparticles: Technological concepts and future applications. *Journal of Nanoparticle Research* 10: 507–517.
52. Metz, K. M. **2015**. Green synthesis of metal nanoparticles via natural extracts: The biogenic nanoparticle corona and its effects on reactivity. *ACS Sustainable Chemistry Engineering* 3: 1610–1617.
52. Schröfel, A.; Kratošová, G.; Šafařík, I.; Šafaříková, M.; Raška, I.; Shor, L. M. **2014**. Applications of biosynthesized metallic nanoparticles – A review. *Acta Biomaterialia* 10: 4023–4042.
54. Lee, Y.; et al. **2012**. A transfer of silver nanoparticles from pregnant rat to offspring. *Toxicology Research* 28: 139–141.
55. Moulton, M. C.; Braydich-Stolle, L. K.; Nadagouda, M. N.; Kunzelman, S.; Hussain, S. M.; Varma. R. S. **2010**. Synthesis, characterization and biocompatibility of 'green' synthesized silver nanoparticles using tea polyphenols. *Nanoscale* 2: 763–770.
56. Jha, K. A.; Prasad, K.; Kulkarni, A. R. **2009**. Synthesis of TiO_2 nanoparticles using microorganisms. *Colloids and Surfaces B: Biointerfaces* 71: 226–229.
57. Mandal, S.; Puniya, A. K.; Singh, K. **2006**. Effect of alginate concentrations on survival of microencapsulated Lactobacillus casei NCDC-298. *International Dairy Journal* 16: 1190–1195.
58. Balamurugan, M.; Saravanan, S.; Soga, T. **2014**. Synthesis of iron oxide nanoparticles by using eucalyptus globulus plant extract. *e-Journal of Surface Science and Nanotechnology* 12: 363–367.
59. Nava, O. J.; et al. **2017**. Influence of Camellia sinensis extract on zinc oxide nanoparticle green synthesis. *Journal of Molecular Structure* 1134: 121–125.
60. Annamalai, A.; Babu, S. T.; Jose, N. A.; Sudha, D.; Lyza, C. V. **2011**. Biosynthesis and characterization of silver and gold nanoparticles using aqueous leaf extraction of Phyllanthus amarus Schum & Thonn. *World Applied Sciences Journal* 13: 1833–1840.
61. Kumar, B.; Smita, K.; Cumbal, L.; Debut. A. **2017**. Green synthesis of silver nanoparticles using Andean blackberry fruit extract. *Saudi Journal of Biological Sciences* 24: 45–50.
62. Lee, H. J.; Lee, G.; Jang, N. R.; Yun, J. H.; Song, J. Y.; Kim, B. S. **2011**. Biological synthesis of copper nanoparticles using plant extract. *Technical Proceedings of the 2011 NSTI Nanotechnology Conference and Expo, NSTI-Nanotech* 1: 371–374.
63. Verma, S. K.; Jha, E.; Panda, P. K.; Thirumurugan, A.; Suar, M. **2019**. Biological effects of green-synthesized metal nanoparticles: A mechanistic view of antibacterial activity and cytotoxicity. *Advanced Nanostructured Materials for Environmental Remediation* 25: 145–171.
64. Mittal, A. K.; Chisti, Y.; Banerjee, U. C. **2013**. Synthesis of metallic nanoparticles using plant extracts. *Biotechnological Advances* 31: 346–356.
65. Rai, M.; Yadav, A.; Gade, A. **2008**. CRC 675 – Current trends in phytosynthesis of metal nanoparticles. *Critical Reviews in Biotechnology* 28: 277–284.
66. Iravani, S. **2011**. Green synthesis of metal nanoparticles using plants. *Green Chemistry* 13: 2638–2650.
67. Nguyen, N. T. T.; Nguyen, L. M.; Nguyen, T. T. T.; Nguyen, T. T.; Nguyen, D. T. C.; Van Tran, T. **2022**. Formation, antimicrobial activity, and biomedical performance of plant-based nanoparticles: A review. *Environmental Chemistry Letters* 20: 2531–2571.

68. Herlekar, M.; Barve, S.; Kumar, R. **2014**. Plant-mediated green synthesis of iron nanoparticles. *Journal of Nanoparticles* 2014: 1–9.
69. Vanaja, M.; Annadurai, G. **2013**. Coleus aromaticus leaf extract mediated synthesis of silver nanoparticles and its bactericidal activity. *Applied Nanoscience (Switzerland)* 3: 217–223.
70. Kumar, V.; Gautam, A.; Guleria, P. **2020**. Platinum nanoparticles: Synthesis strategies and applications. *Nanoarchitectonics* 1: 53–92.
71. Velayutham, K.; et al. **2012**. Evaluation of Catharanthus roseus leaf extract-mediated biosynthesis of titanium dioxide nanoparticles against Hippobosca maculata and Bovicola ovis. *Parasitology Research* 111: 2329–2337.
72. Bachheti, A.; Bachheti, R. K.; Abate, L.; Husen, A. **2021**. Current status of Aloe-based nanoparticle fabrication, characterization and their application in some cutting-edge areas. *South African Journal of Botany* 147: 1058–1069.
73. Sahaya, P.; Kumar, M.; Francis, A. P.; Devasena, T. **2014**. Biosynthesized and chemically synthesized titania nanoparticles: Comparative analysis of antibacterial activity. *Journal of Environmental Nanotechnology* 3: 73–81.
74. Choudhary, B. C.; et al. **2017**. Photocatalytic reduction of organic pollutant under visible light by green route synthesized gold nanoparticles. *Journal of Environmental Sciences (China)* 55: 236–246.
75. Al-Qahtani, K. M. **2017**. Cadmium removal from aqueous solution by green synthesis zero valent silver nanoparticles with Benjamina leaves extract. *Egyptian Journal of Aquatic Research* 43: 269–274.
76. Suárez-Cerda, J.; Alonso-Nuñez, G.; Espinoza-Gómez, H.; Flores-López, L. Z. **2015**. Synthesis, kinetics and photocatalytic study of 'ultra-small' Ag-NPs obtained by a green chemistry method using an extract of Rosa 'Andeli' double delight petals. *Journal of Colloid and Interface Science* 458: 169–177.
77. Gangula, A.; Podila, R.; R. M.; Aranam, L.; Janardhana, C.; Rao, A. M. **2011**. Catalytic reduction of 4-nitrophenol using biogenic gold and silver nanoparticles derived from breynia rhamnoides. *Langmuir* 27: 15268–15274.
78. Saratale, G. D.; et al. **2020**. Investigation of photocatalytic degradation of reactive textile dyes by Portulaca oleracea-functionalized silver nanocomposites and exploration of their antibacterial and antidiabetic potentials. *Journal of Alloys and Compounds* 833: 155083.
79. Devatha, C. P.; Thalla, A. K.; Katte, S. Y. **2016**. Green synthesis of iron nanoparticles using different leaf extracts for treatment of domestic waste water. *Journal of Cleaner Production* 139: 1425–1435.
80. Ozkan, Z. Y.; Cakirgoz, M.; Kaymak, E. S.; Erdim, E. **2018**. Rapid decolorization of textile wastewater by green synthesized iron nanoparticles. *Water Science and Technology* 77: 511–517.
81. Mystrioti, C.; Xanthopoulou, T. D.; Tsakiridis, P. E.; Papassiopi, N.; Xenidis, A. **2016**. Comparative evaluation of five plant extracts and juices for nanoiron synthesis and application for hexavalent chromium reduction. *Science of the Total Environment* 539: 105–113.
82. Wang, T.; Lin, J.; Chen, Z.; Megharaj, M.; Naidu, R. **2014**. Green synthesized iron nanoparticles by green tea and eucalyptus leaves extracts used for removal of nitrate in aqueous solution. *Journal of Cleaner Production* 83: 413–419.
83. Ehrampoush, M. H.; Miria, M.; Salmani, M. H.; Mahvi, A. H. **2015**. Cadmium removal from aqueous solution by green synthesis iron oxide nanoparticles with tangerine peel extract. *Journal of Environmental Health Science and Engineering* 13: 1–7.
84. Davar, F.; Majedi, A.; Mirzaei. A. **2015**. Green synthesis of ZnO nanoparticles and its application in the degradation of some dyes. *Journal of the American Ceramic Society* 98: 1739–1746.
85. Peralta-Videa, J. R.; et al. **2016**. Plant-based green synthesis of metallic nanoparticles: Scientific curiosity or a realistic alternative to chemical synthesis. *Nanotechnology for Environmental Engineering* 1: 1–29.
86. Sinha, T.; Ahmaruzzaman, M. **2016**. Photocatalytic decomposition behavior and reaction pathways of organic compounds using Cu nanoparticles synthesized: Via a green route. *Photochemical and Photobiological Sciences* 15: 1275–1281.
87. Goutam, S. P.; Saxena, G. **2021**. Biogenic nanoparticles for removal of heavy metals and organic pollutants from water and wastewater: Advances, challenges, and future prospects. *Bioremediation for Environmental Sustainability* 623–636.
88. Xiao, X.; et al. **2017**. Self-assembly of complex hollow CuS nano/micro shell by an electrochemically active bacterium Shewanella oneidensis MR-1. *International Biodeterioration and Biodegradation* 116: 10–16.
89. Mittal, A. K.; Thanki, K.; Jain, S.; Banerjee, U. C. **2016**. Comparative studies of anticancer and antimicrobial potential of bioinspired silver and silver-selenium nanoparticles. *Applied Nanomedicine* 1: 1–6.
90. Ahluwalia, S.; Prakash, N. T.; Prakash, R.; Pal, B. **2016**. Improved degradation of methyl orange dye using bio-co-catalyst Se nanoparticles impregnated ZnS photocatalyst under UV irradiation. *Chemical Engineering Journal* 306: 1041–1048.

91. Wang, J.; et al. **2016**. Silver-nanoparticles-modified biomaterial surface resistant to staphylococcus: New insight into the antimicrobial action of silver. *Scientific Reports* 6: 1–16.
92. Hazra, C.; Kundu, D.; Chaudhari, A.; Jana, T. **2013**. Biogenic synthesis, characterization, toxicity and photocatalysis of zinc sulfide nanoparticles using rhamnolipids from Pseudomonas aeruginosa BS01 as capping and stabilizing agent. *Journal of Chemical Technology and Biotechnology* 88: 1039–1048.
93. Furgal, K. M.; Meyer, R. L.; Bester, K.; **2014**. Removing selected steroid hormones, biocides and pharmaceuticals from water by means of biogenic manganese oxide nanoparticles in situ at ppb levels. *Chemosphere* 136: 321–326.
94. Deplanche, K.; et al. **2014**. Catalytic activity of biomass-supported Pd nanoparticles: Influence of the biological component in catalytic efficacy and potential application in 'green' synthesis of fine chemicals and pharmaceuticals. *Applied Catalysis B* 147: 651–665.
95. Quan, X.; Xu, H.; Sun, B.; Xiao, Z. **2018**. Anode modification with palladium nanoparticles enhanced Evans Blue removal and power generation in microbial fuel cells. *International Biodeterioration and Biodegradation* 132: 94–101.
96. Ali, I.; et al. **2019**. Overview of microbes based fabricated biogenic nanoparticles for water and wastewater treatment. *Journal of Environmental Management* 230: 128–150.
97. Meerburg, F.; Hennebel, T.; Vanhaecke, L.; Verstraete, W.; Boon, N. **2012**. Diclofenac and 2-anilinophenylacetate degradation by combined activity of biogenic manganese oxides and silver. *Microbial Biotechnology* 5: 388–395.
98. Chellamuthu, P. et al. **2019**. Biogenic control of manganese doping in zinc sulfide nanomaterial using Shewanella oneidensis MR-1. *Frontiers in Microbiology* 10: 938.
99. Castro, L.; Blázquez, M. L.; Muñoz, J. A.; González, F.; Ballester, A. **2013**. Biological synthesis of metallic nanoparticles using algae. *IET Nanobiotechnology* 7: 109–116.
100. Shankar, P. D.; et al. **2016**. A review on the biosynthesis of metallic nanoparticles (gold and silver) using bio-components of microalgae: Formation mechanism and applications. *Enzyme and Microbial Technology* 95: 28–44.
101. Merida, L. G. R.; Zepka, L. Q. **2015**. Microalgae as nanofactory for production of antimicrobial molecules. *Journal of Nanomedicine and Nanotechnology* 6: 5–8.
102. Luangpipat, T.; Beattie, I. R.; Chisti, Y.; Haverkamp, R. G. **2011**. Gold nanoparticles produced in a microalga. *Journal of Nanoparticle Research* 13: 6439–6445.
103. Senapati, S.; Syed, A.; Moeez, S.; Kumar, A.; Ahmad, A. **2012**. Intracellular synthesis of gold nanoparticles using alga Tetraselmis kochinensis. *Materials Letters* 79: 116–118.
104. Parial, D.; Pal, R. **2015**. Biosynthesis of monodisperse gold nanoparticles by green alga Rhizoclonium and associated biochemical changes. *Journal of Applied Phycology* 27: 975–984.
105. Rajasulochana, P.; Dhamotharan, R.; Murugakoothan, P.; Murugesan, S.; Krishnamoorthy, P. **2010**. Biosynthesis and characterization of gold nanoparticles using the alga Kappaphycus alvarezii. *International Journal of Nanoscience* 9: 511–516.
106. Singaravelu, G.; Arockiamary, J. S.; Kumar, V. G.; Govindaraju, K. **2007**. A novel extracellular synthesis of monodisperse gold nanoparticles using marine alga, Sargassum wightii Greville. *Colloids and Surfaces B: Biointerfaces* 57: 97–101.
107. Velusamy, P.; Kumar, G. V.; Jeyanthi, V.; Das, J.; Pachaiappan, R. **2016**. Bio-inspired green nanoparticles: Synthesis, mechanism, and antibacterial application. *Toxicology Research* 32: 95–102.
108. Keat, C. L.; Aziz, A.; Eid, A. M.; Elmarzugi., N. A. **2015**. Biosynthesis of nanoparticles and silver nanoparticles. *Bioresources and Bioprocessing* 2: 47.
109. Li, S. N.; Wang, R.; Ho, S. H. **2021**. Algae-mediated biosystems for metallic nanoparticle production: From synthetic mechanisms to aquatic environmental applications. *Journal of Hazardous Materials* 420: 126625.
110. Balaraman, P.; et al. **2020**. Phyco-synthesis of silver nanoparticles mediated from marine algae Sargassum myriocystum and its potential biological and environmental applications *Waste and Biomass Valorization* 11: 5255–5271.
111. Bhuyar, P.; Rahim, M. H. A.; Sundararaju, S.; Ramaraj, R.; Maniam, G. P.; Govindan, N. **2020**. Synthesis of silver nanoparticles using marine macroalgae Padina sp. and its antibacterial activity towards pathogenic bacteria. *Beni- Suef University Journal of Basic and Applied Sciences* 9: 3.
112. Lengke, M. F.; Fleet, M. E.; Southam, S. **2007**. Biosynthesis of silver nanoparticles by filamentous cyanobacteria from a silver(I) nitrate complex. *Langmuir* 23: 2694–2699.
113. Momeni, S.; Nabipour, I. **2015**. A simple green synthesis of palladium nanoparticles with Sargassum alga and their electrocatalytic activities towards hydrogen peroxide. *Applied Biochemistry and Biotechnology* 176: 1937–1949.

114. Arsiya, F.; Sayadi, M. H.; Sobhani, S. **2017**. Green synthesis of palladium nanoparticles using Chlorella vulgaris. *Materials Letters* 186: 113–115.
115. Gade, A.; Ingle, A.; Whiteley, C. **2010**. Mycogenic metal nanoparticles: Progress and applications. *Biotechnology Letters* 32: 593–600.
116. Siddiqi, K. S.; Husen, A. **2016**. Fabrication of metal nanoparticles from fungi and metal salts: Scope and application. *Nanoscale Research Letters* 11: 22–24.
117. Dhillon, G. S.; et al. **2012**. Green approach for nanoparticle biosynthesis by fungi: Current trends and applications. *Critical Reviews in Biotechnology* 32: 49–73.
118. Mahanty, S.; Bakshi, M.; Ghosh, S.; Chatterjee, S.; Bhattacharyya, S. **2019**. Green synthesis of iron oxide nanoparticles mediated by filamentous fungi isolated from Sundarban mangrove. *BioNanoScience* 9: 637–651.
119. Durán, N.; Marcato, P. D.; De Souza, G. I. H.; Alves, O. L.; Esposito, E. **2014**. Antibacterial effect of silver nanoparticles produced by fungal process on textile fabrics and their effluent treatment. *Journal of Biomedical Nanotechnology* 3: 203–208.
120. Moustafa, M. T. **2017**. Removal of pathogenic bacteria from wastewater using silver nanoparticles synthesized by two fungal species. *Water Science* 31: 164–176.
121. Paknikar, K. M.; Nagpal, V.; Pethkar, A. V.; Rajwade, J. M. **2005**. Degradation of lindane from aqueous solutions using iron sulfide nanoparticles stabilized by biopolymers. *Science and Technology of Advanced Materials* 6: 370–374.
122. Bhargava, A.; Jain, N.; Khan, M. A.; Pareek, V.; Dilip, R. V.; Panwar, J. **2016**. Utilizing metal tolerance potential of soil fungus for efficient synthesis of gold nanoparticles with superior catalytic activity for degradation of rhodamine B. *Journal of Environmental Management* 183: 22–32.

2 Characterization Tools for Biogenic Nanomaterials

Karina Bano, Sandeep Kaushal, and Ranvijay Kumar

2.1 INTRODUCTION

The convenience of technology in day-to-day applications is increasingly being seen by people, and the term "nanotechnology" has become established in the terminology of modern tools. The discipline of synthesizing nanoparticles with precise morphologies and amazing properties has grown significantly over the last 20 years, becoming a vast area of study [1–3]. The key goal in chemistry is to produce nanomaterials with controlled particle size, structure, and crystallinity that might be employed for various applications. Nanotechnology deals with the production of particles having size from 1 to 100 nm, including the development of several dimensions of nanomaterials, like metal NPs, metal oxide NPs, nanocomposites, bimetallic NPs, semiconductor nanoparticles (SNs). These nanoparticles are different in their morphological size, form, and optical characteristics, and they are excellent in different applications in diverse fields, including anticancer, antimicrobial, and catalytic reactions [4]. A nanometer is approximately 1,00,000 times smaller as compared to a diameter of human hair. The emergent magnetic, optical, electrical, and other properties of the nanomaterials make them novel materials in various fields.

The fabrication of nanomaterials may be generally categorized into two approaches: top-down and bottom-up. Top-down strategies include the nanoscale breaking or cutting of the bulk materials into small particles. The bottom-up methods include building nanoscale materials one atom at a time, one molecule at a time, or one cluster at a time. For synthesis of these nanomaterials (NMs), numerous methods have been reported, which include physical, chemical, and biological methods [5–7]. The conventional methods utilized for NM fabrication are regarded as costly, laborious, and environmentally hazardous [8]. Even though these techniques have produced better NMs, a fundamental understanding of improved manufacturing processes is still necessary. This knowledge could be applied at the industrial and commercial level to produce products that are more powerful, long-lasting, clean, safe, and sophisticated [9]. Additionally, biogenic synthesis methods are simple, enduring, eco-friendly, and affordable; thus, researchers are leaning toward using them to create nanomaterials from various green reagents [10–12]. By using these methods, nanoparticle-sized metal and metal oxides can be synthesized efficiently. The most popular and reported biological agents for the synthesis of nanomaterials involve plant parts, bacteria, fungi, yeast, and waste materials of plants or fruits, as they act as excellent reducing and capping agents [13, 14]. These NMs, synthesized via biogenic techniques, possess fascinating size and morphologies assured via different characterization tools, as shown in Figure 2.1. This chapter summarizes various characterization tools utilized for the analysis of biogenic nanomaterials.

2.2 CHARACTERIZATION OF BIOGENIC NANOMATERIALS

To ensure synthesis of NMs, a physical and chemical study of the material is necessary. The ability of the industry to meet requirements and the adoption of these materials in commercial applications will both be strongly impacted by the validity and dependability of NMs measurement techniques. With the advancement in nanotechnology, more authentic and reliable methods are demanded. Alternate techniques are necessary to characterize NMs to enhance its scope in various fields. The main methods for analyzing biogenic nanomaterials are UV-visible spectroscopy, Fourier transform

DOI: 10.1201/9781003430087-2

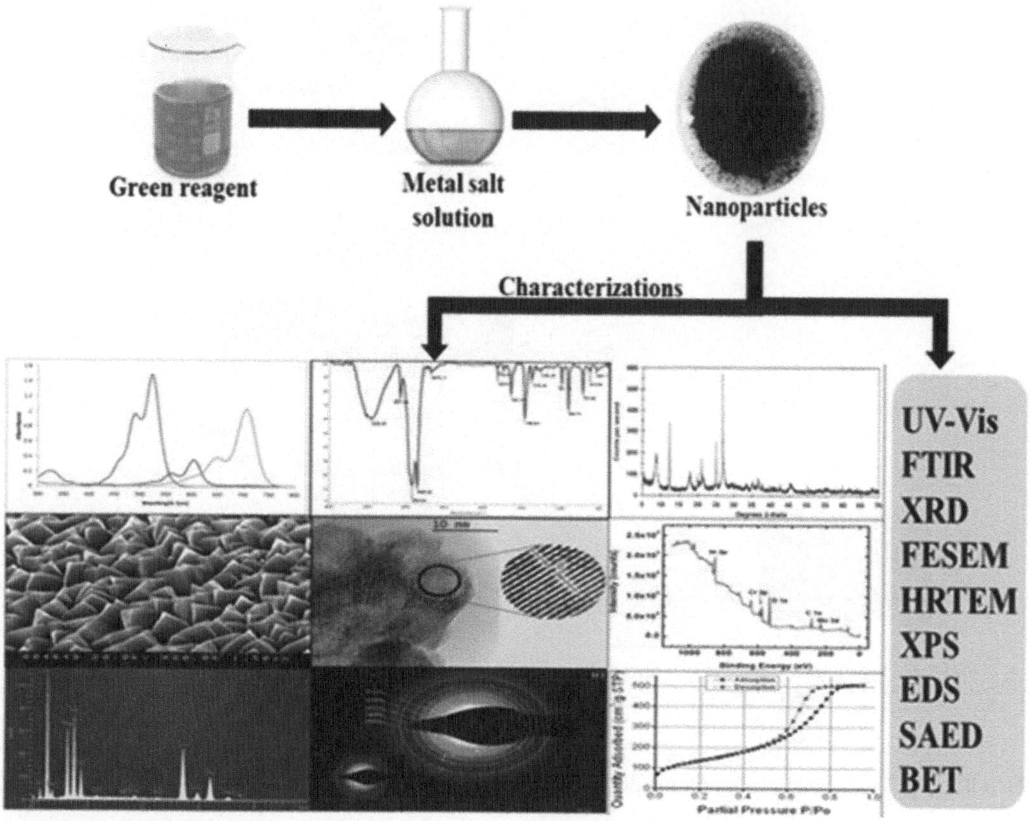

FIGURE 2.1 Representation of general mechanism of synthesis and different characterization tools for nanomaterials (NMs).

infrared (FTIR), TEM, SEM, X-ray diffraction (XRD), X-ray photon emission (XPS), Brunauer-Emmett-Teller (BET), energy-dispersive X-ray (EDX), and zeta potential. UV-visible spectroscopy, FTIR, and XRD analysis are considered to be primary methods to analyze the optical properties, atomic bindings, and crystalline nature of synthesized sample [15–17]. In addition to this, FESEM, HRTEM, and XPS techniques are microscopic methods to analyze size and morphology of NMs [18]. Chemical composition and surface area studies have been performed by EDS and BET techniques.

2.2.1 Electron Spectroscopy–Based Characterization Techniques

2.2.1.1 Scanning Electron Microscopy

The morphology (such as particle sizes and shapes), crystallographic properties, flaws, and structure of nanocrystalline particles are investigated using SEM microscopic techniques. In addition to this, micrometer size and elemental compositions of materials can also be determined by SEM. In this technique, interparticle interactions and sample drying conditions may affect the structure of nanomaterials [19]. Figure 2.2 represents instrumentation of SEM microscope consisting high-energy electron beam, magnetic lens, apertures, and monitor.

SEM involves a highly focused electron beam (consisting a negative charge) falling on the sample surface and provides magnified images of particle shape and orientation. SEM has a resolution of 1 nm and a magnification of 400,000 [20]. Interaction of electron beam with sample surface creates distinct signals that represent the topographic detail and atomic constitution of the examined material area. The input electron beam splits into X-rays, backscatter electrons, secondary electrons, and

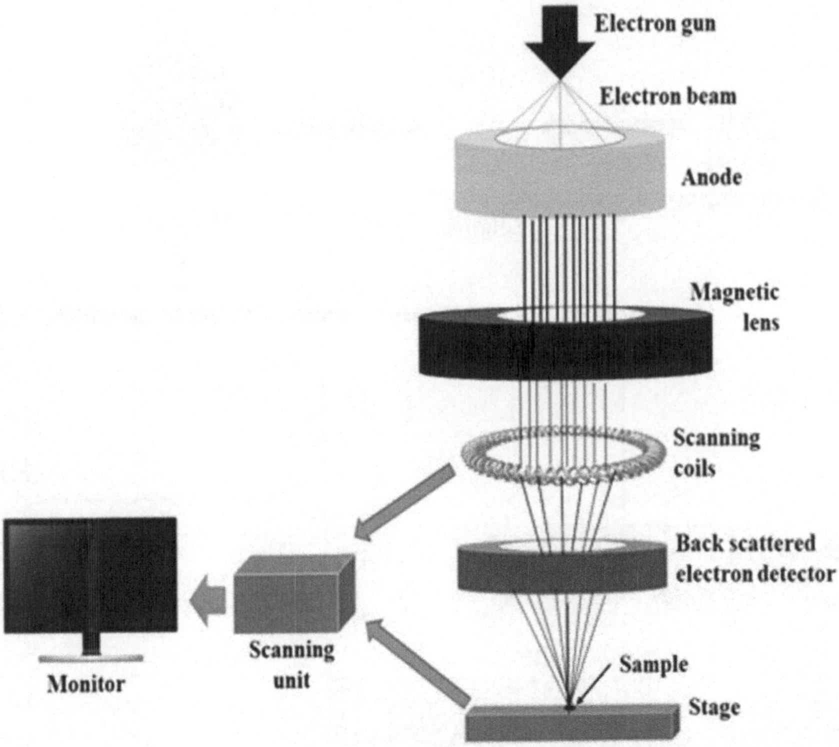

FIGURE 2.2 Schematic representation of instrumentation for FESEM analysis.

Auger electrons. Among all these active charges, secondary electrons play a vital role in determining material morphology and topography.

In this method, along with the excited electrons, X-rays are also emitted, which further investigate the composition of sample, known as EDS technique, which is frequently used in conjunction with the SEM. This combination provides information on composition of material, texture, and weight % of a particular element in sample.

Various NPs, including transition and inert metals, can be comfortably recognized on the surface of a material using EDX analysis. Elements with a low atomic number, on the other hand, are more difficult to identify using EDX. It provides the chemical composition of each element present in the sample, along with the graphical representation.

In SEM technique, samples to be scanned should be electrically conducted to prevent electrostatic charge buildup, which results in electrostatically deformed pictures and distortions. This is accomplished by extremely thin laying of as Au, Au/Pd alloys, or Pt [21]. Moreover, after SEM, FESEM is intended to function at lower vacuum conditions and can be used to see specimens in their "natural" state. This includes the capacity to inspect wet materials, like biological samples [22]. However, the SEM observations have several limitations. During sample preparation, contrasting processes may cause reduction of the nanomaterials, altering their size and form.

2.2.1.2 Transmission Electron Microscopy

An electron microscopy technique commonly utilized for the analysis of NPs is transmission electron microscopy (TEM). At atomic dimensions (1 nm), TEM may provide direct high-resolution images as well as thorough qualitative/quantitative information for nanomaterials. Instrumentation

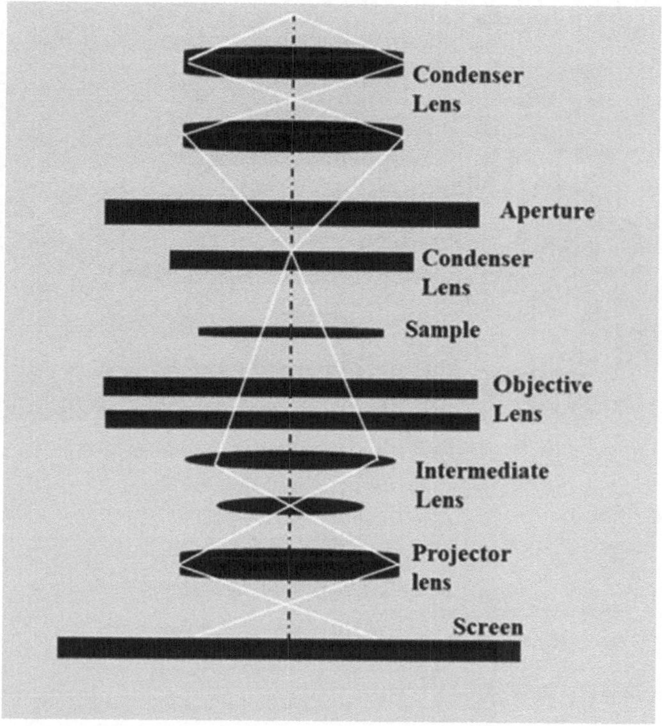

FIGURE 2.3 Instrumentation of transmission electron microscope [23].

of TEM is almost similar to that of SEM, the major difference being the position of sample and results interpreted from the obtained data.

During TEM, inter-electron interactions between the beam and the sample are seen when a high-energy electron ray is guided through an ultrathin material. This, in turn, leads to the formation of unshattered, elastically scattered, or inelastically scattered electrons. The general instrumentation of TEM is shown in Figure 2.3. On the basis of the density of these electrons, they get focused on lenses and then provide electron diffraction pattern or a dark image of varied darkness.

Powders, nanoparticles, and nanotubes have tiny dimensions and seem to be electron-transparent. The coating of a low-concentration sample onto supporting grids produces the result. Cryo-microtomes are used in some circumstances to cut samples under cryogenic conditions. To transmit the electron beam, the TEM is operated in a high vacuum. A high vacuum is necessary to prevent the electron beam from being deflected or attenuated by air molecules. The fluctuating electron concentrations surrounding the atoms within the material deflect the electron beam. Osmium tetroxide is a common staining reagent that is absorbed by unsaturated bonds [20].

HR-TEM provides direct imaging of a samples' crystallographic structure at the atomic level, a useful method for studying material properties on an atomic scale, allowing for the visualization of crystal structures and crystal defects. Kumari et al. reported the utilization of *Terminalia arjuna* bark for the preparation of CuO- and rGO-based CuO/rGO nanocomposite material. Size and morphology of particles were evaluated by FESEM and HRTEM analysis along with EDS mapping. Agglomerates of nanocomposite showed spherical morphology of CuO NPs and fine sheets of GO, as shown in Figure 2.4. Elemental composition was ensured by the presence of sharp peaks assigned to Cu, O, and C elements during EDS mapping [23].

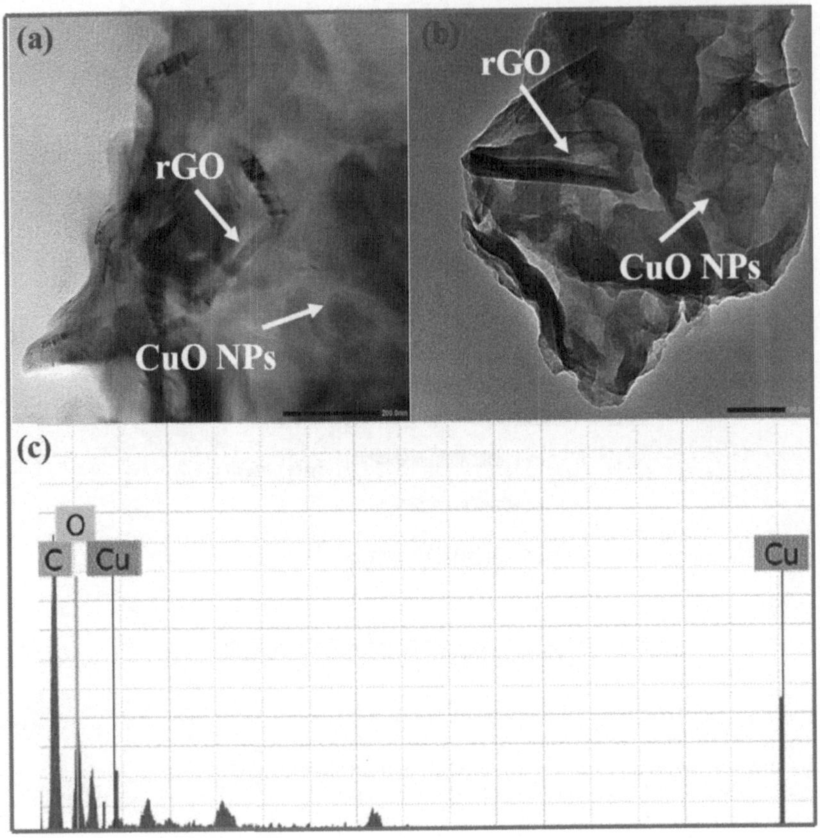

FIGURE 2.4 Representation of (a–b) HRTEM images and (c) EDS of biogenic CuO/rGO nanocomposites.

Although TEM is a strong tool for studying the distribution of particles, level of aggregation, and dispersion in particular NPs, high-vacuum and ultrathin sample sections are needed for this method, which increase the possibility of deformation of sample structure. Furthermore, the striking high-voltage beam has the potential to harm or even destroy the specimen.

2.2.2 X-Ray-Related Characterization Techniques

2.2.2.1 XRD Analysis

Among the most often used methods for characterizing NPs is X-ray diffraction. XRD typically reveals the lattice parameters, phase, crystalline grain size, and crystalline structure. To calculate the particle size and enlargement of the high-intensity peak of an XRD data for a specific material, the most popular Scherrer equation was used. The advantage of using XRD techniques is that they yield results that are statistically representative. These methods are typically applied to powdered samples. The particle composition can be established by comparing the peaks' locations and sizes to reference patterns found in the International Centre for Diffraction Data (JCPDS) database.

Every crystal possesses a well-defined atomic framework which, when subjected to X-rays, generates constructive and destructive interference of the scattered X-ray beam and results in the generation of different diffraction patterns with many distinct spots, or Bragg diffraction peaks. The spectra of crystalline material demonstrate the occurrence of sharp signals at particular 2θ value, whereas the amorphous material illustrates the intensity maximum, which stretches over multiple scattering angles [24].

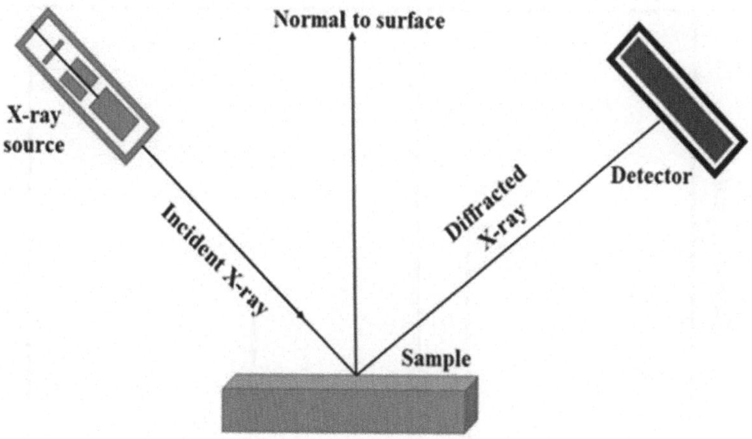

FIGURE 2.5 Basic instrumentation for working of an X-ray diffractometer.

X-rays are electromagnetic energy with a shorter wavelength, produced from electrically charged particles. High voltage is preserved at the interface of two electrodes, which further attract electrons to the metal target. X-rays are created at the impact spot and radiate in all directions. These X-rays directly hit at highly refined powder sample, as shown in Figure 2.5. The detector detects X-rays, processed using a microcontroller. Changing the angle between the source, powder sample, and sensor produces an X-ray or spectrum analyzer scan.

AuNPs synthesized from *Lonicera japonica* were characterized by using the XRD technique, and a crystalline nature was confirmed with particle size in the range of 10–40 nm. Here, green extract was prepared from a flower extract. Alshehri et al. synthesized Fe nanoparticles by using *Hibiscus sabdariffa* extract, and the obtained material was characterized by various techniques. From XRD analysis, highly intense peaks were seen in the range of 20°–70° with all the values fitted in JCPDS card no. 89–0950. Fe NPs comprised purely cubic phase and pure crystalline nature [25].

Singh et al. [26] synthesized NiO NPs by using *Spirogyra* sp. cell culture as green extract and nickel nitrate as precursor salt. The XRD analysis of NiO nanoparticles represents a highly intense peak at $2\theta = 43.3°$ with indices (200) and data matched with JCPDS card number 47–1049, as shown in Figure 2.6.

This face-centered cubic structure possessed average lattice parameter of 4.17 Å. For calculation of crystallite size, Debye–Scherrer approximation was applied, and obtained particle size was 27.7 nm [26]. Although peak broadening is minimal for larger particles, XRD tends to be applicable for nanometer crystals of dimensions under 100–200 nm. On the contrary, the peak widening gets so noticeable for extremely tiny particles having dimensions under 2–3 nm, as peak width is unable to be precisely quantified, though it is occasionally observed.

Overall, XRD tends to be less beneficial for anisotropic and non-uniform nanoparticles; hence, in these situations, it ought to be evaluated with other investigations, like TEM. The component composition of a mixture, its level of crystallinity, and the divergence of a specific element from its optimal content, as well as structure, may all be identified employing XRD patterns [27].

2.2.2.2 X-Ray Photoelectron Spectroscopy

The quantitative spectroscopic surface analysis method known as XPS can be employed to determine a material's elemental structure and electronic and chemical state of each atom with an atomic sensitivity of 0.1%–1% [28]. The photoelectric effect is the basic theory that drives it. With the exception of H and He, this technique provides elemental and chemical information for every element found on the surface of the material being studied [29]. A material exposed to X-rays, which cause electron emission from the elements' surface core orbitals, is examined in an ultra-high vacuum. The measurements of amount of ejected electrons from the material surface produce the XPS

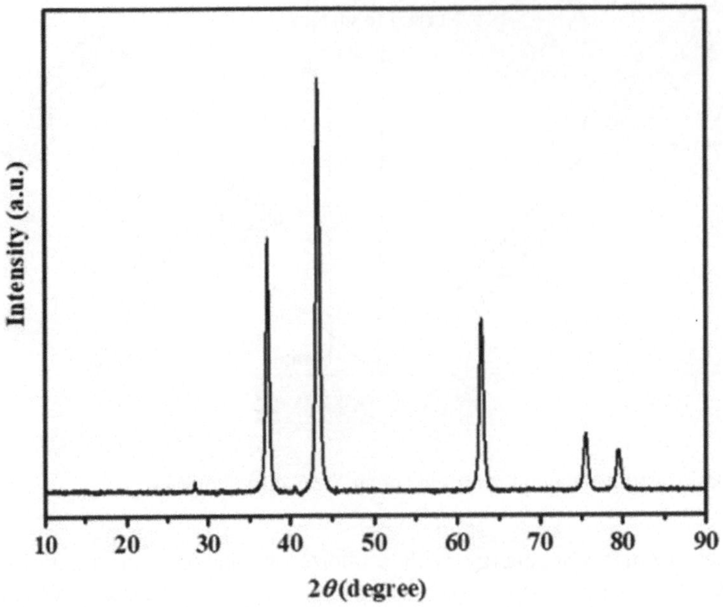

FIGURE 2.6 Powder XRD pattern of cell-mediated NiO NPs [26].

spectra. Kinetic energy can be employed to compute the electrons' binding energy, which represents the oxidation state of the surface elements. Contamination in the sample can be found using XPS as well. The following equation can be utilized to compute the binding energy of the electrons using the kinetic energy:

$$E_{binding} = E_{photon} - (E_{kinetic} + \Phi)$$

Where $E_{binding}$, E_{photon}, and $E_{kinetic}$ are the energy of the emitted electrons, X-ray photon energy, and kinetic energy of the emitted electrons, correspondingly, and Φ is the work function of the electron energy analyzer.

Due to the fact that every substance generates a distinct group of XPS signals at distinctive binding energy levels, the spectrum of energy of the photoelectrons reveals the electronic structures of the atoms. Usually, with the rise in atomic number, the binding energy rises, actuality approximately equivalent to the square of it. In addition, the chemical atmosphere impacts the binding energy, leading to tiny alterations in the location of the XPS signals in the spectrum, referred to as chemical shifts. For instance, Hosny et al. reported the synthesis of platinum NPs via a green synthesis method in which *Polygonum salicifolium* leaves were used to prepare a green extract. Chemical composition was examined with X-ray photoelectron spectroscopy, which revealed the presence of $Pt^0 4f_{7/2}$ and $Pt^0 4f_{5/2}$ orbitals at binding energies of 72.98 and 75.38 eV [30]. Similarly, XPS analysis, peaks at 933.7eV and 953.5eV, was allocated to binding energy of $Cu2p_{3/2}$ and $Cu2p_{1/2}$ [31].

As a consequence, the recognition of the elements and their oxidation state may be deduced from the binding energy spectrum, whereas numerical elemental characteristics is gained by quantifying the peak altitudes. Despite this, the XPS study of a nonconductive substance necessitates specific attention since these materials typically establish undesirable changes in the energy spectrum; XPS is a valuable instrument for the examination of surfaces and provides details which are not possible to obtain by strategies including SEM-EDX.

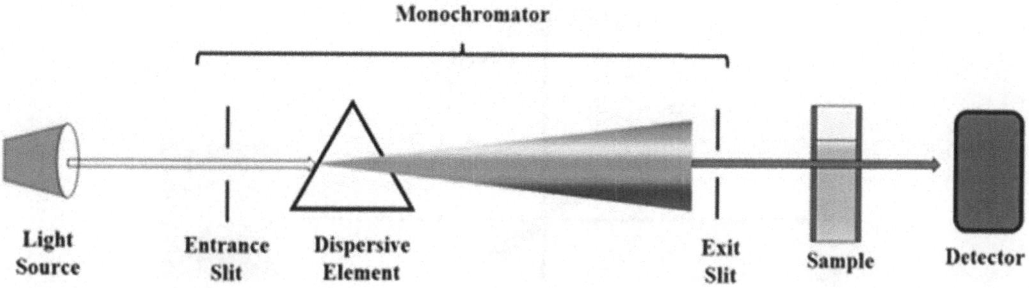

FIGURE 2.7 Representation of instrumentation of UV-visible spectrophotometer.

2.2.3 Absorption-Emission-Based Spectroscopic Characterization Techniques

2.2.3.1 UV-Visible Spectroscopy

UV-visible spectrophotometer analyzes the absorbance of a sample with respect to wavelength of incident light energy. A very low-concentration liquid sample was put into 1 cm wide quartz cuvettes. The fundamental working principle of a spectrophotometer is the difference in the intensity of light that strikes the material and that is transmitted through it. As shown in Figure 2.7, some light may be captured while the rest of the light is transmitted through a solution when a light beam passes through it. *Transmittance* is the measure of incident-light-to-transmitted-light ratio. The maximum wavelength of the sample was calculated using a plot of wavelength vs. absorption, and this wavelength was then used to derive bandgap energy values and potential energy values of nanomaterials.

Baruah et al. synthesized biogenic ZnO NPs by using *Livistona jekinsiana*, which was reported for the first time, and examined the chemical properties via UV-visible spectroscopy. The highest intensity toward absorption was observed at 332 nm. From Tauc plot calculations, bandgap energy value was found to be 3.73 eV. ZnO has been employed for photocatalytic and antibacterial activity based on these optical characteristics [32].

Mani et al. prepared silver NPs by using butter fruit extract, which includes various vitamins, minerals, and phytochemicals, and it showed excellent stabilizing results. Synthesized NPs displayed the size of particles in the range of 20–50 nm, and surface plasmon resonance was detected through UV-visible analysis, which shows highly intense peak at 420 nm. These results confirmed the successful formation of AgNPs [33]. Similarly, Rjakumar et al. also described the fabrication of silver NPs employing extract of *Persea americana* seed. During the addition of green extract into the salt solution, the yellow-colored solution turned reddish-brown; UV-visible spectra revealed the formation of SPR band at 430 nm due to the presence of free electron in plasmon resonance, and this property of metals distinguishes them from other metal NPs [34].

2.2.3.2 Fourier Transform Infrared Spectroscopy

FTIR investigation is employed to discover various functional groups at particular wave number and pattern of absorption bands. A light lamp, specimen cell, detector, amplifier, and processor are the main components of an FTIR spectrometer, as shown in Figure 2.8 When transmitted light reaches the detector, the amplifier enhances the signal, transforms it to a digital signal, and then passes it to the computer. The sample converts the absorbed energy of the radiation into vibrational energy. The detector's final signal has a spectrum that normally occupies a range from 4,000 to 400 cm^{-1} and signifies the molecular signature of the samples.

Changes in the distinctive pattern of absorption bands indicate the composition of a given material and unknown substances, locate contaminants, detect breakdown and oxidizing behavior, and detect additives. In FTIR, a sample transfers the energy from the incoming radiation into vibrational or rotational energy, and the output signal is represented in the form of a spectrum (4,000

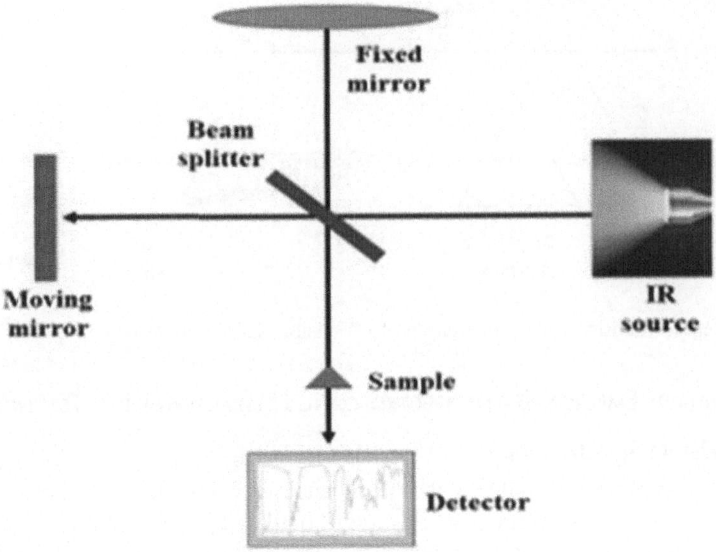

FIGURE 2.8 Representation of instrumentation of FTIR.

to 400 cm^{-1}) that signifies the molecular fingerprint of the samples. FTIR is a powerful method for chemical recognition because each molecule has a unique fingerprint.

Bano et al. has fabricated CuO/SnO_2 heterojunction material employing papaya fruit extract. Characterization of NPs was performed with UV, FTIR, FESEM, and HRTEM analysis, which ensures synthesis of an interface between two metal oxides. FTIR analysis confirmed the presence of Cu-O and Sn-O bonds as vibrations bands appeared at 677 cm^{-1} and 660 cm^{-1}, respectively, as shown in Figure 2.9. Upon mixing of CuO and SnO_2, new electrostatic interactions were generated and, corresponding to these new peaks, appeared at 1,653, 2,351, and 2,859 cm^{-1} [35].

2.2.3.3 Photoluminescence Spectroscopy (PL)

The process through which light energy promotes the emission of a photon from any substance is known as photoluminescence spectroscopy. It is a way of material testing that is non-contact and non-destructive. Adinarayana and his coworkers studied sensing of Fe(III) ions with the help of green synthesized NPs via photoluminescence technique. For this experiment, solutions with different concentrations of Fe ions were prepared, and Si NPs were added in each solution and PL was recorded. Being a fluorescence quencher, Si NPs showed a decline in intensity upon an increment of Fe ions. In this way, we found out Si NPs to be best fluorescents [36]. Similarly, Achadu et al. synthesized carbon dots to analyze Fe(III) ions by using fluorescence analysis. Here, wasted agar was used to synthesize carbon NPs. In another experiment, approximately 0.15 M of Fe^{3+} was examined in spiked human serum and water; thus, the FL nanosensor based on ag-oxCDs was used to demonstrate the practical applicability of the technology for statistical monitoring of Fe(III) [37].

2.2.4 Adsorption-Based Characterization Method

BET study is a physical characterization method which gives quantitative information about solid material surface area and porosity distribution. In this method, nitrogen or any other inert gas has been adsorbed and desorbed on the surface of a solid material under high-temperature pressure conditions. On the basis of BET theory, surface area of sample has been calculated at different points by adsorbing nitrogen gas molecules. Here are the few studies on BET, including Guo et al., who concluded that biogenic-synthesized iron silicide NPs showed amazing activity toward cancer

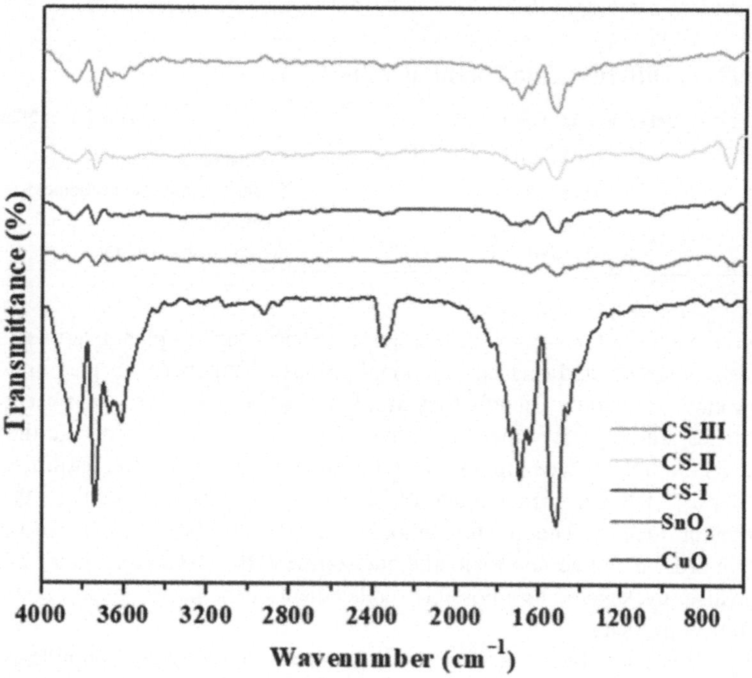

FIGURE 2.9 FTIR spectra of biogenic heterojunctions.

Source: Reprinted with permission from [35]. Copyright Elsevier, 2023.

therapy. From BET analysis, it was found that FeSi exhibited type IV BET isotherm, which suggested the formation of mesoporous structure. FeSi NPs exhibited surface area of 90 m^2/g and the pore volume of 0.21 cm^3/g [38]. Similarly, Lingamdinne et al. reported biogenic fabrication of ferrous oxide NPs utilized for the adsorption of toxic metal ions. When they compared the surface area of NPs by using BET analysis, results revealed greater surface area of biogenic Fe NPs (122.54m^2/g) as compared to other Fe NPs which were synthesized without green extract [39].

2.2.5 Nanoparticle Surface Charge Analysis

One of the key variables that affects how a nanomaterial behaves is typically thought to be the surface charge of it. The dispersion durability and agglomeration phase of the NPs are impacted by the surface charge. Nanoparticles that have a small positive or negative surface charge frequently agglomerate and precipitate, whereas particles with large surface charges have a tendency to resist one another and generate an enduring colloidal dispersion. A colloidal solution's surface charge of nanoparticles is determined via zeta potential assessments [18]. The following table shows the behavior of NPs in different zeta potential values.

Finally, zeta potential is a characteristic which is influenced by both nanoparticle itself and conditions like pH and ionic strength of the solution. Hence, a precise and repeatable zeta potential evaluation in dilute suspension can't replicate the accurate value in a concentrated suspension.

2.3 CONCLUSION

In comparison to conventional physically and chemically synthesized nanoparticles, the green synthesis approach decreases environmental toxicity by eliminating hazardous by-products and

TABLE 2.1
Behavior of NPs in Different Zeta Potential Values

Sr No.	Zeta Potential Range (mV)	Behavior of NPs
1	−100 to +100	Normal range
2	−30 to +30	Strongly anionic and cationic, with greater stability
3	−25 to +25	Agglomerates due to interparticle interactions
4	−10 to +10	Approximately neutral

also reducing the potential toxicity to various living species, such as plants, animals, and humans. Factors such as size, shape, and surface functional groups of nanoparticles play important roles in the fact of how environmentally friendly they are. It should be highlighted that because biosynthesized nanoparticles are biogenic rather than physically or chemically produced, the biomolecules and biochemicals on their surface are less harmful and less abundant. In contrast, biosynthesized nanoparticles have functional groups made of biochemicals and biomolecules, which ultimately lower their potential toxicity. The green technology has recently been introduced into the biosynthetic method in order to create non-toxic and environmentally sustainable nanoparticles. Further, solar energy is also investigated as a possible replacement of the heating step in the production of biosynthesized nanoparticles.

Additionally, efforts have been made to isolate the precise biomolecule, peptide, or plant chemicals that are the main reason behind the nanoparticle formation, using a variety of characterization methods. These particular biomolecules were recently utilized as stabilizing or reducing agents after their isolation to boost the efficiency of nanoparticle formation and increase their degree of stability via some additional processes, like pH change or the introduction of bio-templates. For the development of particular phytochemicals that can be harvested for the mass production of nanoparticles, strategies such as genetic engineering and plant tissue culture to create hybrid plants will be helpful. Additionally, these laboratory-based green biosynthesis techniques will aid in resolving problems with biological exploration and biological piracy without harming the natural ecosystem.

The function of various approaches for the characterization of materials having nanoscale particle size was emphasized in this chapter. Through our deep analysis of nanoparticle characterization methods, we discussed the advantages and disadvantages of each technique in addition to a demonstration of to how to utilize these techniques positively in combination or as a complement to each other. It is common that we need to utilize multiple techniques to thoroughly provide the complete idea about various features associated with a nanomaterial.

Our evaluation will serve as a solid reference, assisting researchers to comprehend the discussed topic better by showing the role of each technique in a comparable manner. Researchers will thus be assisted in selecting the most appropriate methodologies for their characterization and in being able to evaluate their use more precisely.

In reality, the scientific community faces difficulties in further enhancing the precision and clarity of several procedures. Finally, we believe that thorough reading of this evaluation will enable readers to choose if valuable techniques require additional technical advancements.

REFERENCES

1. Mustafa, F.; Andreescu, S. Nanotechnology-based approaches for food sensing and packaging applications, *RSC Adv.* **2020**, 10, 19309–19336.
2. Liu, Y.; Shi, L.; Su, L. et al. Nanotechnology-based antimicrobials and delivery systems for biofilm-infection control, *Chem Soc Rev.* **2019**, 48, 428–446.
3. Kargozar, S.; Baino, F.; Hamzehlou, S.; Hamblin, M. R.; Mozafari, M. Nanotechnology for angiogenesis: Opportunities and challenges, *Chem Soc Rev.* **2020**, 49, 5008–5057.

4. Baig, N.; Kammakakam, I.; Falath, W. Nanomaterials: A review of synthesis methods, properties, recent progress, and challenges, *Mater Adv.* **2021**, 2, 1821–1871.
5. Yang, M.; Zhu, H.; Zheng, Y. et al. One-step chemical vapor deposition fabrication of Ni@NiO@graphite nanoparticles for the oxygen evolution reaction of water splitting, *RSC Adv.* **2022**, 12, 10496–10503.
6. Zhang, M.; Caldwell, T.; Hector, A. L.; Garcia-Araez, N.; Falvey, J. Solvothermal synthesis of nanoscale $BaTiO_3$ in benzyl alcohol–water mixtures and effects of manganese oxide coating to enhance the PTCR effect, *Dalton Trans.* **2023**, 52, 297–307.
7. Parashar, M.; Shukla, V. K.; Singh, R. Metal oxides nanoparticles via sol–gel method: A review on synthesis, characterization and applications, *J Mater Sci: Mater Electron.* **2020**, 31, 3729–3749.
8. Hasan, M.; Yang, W.; Ju, Y. et al. Biocompatibility of iron carbide and detection of metals ions signalling proteomic analysis via HPLC/ESI-Orbitrap, *Nano Res.* **2017**, 10, 1912–1923.
9. Singh, Y.; Kaushal, S.; Sodhi, R. S. Biogenic synthesis of silver nanoparticles using cyanobacterium Leptolyngbya sp. WUC 59 cell-free extract and their effects on bacterial growth and seed germination, *Nanoscale Adv.* **2020**, 2, 3972.
10. Siddiqi, K. S.; Husen, A.; Rao, R. A. K. A review on biosynthesis of silver nanoparticles and their biocidal properties. *J Nanobiotechnol.* **2018**, 16.
11. Shreyash, N.; Bajpai, S.; Khan, M. A.; Vijay, Y.; Tiwary, S. K.; Sonker, M. Green synthesis of nanoparticles and their biomedical applications: A review, *ACS Appl Nano Mater.* **2021**, 4, 11428–11457.
12. Kaushal, S.; Kumar, A.; Bains, H.; Singh, P. Photocatalytic degradation of tetracycline antibiotic and organic dyes using biogenic synthesized CuO/Fe_2O_3 nanocomposite: Pathways and mechanism insights, *Environ Sci Pollut Res.* **2023**, 30, 37092–37104.
13. Mittal, A. K.; Chisti, Y.; Banerjee, U. C. Synthesis of metallic nanoparticles using plant extracts, *Biotechnol Adv.* **2013**, 31, 346–356.
14. Aswathi, V. P.; Meera, S.; Maria, C. G. A.; Nidhin, M. Green synthesis of nanoparticles from biodegradable waste extracts and their applications: A critical review, *Nanotechnol Environ Eng.* **2022**, 8, 377–397.
15. Doyen, M.; Bartik, K.; Bruylants, G. UV–Vis and NMR study of the formation of gold nanoparticles by citrate reduction: Observation of gold–citrate aggregates, *J Colloid Interface Sci.* **2013**, 399, 1–5.
16. Rogero, C. B.; Brimaud, S.; Gullon, J. S. et al. Ethanol oxidation on shape-controlled platinum nanoparticles at different pHs: A combined in situ IR spectroscopy and online mass spectrometry study, *J Electroanal Chem.* **2016**, 763, 116–124.
17. Upadhyay, S.; Parekh, K.; Pandey, B. Influence of crystallite size on the magnetic properties of Fe_3O_4 nanoparticles, *J Alloys Compd.* **2016**, 678, 478–485.
18. Mourdikoudis, S.; Pallares. R. M.; Thanh, N. T. K. Characterization techniques for nanoparticles: Comparison and complementarity upon studying nanoparticle properties, *Nanoscale* **2018**, 10, 12871–12934.
19. Sun, Z.; Shi, J.; Wang J.; Jiang M.; Wang, Z.; Bai, X.; Wang, X. A deep learning-based framework for automatic analysis of the nanoparticle morphology in SEM/TEM images Cells, *Nanoscale* **2022**, 14, 10761–10772.
20. Shanks, R. A. Chapter 2 – Characterization of nanostructured materials, *Nano Structured Polymer Blends* **2014**, 15–31.
21. Barhoum, A.; Garcia-Betancourt, M. L. Physicochemical characterization of nanomaterials: Size, morphology, optical, magnetic, and electrical properties (Chapter 10). In *Emerging Applications of Nanoparticles and Architecture Nanostructures.* Elsevier, Amsterdam, The Netherlands, **2018**, 279–304.
22. Lin, P. C.; Lin, S.; Wang, P. C.; Sridhar, R. Techniques for physicochemical characterization of nanomaterials, *Biotechnol Adv.* **2014**, 32, 711–726.
23. Kumari, V.; Kaushal, S.; Singh, P. Green synthesis of a CuO/rGO nanocomposite using a Terminalia arjuna bark extract and its catalytic activity for the purification of water, *Mater Adv.* **2022**, 3, 2170–2184.
24. Lamas, D. G.; Oliveira, N. M.; Kellermann, G.; Craievich, A. F. X-ray diffraction and scattering by nanomaterials (Chapter 5). In *Nanocharacterization Techniques.* William Andrew, Norwich, England, **2017**, 111–182.
25. Alshehri, A.; Malik, M. A.; Khan, Z.; Al-Thabaiti, S. A.; Hasan, N. Biofabrication of Fe nanoparticles in aqueous extract of Hibiscus sabdariffa with enhanced photocatalytic activitie, *RSC Adv.* **2017**, 7, 25149–25159.
26. Singh, Y.; Sodhi, R. S.; Singh, P. P.; Kaushal, S. Biosynthesis of NiO nanoparticles using Spirogyra sp. cell-free extract and their potential biological applications, *Mater Adv.* **2022**, 3, 4991–5000.

27. Titus, D.; Samuel, E. J. J.; Roopan, S. M. Nanoparticle characterization techniques (Chapter 12). In *Green Synthesis, Characterization and Applications of Nanoparticles*. Elsevier, Amsterdam, The Netherlands, **2019**, 303–319.
28. Ohara, T. A. S.; Ida, T.; Yashima, M.; Mikayama, T.; Abe, H.; Setsuhara, Y.; Nogi, K.; Miyahara, M.; Kaneko, K.; Ohtomo, A. Characterization methods for nanostructure of materials (Chapter 5). In *Nanoparticle Technology Handbook*. Elsevier, Amsterdam, The Netherlands, **2018**, 255–300.
29. Kumar, A.; Dixit, C. K. Methods for characterization of nanoparticles. In *Advances in Nanomedicine for the Delivery of Therapeutic Nucleic Acids*. Woodhead Publishing, Darya Ganj, India, **2017**, 43–58.
30. Hosny, M.; Fawzy, M.; El-Fakharany, E. M. et al. Biogenic synthesis, characterization, antimicrobial, antioxidant, antidiabetic, and catalytic applications of platinum nanoparticles synthesized from Polygonum salicifolium leaves, *J Envion Chem Engg*. **2022**, 10, 106806.
31. Shaheen, T. I.; Fouda, A.; Salem, S. S. Integration of cotton fabrics with biosynthesized CuO nanoparticles for bactericidal activity in the terms of their cytotoxicity assessment, *Ind Eng Chem Res*. **2021**, 60, 1553–1563.
32. Bekru, A. G.; Tufa, L. T.; Zelekew, O. A.; Goddati, M.; Lee, J.; Sabir, F. K. Green synthesis of a CuO–ZnO nanocomposite for efficient photodegradation of methylene blue and reduction of 4-nitrophenol, *ACS Omega*. **2022**, 7, 30908–30919.
33. Baruah, R.; Yadav, A.; Das, A. M.; Jekinsiana, L. Fabricated ZnO nanoparticles and their detrimental effect towards anthropogenic organic pollutants and human pathogenic bacteria, *Spectrochim Acta A: Mol Biomol Spectrosc*. **2021**, 251, 119459.
34. Mani, R.; Vijayakumar, P.; Dhas, T. S. et al. Synthesis of biogenic silver nanoparticles using butter fruit pulp extract and evaluation of their antibacterial activity against Providencia vermicola in Rohu, *J King Saud Univ Sci*. **2022**, 34, 101814.
35. Bano, K.; Kaushal, S.; Kumar, A.; Singh, P. Sunlight-driven photocatalytic degradation of 4-nitrophenol and adsorptive removal of Mn (II) ions from industrial wastewater by biogenic synthesized CuO/SnO_2 heterojunction, *Mater Today Chem*. **2022**, 26, 101193.
36. Adinarayana, T. V. S.; Mishra, A.; Singhal, I.; Reddy, D. V. R. Facile green synthesis of silicon nanoparticles from equisetum arvense for fluorescence based detection of Fe(III) ions, *Nanoscale Adv*. **2020**, 2, 4125–4132.
37. Achadu, O. J.; Elizur, G. L.; Boye, T. G. E.; Park, E. Y. Green synthesis of carbon dots using expired agar for a label-free fluorescence signal-amplified detection of ferric ion utilizing oxalate functionalization, *Mater Adv*. **2022**, 3, 6307–6315.
38. Xu, Z.; Guo, X.; Wang, S. et al. Green synthesis of mesoporous and biodegradable iron silicide nanoparticles for photothermal cancer therapy, *J Mater Chem B*. **2023**, 11, 4065–4075.
39. Lingamdinne, L. P.; Chang, Y. Y.; Yang, J. K.; Singh, J.; Choi, E. H.; Shiratani, M.; Koduru, J. R.; Attri, P. Biogenic reductive preparation of magnetic inverse spinel iron oxide nanoparticles for the adsorption removal of heavy metals, *J Chem Eng*. **2017**, 307, 74–84.

3 Applications of Biogenic Nanoparticles in Health

Ragaa A. Hamouda, Badr E. El Bialy, and Muhammad A. Abuelmagd

3.1 INTRODUCTION

Phytomedicines are widely applied by a greater number of people who need remedies for health protection and diseases treatment with low or even no adverse effects. Recently, people from both low- and high-income countries have paid progressed attention toward eco-friendly and effective biologically based products for treatment and prevention of infectious diseases [1].

Nano-biogenic technology has an ability to combat various human diseases and deals with the application of nanomaterials in targeting cell- or tissue-specific clinical intervention. The engineered nanoparticles can react at the molecular level, utilizing synthesized nano-devices with high specificity, achieving maximum and the most desirable therapeutic potency with insignificant adverse effects. The toxic nature of chemicals has restricted their usage in the biomedical sectors of chemically fabricated nanoparticles [2].

The building units of the living organisms are cells, which have a dimension of about 10 μm; the cellular components are also smaller and have the size of less than a micron, such as proteins, which have a typical size of about 5 nm (comparable size to smallest synthesized nanoparticles) [3]. This comparable nanoscaled size offers an ability to utilize biogenic nanoparticles as nanoprobes that help us to diagnose, treat, and prevent diseases at the cellular and molecular level without the excessive need of much interference [3]. The NPs' rational size in the nanometer scale makes it easy to deliver into human body cells and efficient for specific natures of cell-targeting therapy, for instance, drug delivery, and the competent detection of diseases [4]. Nanoscaled structures can be fabricated using different biological, physical, and chemical methods; however, the development of biogenic or green synthesized nanoparticles (bio-based nanoparticles) has gained significant worldwide attention in medical science and the health sector [5]. Although chemical and physical techniques in nanotechnology produce nanoparticles with definite morphology and unique properties, these techniques have the problems of being expensive and potentially unsafe to the environment and human health [1]. Biofabricated NPs have been utilized in a wide range of medical implementations that may be in the diagnosis or treatment fields, as they can be utilized as a strategic system for delivering conventional drugs, cancer treatment, antibacterial agents, vaccines, gene therapy, recombinant proteins, and DNA analysis, enhancing the reaction rates, magnetic resonance imaging (MRI), and biosensors [6]. Recent techniques have been focused on developing new green/natural chemistry routes of metal nanoparticles fabrication with the advantage of using biological products, such as plant, bacterial, algal, or even fungal extract, and avoiding toxic chemical reducing agents. The biogenic fabrication of nanoparticles is trustworthy and ecologically friendly, in which toxic chemicals are avoided in the synthesis process [7]. Biologically fabricated nanoparticles have also received a certain consideration, as the biological components contributed in the fabrication of nanoparticles act in dual functional tasks, as ligands for reducing metal salts as silver nitrate to produce silver nanoparticles and capping of generated nanoparticles, making these nanoparticles more fit for medical application [8].

DOI: 10.1201/9781003430087-3

3.2 BIOMEDICAL APPLICATIONS

3.2.1 Antifungal Activities

Several studies have reported the use of biogenic nanoparticles in antifungal activities. Biosynthesized silver nanoparticles by the filtrate of *Trichoderma viride* (MTCC 5661) had antifungal activities against pathogenic fungi *Fusarium oxysporum* and *Alternaria brassicicola* [9]. Biosynthesized silver, zinc oxide, and copper nanoparticles exhibit antifungal activities against pathogenic fungi, such as *C. neoformans* and *C. albicans* species [10]. The biogenic Ag-NPs, by callus extract, displayed antifungal activity against *Candida albicans*, *Candida nonalbicans*, and *Candida tropicalis* [11]. Biosynthesized silver nanoparticles by *Malva parviflora* effectively diminish the mycelial growing of *Helminthosporium rostratum*, *Alternaria alternata*, *Fusarium solani*, and *Fusarium oxysporum* [12]. Silver nanoparticles synthesized by *Aspergillus niger* had antifungal activities against some pathogenic fungi [13]. Biofabricated silver nanoparticles by *Allium ampeloprasum* leaves extract showed antifungal activities against *Aspergillus Flavus*, *A. parasiticus*, *A. niger*, and *A. ochraceus* [14]. Biogenic ZnONPs by *Allium sativum* showed antifungal effect against *Cercospora* species [15].

Green synthesized selenium nanoparticles using *Melia azedarach* leaves extract have antifungal activity *in vitro* against *Fusarium mangiferae* that affect mango plants [16]. Nanocellulose, derived from cellulose (glucose polymer) from plant cell, had low antimicrobial activities but, when conjugated with allicin, had good antifungal effects against *Candida albicans*, *Aspergillus niger*, as well as antibacterial effects against *Escherichia coli* and *Staphylococcus aureus* [17–19]. Synthesized gold nanoparticles using culture filtrate of lactic acid bacteria exhibited an efficient antifungal activity against *Aspergillus flavus* and *Aspergillus ochraceus* [20]. Also, bacterial extracellular polysaccharides have been used in bio-fabrication and stabilizing of metal NPs to be used in different medical purposes, such as antimicrobials, antioxidants, and hybrid drug delivery systems [21]. Moreover, many studies have proven the antifungal effects of the biosynthesized nanoparticles against various pathogenic fungi, as presented in Table 3.1.

3.2.2 Antibacterial Activities

The long-term use and misuse of antibiotics result in the development of the multidrug-resistant pathogenic bacterial strains and reduce the efficiency of available antibiotics [33]. In the last years, there has been an extraordinary need to synthesize agents that have the ability to overcome microbial resistance; hence, the biogenic synthesis of nanoparticles became an alternative strategy to overcome this limitation [34]. Biogenic palladium nanoparticles (Pd-NPs) derived from plant extracts have been reported to be a valuable nanocatalyst in different catalytic reactions due to their cost-effectiveness, reduced noxious chemicals, excellent activities, and selectivity [35]. Green synthesized Pd-NPs reveal a prominent antimicrobial activity as a result of their infinitesimal particle size that gives synthesized particles a large surface area that results in a strong lethal activity due to more effective interaction with cell membranes of the pathogenic bacteria [35]. *Moringa oleifera* peel–capped Pd-NPs has been reported to have good antibacterial effect against pathogenic bacteria, such as *Staphylococcus aureus* [36]. Biologically synthesized AgNPs using leaf extract of *Datura stramonium* proved considerable antibacterial activity, DNA-cleaving action, and had antioxidant activities [37]. Biogenic iron nanoparticles by different plants were shown to be toxic to many bacterial strains [38]. The biogenic AuNPs using *Cladosporium cladosporioides* showed antipathogenic microorganisms [39]. Biosynthesized copper (Cu) NPs with *Hagenia abyssinica* had been reported as antibacterial agent [40]. Ag-NPs and ZnO-NPs synthesis by marine alga *Ulva fasciata* has potential synergistic effect with important antibiotics against some pathogenic bacteria [41]. *Libidibia ferrea* leaves loaded in bacterial nanocellulose (BNC-Lf) displayed antibacterial potentiality against *P.*

TABLE 3.1
Antifungal Activities of Biogenic Nanoparticles

Nanoparticle	Reducing Agents	Size (nm)	Fungi	References
Silver (Ag)	*Caesalpinia ferrea* seeds extracts (plant)	30–50	*Candida kruzei*, *C. albicans*, *C. guilliermondii*, and *C. glabrata*	[22]
	Aspergillus sydowii filtrates (fungi)	12±2	*Aspergillus* spp., *Fusarium* spp., *Candida* spp., *Cryptococcus neoformans*, and *Sporothrix schenckii*	[23]
	Penicillium chrysogenum and *Aspergillus oryzae*	19–60	*Trichophyton rubrum*	[24]
	Alcohol extract of tulsi leaves	2–7	Candida	[25]
	Nocardiopsis spp. MBRC-1	30–90	*A. brasiliensis*,	[26]
	Alternaria alternate	15–30	*Aspergillus niger* and *candida albicans*	[27]
	Sinapis arvensis	1–35	*Neofusicoccum parvum*	[28]
	Geranium leaves	16–40	*Fusarium oxysporum*	[29]
Iron oxide (FeO)	*Platanus orientalis*	78–80	*Aspergillus niger* and *Mucor piriformis*	[30]
	Borassus flabellifer seed coat	35	*A. niger* and *C. albicans*	[31]
Zinc oxide (ZnO)	*Camellia Sinensis*	30–40	*Aspergillus flavus*, *A. fumigatus*, and *Penicillium* sp.	[32]

aeruginosa, *S. aureus*, and *E. coli* [42]. Nanocellulose–chitosan biocomposites with a ratio of 10:2 displayed antibacterial action against *E. coli* [43]. Chitosan/bacterial nanocellulose/ZnO showed antibacterial activity [44]. Cellulose in nanosize has low antibacterial and antifungal effect but, when blended to melanin, has manifested high antimicrobial effects versus *C. albicans*, *A. niger*, *S. aureus*, and *E. coli* [45]. Alcohol-free hand sanitizer prepared with nanocellulose provides long permanent antibacterial effect without skin dehydration [46]. Figure 3.1 illustrates the mechanisms of antibacterial potentiality of some metal nanoparticles. Table 3.2 demonstrates some applications of biogenic nanoparticles as antibacterial activities.

3.2.3 Antiviral Activities

Nanoparticles synthesized by different methods possess antiviral activities, but the adequate mechanisms through which nanoparticles can destroy viruses remain uncertain, so many studies suggest various mechanisms. The suggested mechanisms are: obstructing the infection by either the virus inside the cells or by directly disabling the viruses [55]. This is achieved via attaching to the proteins in viral coat and obstructing structural interactions due to the large surface area of nanoparticles [56]. Moreover, nanoparticles can be attached with the viral surface, leading to damage of the viral genomic, and also act as inhibitors of the nucleocapsids [57]. The green synthesized Ag-NPs by *Malephora lute* and *Lampranthus coccineus a* were able to control HSV-1 virus, HAV-10 virus, and CoxB4 virus infectivity [58]. Aqueous extract of *Syzygium aromaticum* used for green synthesis of silver nanoparticles (Ag-NPs) is an effective tool against the Newcastle virus [59]. The biosynthesized Ag-NPs using *Phyllanthus niruri*, *Tinospora cordifolia*, and *Andrographis paniculata* had antiviral activity against Chikungunya virus [60]. Figure 3.2 shows a comparison between the treated and nontreated virus with nanoparticles.

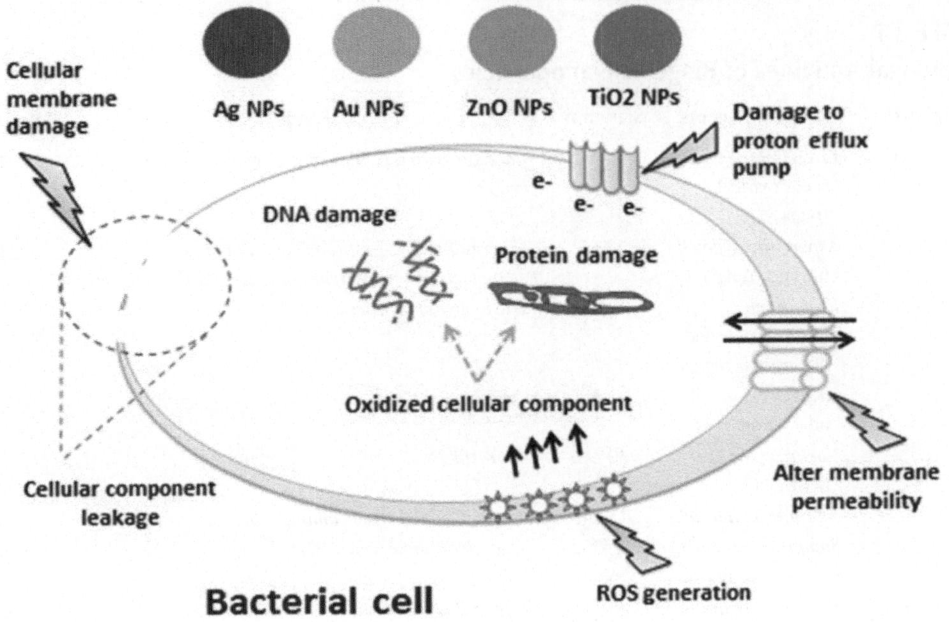

FIGURE 3.1 Schematic figure illustrating the antibacterial potentiality of some metal nanoparticles.

TABLE 3.2
Antibacterial Activities of Biogenic Nanoparticles

Nanoparticle	Reducing Agents	Size (nm)	Bacteria	References
Silver (Ag)	*Nocardiopsis* sp. MBRC-1 filtrates	30–90	*Bacillus subtilis*	[25]
	Alternaria alternata	15–30	*Escherichia coli* and *Staphylococcus aureus*	[26]
Ag	*Beauveria bassiana*	10–50	*E. coli* and *S. aureus*	[47]
Iron oxide (FeO)	*Borassus flabellifer* seed coat	35	*E. coli* and *Shigella*	[30]
Zinc oxide (ZnO)	*Pseudomonas putida*	25–45	*Acinetobacter baumannii* (MCC 2366), *Pseudomonas oleovorans* (MCC 2566), *Pseudomonas otitidis* (MCC 2509), *Enterococcus faecalis* (MCC 2041), and *Bacillus cereus* (MCC 2039)	[48]
	Coccinia abyssinica	10.4	*Staphylococcus aureus*, *Bacillus coagulans*, and *Sphingomonas paucimobilis*, *S. typhimurium*, and *Shigella dysenteriae*	[49]
	Aeromonas hydrophila	50–70	*Pseudomonas aeruginosa* and *Aspergillus flavus*	[50]
Platinum (Pt)	*Padina gymnospora*		*Escherichia coli*, *Lactococcus lactis*, *Klebsiella pneumoniae*	[51]
Copper oxide (CuO)	*Phyllanthus amarus*	20	*S. aureus* and *E. coli*	[52]
Cadmium sulfide (CdS)	*Escherichia coli*	2–5	*E. coli* strain BW25113	[53]
Titanium oxide (TiO_2)	*Aspergillus flavus*	60–74	*S. aureus*	[54]

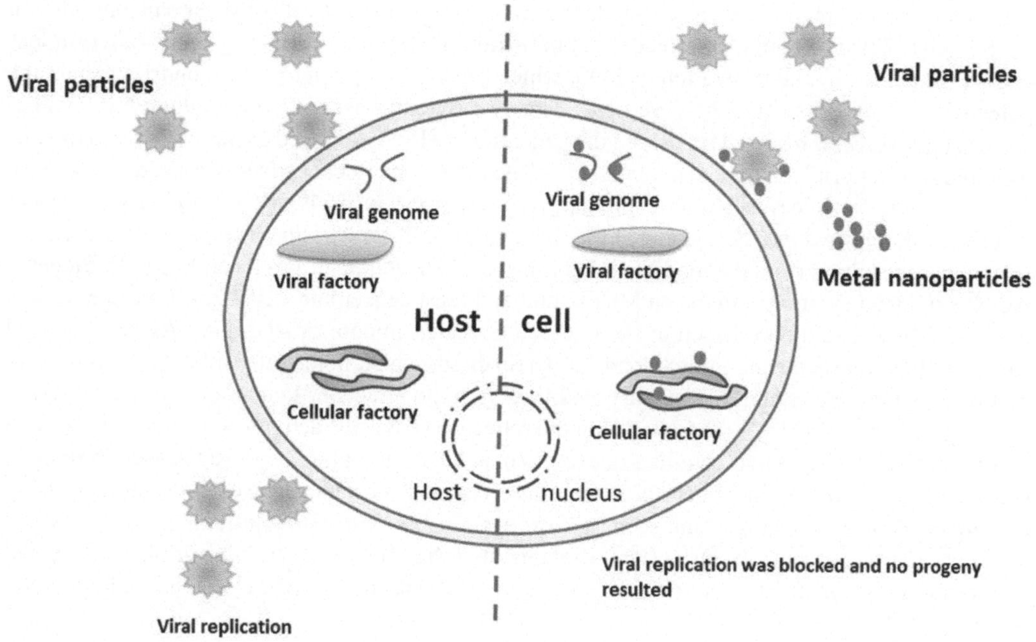

FIGURE 3.2 Schematic figure illustrating the antiviral potentiality of metal nanoparticles.

3.2.4 Biogenic Nanoparticles in Cancer Therapy

Cancer is a terrible worldwide disease that causes major health risks and high death percentage among diseased persons [61]. The existing technologies for the treatment of cancer include surgery, chemotherapy, and radiation, which have enormous side effects, such as bone marrow and immune system regressions and multiorgan affections, such as cardiomyopathy, hepatorenal syndrome, neurotoxicity, and gastrointestinal tract damage. However, an advanced stage of cancer is not sensitive to conventional therapies [62]. Chemotherapy uses chemicals to stop or kill the progression of cancer cells [63]. The chemotherapeutic agents used in chemotherapy attach to uncontrolled, fast-growing cancer cells and also attack fast-growing healthy cells. This undesired attack and developing of multidrug resistance result in the non-effectiveness of traditional chemotherapy [64]. The restrictions of traditional chemotherapeutic agent have caused the progress toward alternative smart nanocarrier-based drug delivery systems that have the capability to deliver drugs to specific targeted sites [65].

Several biosynthesized metals or metal oxide nanoparticles have antimicrobial and anticancer activities via their cytotoxic effects [66]. The main strategies for the anticancer potentiality of biologically synthesized nanoparticles are mediated by the cytotoxic effects of these nanoparticles. Nanoparticles can generate reactive oxygen species (ROS) or act themselves as ROS, leading to oxidants/antioxidants imbalance or oxidative stress. The consequences of oxidative stress are lipids, proteins, and DNA oxidative damage and activation of the apoptotic or necrotic pathways of the cancer cells. The interference with cell chemistry functions, peroxidation of cellular and mitochondrial membranes, causes cell permeability disturbance and mitochondrial dysfunction [67–68]. Moreover, the activation of p38 mitogen-stimulated protein kinase (p38 MAPK) and caspase-3 at gene expression produces an outcome of apoptotic reaction to nanoparticles [69]. The toxic effects of nanoparticles on cells depend on their shape, size, and nature of coating agent, as well as their concentration and composition, as concluded from the study of the effects of silver nanoparticles on cancer cells [70]. The apoptotic effect of copper oxide NPs synthesized using *Ficus religiosa* leaf extract is induced by the generation of ROS, which causes disruption of mitochondrial membrane potential in A549 cells [71]. The dose-dependent cytotoxicity effect of the palladium NPs synthesized using *Evolvulus alsinoides* induced the generation of ROS, followed by autophagy and impairment of mitochondrial membrane potential [72]. The cytotoxic effect of silver NPs synthesized from *Panax ginseng* fresh leaves extract exhibited oxidative stress in A549, MCF7, and HepG2 cancer cell lines. Moreover, silver NPs changed the shape of the cell nucleus and increased the percentage of apoptosis resulting from the modification of the p38 MAPK/p53 in A549 cell line [73]. Biogenic Ag-NPs-induced cytotoxic effects on MCF-7 human breast cancer and RAW 264.7 murine macrophage cell lines could be referred to the release of greater amount of ROS [74]. Ag-NPs induced cancer cell apoptosis through mitochondrial dysfunction, which potentially inhibits the proliferation of cells [75]. He et al. [76] demonstrated the cytotoxic effects of longan peel powder–synthetized AgNPs on H1299 cells correlated with the reticence of NF-κB activity, a decline in bcl-2, and an enhance in caspase-3 and continue in expressions. Khalifa et al. [77] proved reduction in EAC cell viability *in vitro* by increment biogenic synthesized silver nanoparticles concentration using blue-green algae as reducing agent, such as *Nostoc muscorum, Calothrix marchica*, and *Anabaena oryzae*, using trypan blue exclusion test. Biofabricated Ag-NPs using *Turbinaria turbinata* could control the progress of Ehrlich cell solid tumor in mice via their potent cytotoxic and apoptotic actions [78].

Additionally, gold nanoparticles synthesized using *Scutellaria barbata*–treated pancreatic cancer cell lines (PANC-1) caused up-regulation in the expression of Bax, caspase-3, and caspase-9 and down-regulated the expression of Bid and Bcl-2 [79]. Magnetic nanoparticles like maghemite (Fe_2O_3) and magnetite (Fe_3O_4) (magnetite) have several applications in targeting cancer treatment (magnetic hyperthermia), stem cell sorting and manipulation, and magnetic resonance imaging (MRI) [80].

Hsu et al. [81] reported that carbon dots prepared from green tea exhibited inhibition of activity of cancer cell lines, including MCF-7, MDA-MB-231, and HeLa (human cervical carcinoma) cells. They found more reduction of cell viability with increasing the concentration of carbon dots due to the generation of more H_2O_2 and ROS. Biogenic Ag-NPs by leaf extract of *Nathophodytes foetida* (medicinal plant) demonstrated a good cytotoxic effect on human leukemic cells (K562) [82]. ZnONPs biosynthesized by *Albizia lebbeck* stem bark showed antioxidant, antimicrobial, and cytotoxic activity against human breast cancer cell lines [83]. Hamouda et al. [84] reported Au/cellulose nanocomposites mediated by green alga *Chlorella vulgaris* possess antiproliferative effects against cancer cells by increasing the relative expression of p53 gene and decreased that of the Raf-1 gene; the activation of the p53 gene and inhibition of the Raf-1 gene are the most likely mechanisms which they regulate cell cycles. Table 3.3 shows the anticancer activities of biogenic nanoparticles.

3.2.5 Biogenic Nanoparticles in Drug Delivery

Drugs are administered mainly orally or by inhalation, dermal application, or different route of injection to produce a systemic pharmacological effect, where the oral/swallowing is the most common route of drug administration to patients. Then, drugs are absorbed via the mucous membranes of the gastrointestinal tract or respiratory tract or via the skin toward the systemic circulation. However, the poor mucosal absorption and low organ bioavailability of some drugs of high molecular weights, or that of protein and peptide in nature, are problems that result in low therapeutic potency [93]. The use of NPs in cancer treatment as carriers for chemotherapeutic drugs helps in increasing their solubility and half-life and hence increasing their bioavailability [94]. NPs can increase drug accumulation in the cancer tissues by passive targeting via the enhanced permeability and retention mechanism due to leaky endothelium and deficiency of the lymphatic drainage in

TABLE 3.3
Anticancer Activities of Biogenic Nanoparticles

Nanoparticle	Reducing Agents	Size (nm)	Cancer Types	References
Silver (Ag)	*Aspergillus sydowii* filtrates (fungi)	12±2	HeLa cells and MCF-7 cells	[23]
	Nocardiopsis sp. MBRC-1 filtrates	30–90	HeLa cancer cell lines	[26]
	Heliotropiumbacciferum plant	15	MCF-7 and HCT-116	[85]
	Klebsiella oxytoca DSM 29614	–	Human breast	[86]
	Cuminum cyminum L	–	MCF-7	[87]
	Bacillus tequilensis and *Calocybe indica*	20	Human breast cancer cells	[73]
Zinc oxide (ZnO)	*Annona muricata*	80	A549 and MOLT4 cells	[88]
	Gracilaria edulis	4.04 ± 1.81	SiHa cells	[89]
	Pandanus odorifer leaf extract	90	n MCF-7, HepG2, and A-549 cells	[90]
Gold (Au)	*Muntingia calabura*	27	Hep2 cells	[91]
	Brazilian red propolis extract	–	T24 and PC-3 cells	[92]

cancer tissue [95]. Moreover, by active targeting, NPs–anticancer drug combination can differentiate cancer cell from healthy cell and improve the efficiency of the therapy using ligands that can target specific cancer sites [96]. A biosynthesized AuNP used as a drug carrier in therapeutic appliances to develop drug delivery, doxorubicin, released from Bio-AuNPs, can be more efficient anticancer agent than free DOX [97].

Nanoparticles are altered to become hydrophobic, which lengthens the duration that medications are in use and improves their penetration and accumulation in tumors [98]. For decades, organic nanoparticles have been extensively studied and shown to contain a wide variety of components. The first nanoscale medication licensed for clinical use is called a liposome [99]. Several antitumor medications, including doxorubicin and paclitaxel, among other chemotherapeutic medicines, can be delivered *in vivo* using liposomes [100]. Another type of NP is polymer-based, which has unique structural arrangements for drug delivery created by various monomers [101]. Gold nanoparticles (AuNPs), carbon nanotubes (CNTs), quantum dots, magnetic nanoparticles (MNPs), and silica nanoparticles (SNPs) are some of the inorganic NPs that have been examined.

The most extensively researched inorganic NPs are AuNPs, and mixed monolayer-protected clusters with a gold core are thought to be a viable drug delivery system candidate [102]. Surface-functionalized AuNPs have been shown to increase drug accumulation in tumors and to overcome drug resistance, and the gold core is inert and non-toxic [103]. Drug delivery systems made on magnetic NPs are frequently coated with organic compounds like polymers and fatty acids [104]. Lipid–polymer hybrid nanoparticles (NPs), which contain a lipid shell over a polymeric core, have been shown to be an effective drug delivery platform for the treatment of pancreatic cancer [105], breast cancer [106], and metastatic prostate cancer [107]. A liposome–silica hybrid (LSH) nanoparticle was created and demonstrated to be effective at delivering pharmaceuticals to destroy prostate and breast cancer cells. The LSH nanoparticle has a silica core and a lipid bilayer around it [108].

3.2.6 Biogenic Nanoparticles as Antioxidant Activity

Karunakaran et al. [109] reviewed that the green synthetized iron oxide nanoparticles are more effective against gram-negative strains than the gram-positive strains and had high antioxidant activity by chelation of free radicals using 1,1-diphenyl-1,2-picrylhydrazyl (DPPH). The biosynthesized Ag-NPs, by using *Couroupita guianensis* flower extract as a reducing and capping agent, showed significant free radical scavenging activity through DPPH antioxidant assay [110]. AuZnO core–shell nanocomposites, synthesized by *Hibiscus sabdariffa* plant, displayed antibacterial and antibiofilm properties effective against *Staphylococcus aureus* and methicillin-resistant *Staphylococcus haemolyticus*, through the release of reactive oxygen species [1]. PVP/PtNPs nanocomposite synthesized by brown alga *Padina gymnospora* showed an enhanced scavenging activity against DPPH, superoxide, nitric oxide, and hydroxyl radicals [51].

Synthesized nanoparticles using dry leaf extract and its active component have antioxidant-reducing agents that pass into the extract from the plant, the activity of which is determined by the nature of the plant [111]. Indeed, the antioxidant activity of the nanoparticles depends on the chemical composition of the mediated extract with the increment of the nanoparticles concentration, whereas the nanoparticles exhibit high scavenging activity if the biogenic extract is rich in flavonoids and phenolic compounds [112].

3.2.7 Diabetes Mellitus

Diabetes mellitus (DM) is a chronic metabolic disorder characterized by increment in blood glucose level (hyperglycemia), in which the body is not able to synthesize enough insulin from pancreatic β-cells or counteract insulin resistance [113]. Human sedentary lifestyle makes people more susceptible to metabolic disorders, such as obesity, hypertension, and diabetes. Diabetes mellitus affects millions of humans and is expected to affect about 592 million people all over the world by 2035 [114]. Diabetes mellitus is classified into major groups: DM type I, an autoimmune disease, results as a consequence to pancreatic β-cell failure (insulin-producing cells); meanwhile, type II DM results from peripheral resistance to insulin and may be associated with other genetic and physiologic factors (non-insulin-dependent DM) [115]. The treatment of DM type I includes different therapeutic strategies, such as subcutaneous delivery of exogenous insulin in the fatty layer (less-vascularized tissue to reduce insulin absorption rate) [116]. Subcutaneous insulin delivery is often painful and can induce inflammation and allergic reaction at the injection area as well as increased dose of insulin, causing lowering in glucose level in the bloodstream (hypoglycemia) that threats patient life [116]. So alternative routes have been developed to facilitate insulin delivery and to overcome this limitation, as intranasal, oral, and pulmonary could be good alternative routes; however, these routes face some limitations, such as rapid degradation of insulin in gastric juice in oral delivery [117]. In recent years, nanotechnology has emerged as an alternative route for the synthesis of antidiabetic drugs, including insulin administration via oral and nasal tracks, with monitoring the optimum dose and exposure time due to their small size and large-surface-area-to-volume ratio [116, 117]. Nanocarriers loaded with insulin have a mucoadhesive layer formed from polymers such as chitosan that lengthen their existence in the intestine, penetrate the mucous layer, then become unstable in the junctions between epithelial cells and broken as a result of pH; the released insulin seeps to the paracellular pathway and, subsequently, to the bloodstream [118].

Recently, green/natural polymers have been developed for use as nanocarrier for oral insulin delivery, such as alginate [119]. Alginate is a hetero-polymer that is usually extracted from brown algae containing 1,4 linked β-D-mannuronate and α-L-guluronate as the basic components [120]. Alginate is a biodegradable material used in various medical application due to its favorable properties, such as low toxicity and its biocompatibility, as it is dissolved slowly when its cross-linking agents are released and exchange reaction occurs with cations found in the body fluids [121]. Chitosan is a cationic biopolymer with unique intrinsic properties that can be used in nanotechnological application to synthesize chitosan-based materials due to its biodegradability, biocompatibility, antibacterial and inert nature, and high thermal stability [122]. Chitosan has great features making it a perfect candidate to be used as nanomaterial template for drug delivery, such as its cationic nature and the presence of primary amino groups which enhance permeation, mucoadhesion, in situ gelations, as well as its controlled drug release [123]. Li et al. [124] developed an effective system to deliver insulin orally using sodium alginate and chitosan nanoparticles with size about 150 nm; firstly, chitosan

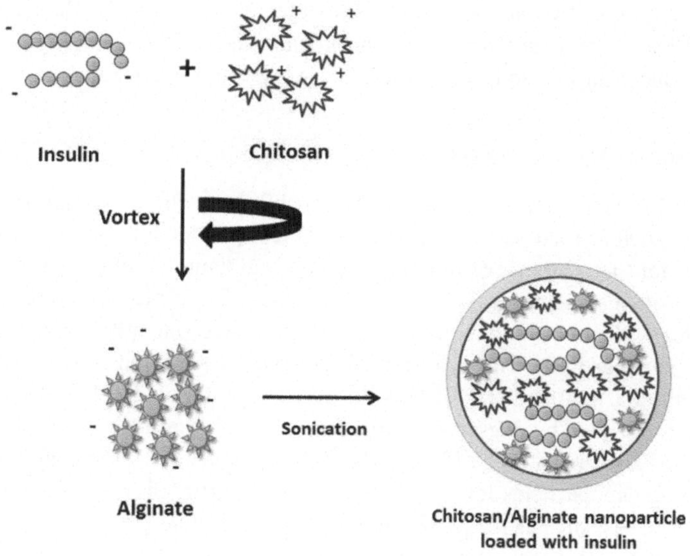

FIGURE 3.3 Schematic figure illustrating the preparation of alginate/chitosan nanoparticles loaded with insulin.

nanoparticles were prepared using sodium tripolyphosphate solution, followed by insulin trapping, then alginate microbeads were prepared and synthesized chitosan nanoparticles were dripped using syringe to perform alginate microbeads containing chitosan nanoparticles. Li et al. [124] also tested the release of inulin and demonstrated its pH when released; whereas acidic pH values delay insulin release, neutral pH, meanwhile, enhances insulin release. This delivery system demonstrated an ability to reduce blood glucose for a period lasting 96 h in the treated diabetic rats. Figure 3.3 illustrates the preparation of alginate/chitosan nanoparticles loaded with insulin.

Shanker et al. [125] reported that the oral administration of three synthesized metal nanoparticles (silver nanoparticles, zinc oxide nanoparticles, and cerium oxide nanoparticles) mediated by *Momordica charantia* (nutritious plant) has an antihyperglycemic potentiality in the pretreated Wistar rats using streptozotocin, which enhances the destruction of β-pancreatic cells. Shanker et al. [125] also reported that both of the synthesized silver and zinc oxide nanoparticles had more potent antidiabetic activity compared to *Momordica charantia*, and cerium oxide nanoparticles, as well as the three synthesized nanoparticles, illustrated no toxic effect in the experimented animals.

3.2.8 Wound Healing

Wound healing is one of the main health problems in diabetic and non-diabetic patients around the world. The pathogenic bacteria play a main role in hindered wound healing; this is related to interaction in the wound area. Biosynthesized copper nanoparticles by *Pseudomonas aeruginosa* were safe on normal cell lines and improved wound healing in rats [126]. Silver and gold nanoparticles bio-synthesis by *Coleus forskohlii* root extract accelerated the wound healing process [127]. Gold nanoparticles biosynthesis by aqueous extract of *Acalypha indica* improved *in vivo* wound healing efficiency [128]. Biosynthesis of silver nanoparticles with *Melia azedarach* (MA-AgNPs) played a significant role in wound healing activity [129]. TiO_2-NPs biosynthesis using *Eichhornia crassipes* extract was evaluated for wound healing activity [130]. The combination of antibacterial nanoparticles like silver nanoparticles, gold nanoparticles, copper nanoparticles, etc. with polymeric matrix could capably hinder bacterial growth and, at the same time, close the healing process of a wound [131]. ZnO-NPs gel biosynthesized by aqueous leaves of *Barleria gibsoni* had efficacy in treating both gram-positive and gram-negative infected burns [132]. Bacterial nanocellulose (BNC) produced by *Gluconacetobacter sucrofermentans* B-11267 is a highly antibacterial agent against *Staphylococcus aureus* and can be used as a wound dressing [133].

3.2.9 Radiolabeled Nanoparticles

Biogenic radiolabeled nanoparticles demonstrated capability for clinical diagnosis and treatment in different diseases, such as rheumatoid arthritis and cancer [134]. Chitosan was utilized as a biodegradable polymer for the synthesis of radiolabeled chitosan nanoparticles; different diagnostic and therapeutic radioisotopes, including 177Lu, 68Ga, 153Sm, 90Y, 166Ho, and 64Cu, were utilized to perform a complex with chitosan, then ionic gelation technique was followed to synthesis of radiolabeled chitosan nanoparticles. The synthesized nanoparticles demonstrated an efficient method in cancer imaging and treatment [135]. Poly lactic-co-glycolic acid (PLGA) is one of the most attractive biocompatible and biodegradable polymers extensively used to construct devices for drug delivery. PLGA was also approved by FDA [136]. PLGA was used as a template for the synthesis of drugs encapsulating nanoparticles for treatment of some arthritic diseases after being loaded with anti-inflammatory drugs, such as methotrexate and hyaluronic acid [134].

3.3 TOXIC EFFECTS OF BIOGENIC NANOPARTICLES

The high dose of delivered nanoparticles induces an adverse immunological disturbance as it interacts with cells in the immune system as well as generates reactive oxygen species which cause damage for metabolic pathways [137]. Excess negatively charged nanoparticles induced the rapid hemolysis of red blood cells [138]. Some of the synthesized nanoparticles are considered as carcinogenic agents because they generate reactive oxygen species which cause damage to DNA, mutation and cell cycle inhibition, and reduced cell viability [137]. The biogenic nanoparticles synthesis by algae have different shapes, sizes, and activities according to Figure 3.4. The biologically synthesized CuO-NPs using *Ulva fasciata* macro green alga used to treat rats at 500 mg/kg bw induced leukocytosis, hepatorenal degenerative changes, and increased P53 mRNA and caspase-3 protein expressions in hepatic tissues [138].

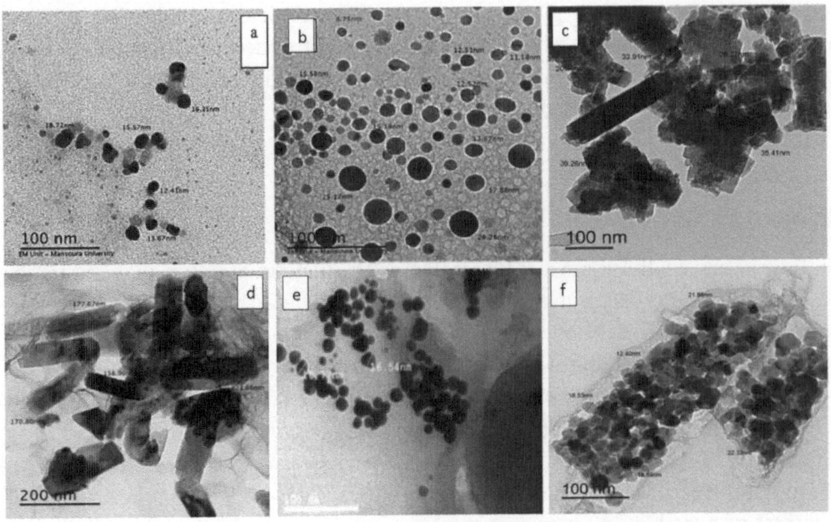

FIGURE 3.4 TEM images of **a)** AgNPs synthesized by *Corallina elongate* [139]; and **b)** by *Gelidium amansii* [139]; **c)** ZnO-NPs synthesized by *Fucus vesiculosus* [120]; and **d)** *Ulva fasciata* [140]; **e)** Au-NPs synthesized by *Chlorella vulgaris* [84]; **f)** CuO-NPs synthesized by *Ulva fasciata* [138].

3.4 CONFLICT OF INTEREST

The authors declare no conflict of interest.

REFERENCES

1. Khan, A. U.; Malik, N.; Khan, M.; Cho, M. H.; Khan, M. M. Fungi-assisted silver nanoparticle synthesis and their applications. *Bioprocess and Biosystems Engineering* **2018**, 41(1), 1–20.
2. Rahimirad, A.; Afshin, J.; Hamid, M.; Navideh, A.; Hoda, J. M. Biological approach in nano-biotechnology – screening of four food pathogenic bacteria extracts ability in extracellular biosynthesis of gold nanoparticles. *Biologia* **2020**, 75(4), 619–625.
3. Taton, T. A. Nanostructures as tailored biological probes. *Trends in Biotechnology* **2002**, 20, 277–279.
4. Saxena, S. K.; Nyodu, R.; Kumar, S.; Maurya, V. K. Current advances in nanotechnology and medicine. S. K. Saxena, S. M. P. Khurana (eds.). Springer Nature Singapore Pte Ltd. *NanoBioMedicine* **2020**. https://doi.org/10.1007/978-981-32-9898-9_1
5. Santhoshkumar, J.; Rajeshkumar, S.; Kumar, S. V. Phyto-assisted synthesis, characterization and applications of gold nanoparticles – A review. *Biochemistry and Biophysics Reports* **2017**, 11, 46–57.
6. Li, X.; Xu, H.; Chen, Z. S.; Chen, G. Biosynthesis of nanoparticles by microorganisms and their applications. *Journal of Nanomaterials* **2011**, 2011, 16, Article ID 270974.
7. Karnani, R. L.; Chowdhary, A. Biosynthesis of silver nanoparticle by eco-friendly method. *Indian Journal of Nanoscience* **2013**, 1(1), 25–31.
8. Li, F.; Lu, J.; Kong, X.; Hyeon, T.; Ling, D. Dynamic nanoparticle assemblies for biomedical applications. *Advanced Materials* **2017**, 29(14), 1605897.
9. Kumari, M.; Giri, V. P.; Pandey, S.; Kumar, M.; Katiyar, R.; Nautiyal, C. S.; Mishra, A. An insight into the mechanism of antifungal activity of biogenic nanoparticles than their chemical counterparts. *Pesticide Biochemistry and Physiology* **2019**, 157, 45–52.
10. Ashajyothi, C.; Prabhurajeshwar, C.; Handral, H. K. Investigation of antifungal and anti-mycelium activities using biogenic nanoparticles: An eco-friendly approach. *Environmental Nanotechnology, Monitoring & Management* **2016**, 5, 81–87.
11. Netala, V. R.; Kotakadi, V. S.; Domdi, L.; Gaddam, S. A.; Bobbu, P.; Venkata, S. K.; Tartte, V. Biogenic silver nanoparticles: Efficient and effective antifungal agents. *Applied Nanoscience* **2016**, 6(4), 475–484.
12. Al-Otibi, F.; Perveen, K.; Al-Saif, N. A.; Alharbi, R. I.; Bokhari, N. A.; Albasher, G.; Al-Mosa, M. A. Biosynthesis of silver nanoparticles using *Malva parviflora* and their antifungal activity. *Saudi Journal of Biological Sciences* **2021**, 28(4), 2229–2235.
13. Al-Zubaidi, S.; Al-Ayafi, A.; Abdelkader, H. Biosynthesis, characterization and antifungal activity of silver nanoparticles by *Aspergillus niger* isolate. *Journal of Nanotechnology Research* **2019**, 1(1), 23–36.
14. Al-Zahrani, S. S.; Al-Garni, S. M. Biosynthesis of silver nanoparticles from *Allium ampeloprasum* leaves extract and its antifungal activity. *Journal of Biomaterials and Nanobiotechnology* **2019**, 10(1), 11.
15. Patino-Portela, M. C.; Arciniegas-Grijalba, P. A.; Mosquera-Sanchez, L. P.; Sierra, B. E. G.; Munoz-Florez, J. E.; Erazo-Castillo, L. A.; Rodriguez-Paez, J. E. Effect of method of synthesis on antifungal ability of ZnO nanoparticles: Chemical route vs green route. *Advances in Nano Research* **2021**, 10(2), 191–210.
16. Shahbaz, M.; Akram, A.; Raja, N. I.; Mukhtar, T.; Mehak, A.; Fatima, N.; Ajmal, M.; Ali, K.; Mustafa, N.; Abasi, F. Antifungal activity of green synthesized selenium nanoparticles and their effect on physiological, biochemical, and antioxidant defense system of mango under mango malformation disease. *PLoS One* **2023**, 18(2), e0274679.
17. Park, S.O.; Baker, J.; Himmel, M.; Parrilla, P.; Johnson, D. Research cellulose crystallinity index: Measurement techniques and their impact on interpreting cellulase performance. *Biotechnology for Biofuels* **2010**, 3, 10.
18. Loelovich, M. Optimal conditions for isolation of nano-crystalline cellulose particles. *Journal of Nanoscience and Nanotechnology* **2012**, 12, 9–13.
19. Pandey, J. K.; Nakagaito, A. N.; Takagi, H. Fabrication and applications of cellulose nanoparticle-based polymer composites. *Polymer Engineering & Science* **2012**, 53, 1–8.
20. Sivaraj, A.; Kumar, V.; Kalyanasundaram, R.; Kasivelu, G. Biogenic production of gold nanoparticles using lactic acid bacteria and their anti-mycobacterial activity. *Research Journal of Pharmacy and Technology* **2020**, 13(9), 4391–4394.

21. Escárcega-González, C. E.; Garza-Cervantes, J. A.; Vázquez-Rodríguez, A.; Morones-Ramírez, J. R. Bacterial exopolysaccharides as reducing and/or stabilizing agents during synthesis of metal nanoparticles with biomedical applications. *International Journal of Polymer Science* **2018**, 2018, ID 7045852.
22. Soares, M. R.; Corrêa, R. O.; Stroppa, P. H. F.; Marques, F. C.; Andrade, G. F.; Corrêa, C. C.; Raposo, N. R. Biosynthesis of silver nanoparticles using *Caesalpinia ferrea* (Tul.) Martius extract: Physicochemical characterization, antifungal activity and cytotoxicity. *PeerJ* **2018**, 6, e4361.
23. Wang, D.; Xue, B.; Wang, L.; Zhang, Y.; Liu, L.; Zhou, Y. Fungus-mediated green synthesis of nano-silver using *Aspergillus sydowii* and its antifungal/antiproliferative activities. *Scientific Reports* **2021**, 11(1), 1–9.
24. Pereira, L.; Dias, N.; Carvalho, J.; Fernandes, S.; Santos, C.; Lima, N. Synthesis, characterization and antifungal activity of chemically and fungal-produced silver nanoparticles against *T Richophyton Rubrum*. *Journal of Applied Microbiology* **2014**, 117(6), 1601–1613.
25. Khatoon, N.; Mishra, A.; Alam, H.; Manzoor, N.; Sardar, M. Biosynthesis, characterization, and antifungal activity of the silver nanoparticles against pathogenic Candida species. *BioNanoScience* **2015**, 5(2), 65–74.
26. Manivasagan, P.; Venkatesan, J.; Senthilkumar, K.; Sivakumar, K.; Kim, S. K. Biosynthesis, antimicrobial and cytotoxic effect of silver nanoparticles using a novel Nocardiopsis sp. MBRC-1. *BioMed Research International* **2013**, 2013, 9, ID 287638. https://doi.org/10.1155/2013/287638.
27. Majeed, S. Biosynthesis and characterization of nanosilver from Alternata alternaria and it antifungal and antibacterial activity in combination with fluconazole and gatifloxacin. *Biomedical and Pharmacology Journal* **2017**, 10(4), 1709–1714.
28. Khatami, M.; Pourseyedi, S.; Khatami, M.; Hamidi, H.; Zaeifi, M.; Soltani, L. Synthesis of silver nanoparticles using seed exudates of *Sinapis arvensis* as a novel bioresource, and evaluation of their antifungal activity. *Bioresources and Bioprocessing* **2015**, 2(1), 1–7.
29. Shankar, S. S.; Ahmad, A.; Sastry, M. Geranium leaf assisted biosynthesis of silver nanoparticles. *Biotechnology Progress* **2003**, 19(6), 1627–1631.
30. Devi, H. S.; Boda, M. A.; Shah, M. A.; Parveen, S.; Wani, A. H. Green synthesis of iron oxide nanoparticles using *Platanus orientalis* leaf extract for antifungal activity. *Green Processing and Synthesis* **2019**, 8(1), 38–45.
31. Sandhya, J.; Kalaiselvam, S. Biogenic synthesis of magnetic iron oxide nanoparticles using inedible *Borassus flabellifer* seed coat: Characterization, antimicrobial, antioxidant activity and *in vitro* cytotoxicity analysis. *Materials Research Express* **2020**, 7(1), 015045.
32. Senthilkumar, S. R.; Sivakumar, T. Green tea (*Camellia sinensis*) mediated synthesis of zinc oxide (ZnO) nanoparticles and studies on their antimicrobial activities. International *Journal of Pharmacy and Pharmaceutical Sciences* **2014**, 6(6), 461–465.
33. Sharma, D.; Shandilya, P.; Saini, N. K.; Singh, P.; Thakur, V. K.; Saini, R. V.; Saini, A. K. Insights into the synthesis and mechanism of green synthesized antimicrobial nanoparticles, answer to the multidrug resistance. *Materials Today Chemistry* **2021**, 19, 100391.
34. Alavi, M. Bacteria and fungi as major bio-sources to fabricate silver nanoparticles with antibacterial activities. *Expert Review of Anti-Infective Therapy* **2022**, (just-accepted). https://doi.org/10.1080/14787210.2022.2045194
35. Nasrollahzadeh, M.; Sajjadi, M.; Dadashi, J.; Ghafuri, H. Pd-based nanoparticles: Plant-assisted biosynthesis, characterization, mechanism, stability, catalytic and antimicrobial activities. *Advances in Colloid and Interface Science* **2020**, 276, 102103.
36. Surendra, T. V.; Roopan, S. M.; Arasu, M. V.; Al-Dhabi, N. A.; Rayalu, G. M. RSM optimized Moringa oleifera peel extract for green synthesis of *M. oleifera* capped palladium nanoparticles with antibacterial and hemolytic property. *Journal of Photochemistry and Photobiology B: Biology* **2016**, 162, 550–557.
37. Mousavi-Khattat, M.; Keyhanfar, M.; Razmjou, A. A comparative study of stability, antioxidant, DNA cleavage and antibacterial activities of green and chemically synthesized silver nanoparticles. *Artificial Cells, Nanomedicine, and Biotechnology* **2018**, 46, 1022–1031.
38. Naseem, T.; Farrukh, M. A. Antibacterial activity of green synthesis of iron nanoparticles using *Lawsonia inermis* and *Gardenia jasminoides* leaves extract. *Journal of Chemistry* **2015**, 912342.
39. Joshi, C. G.; Danagoudar, A.; Poyya, J.; Kudva, A. K.; Dhananjaya, B. L. Biogenic synthesis of gold nanoparticles by marine endophytic fungus-*Cladosporium cladosporioides* isolated from seaweed and evaluation of their antioxidant and antimicrobial properties. *Process Biochemistry* **2017**, 63, 137–144.

40. Ananda Murthy, H. C.; Desalegn, T.; Kassa, M.; Abebe, B.; Assefa, T. Synthesis of green copper nanoparticles using medicinal plant *Hagenia abyssinica* (Brace) JF. Gmel. leaf extract: Antimicrobial properties. *Journal of Nanomaterials* **2020**, 2020, 3924081.
41. Abo-Shama, U. H.; El-Gendy, H.; Mousa, W. S.; Hamouda, R. A.; Yousuf, W. E.; Hetta, H. F.; Abdeen, E. E. Synergistic and antagonistic effects of metal nanoparticles in combination with antibiotics against some reference strains of pathogenic microorganisms. *Infection and Drug Resistance* **2020**, 13, 351.
42. Jozala, A. F.; Santos, J. R.; dos Santos, G. R.; Viroel, F. J. V.; Pickler, T. B.; dos Santos, C. A.; Nesiyama, T. N. G. *Libidibia ferrea* loaded in bacterial nanocellulose: Evaluation of antimicrobial activity and wound care. *Brazilian Journal of Development* **2020**, 6(2), 6212–6226.
43. Nurfadlilah, L.; Hidayah, R. Antibacterial activity of sugarcane bagasse nanocellulose biocomposite with chitosan against *Escherichia coli*. *Journal Kimia Valensi* **2021**, 7(1), 28–37.
44. Jabłońska, J.; Onyszko, M.; Konopacki, M.; Augustyniak, A.; Rakoczy, R.; Mijowska, E. Fabrication of paper sheets coatings based on chitosan/bacterial nanocellulose/ZnO with enhanced antibacterial and mechanical properties. *International Journal of Molecular Sciences* **2021**, 22(14), 7383.
45. Jebali, A.; Hekmatimoghaddam, S.; Behzadi, A.; Rezapor, I.; Mohammadi, B. H.; Jasemizad, T.; Sayadi, M. Antimicrobial activity of nanocellulose conjugated with allicin and lysozyme. *Cellulose* **2013**, 20(6), 2897–2907.
46. Sun, B.; Kc, B. S.; Faruk, O.; Tjong, J.; Sain, M.; Zhang, M.; Kong, F. Antibacterial cellulose based biomedical hand sanitizer-current state and future trends. *Biomedical Journal of Scientific & Technical Research* **2019**, 24(1), 17927–17933.
47. Tyagi, S.; Tyagi, P. K.; Gola, D.; Chauhan, N.; Bharti, R. K. Extracellular synthesis of silver nanoparticles using entomopathogenic fungus: Characterization and antibacterial potential. *SN Applied Sciences* **2019**, 1, 1–9.
48. Jayabalan, J.; Mani, G.; Krishnan, N.; Pernabas, J.; Devadoss, J. M.; Jang, H. T. Green biogenic synthesis of zinc oxide nanoparticles using *Pseudomonas putida* culture and its *in vitro* antibacterial and antibiofilm activity. *Biocatalysis and Agricultural Biotechnology* **2019**, 21, 101327.
49. Safawo, T.; Sandeep, B. V.; Pola, S.; Tadesse, A. Synthesis and characterization of zinc oxide nanoparticles using tuber extract of anchote (*Coccinia abyssinica* (Lam.) Cong.) for antimicrobial and antioxidant activity assessment. *OpenNano* **2018**, 3, 56–63.
50. Jayaseelan, C.; Rahuman, A. A.; Kirthi, A. V.; Marimuthu, S.; Santhoshkumar, T.; Bagavan, A.; Rao, K. B. Novel microbial route to synthesize ZnO nanoparticles using *Aeromonas hydrophila* and their activity against pathogenic bacteria and fungi. *Spectrochimica Acta Part A: Molecular and Biomolecular Spectroscopy* **2012**, 90, 78–84.
51. Ramkumar, V. S.; Pugazhendhi, A.; Prakash, S.; Ahila, N. K.; Vinoj, G.; Selvam, S.; Rajendran, R. B. Synthesis of platinum nanoparticles using seaweed *Padina gymnospora* and their catalytic activity as PVP/PtNPs nanocomposite towards biological applications. *Biomedicine & Pharmacotherapy* **2017**, 92, 479–490.
52. Acharyulu, N. P. S.; Dubey, R. S.; Swaminadham, V.; Kollu, P.; Kalyani, R. L.; Pammi, S. V. N. Green synthesis of CuO nanoparticles using *Phyllanthus amarus* leaf extract and their antibacterial activity against multidrug resistance bacteria. *International Journal of Engineering Research & Technology* **2014**, 3(4), 639–641.
53. Órdenes-Aenishanslins, N.; Anziani-Ostuni, G.; Monrás, J. P.; Tello, A.; Bravo, D.; Toro-Ascuy, D.; Pérez-Donoso, J. M. Bacterial synthesis of ternary CdSAg quantum dots through cation exchange: Tuning the composition and properties of biological nanoparticles for bioimaging and photovoltaic applications. *Microorganisms* **2020**, 8(5), 631.
54. Rajakumar, G.; Rahuman, A. A.; Roopan, S. M.; Khanna, V. G.; Elango, G.; Kamaraj, C.; Velayutham, K. Fungus-mediated biosynthesis and characterization of TiO_2 nanoparticles and their activity against pathogenic bacteria. *Spectrochimica Acta Part A: Molecular and Biomolecular Spectroscopy* **2012**, 91, 23–29.
55. Morris, D.; Ansar, M.; Speshock, J.; Ivanciuc, T.; Qu, Y.; Casola, A.; Garofalo, R. P. Antiviral and immunomodulatory activity of silver nanoparticles in experimental RSV infection. *Viruses* **2019**, 11, 732.
56. Khandelwal, N.; Kaur, G.; Kumar, N.; Tiwari, A. Application of silver nanoparticles in viral inhibition: A new hope for antivirals. *Digest Journal of Nanomaterials and Biostructures DJNB* **2014**, 9, 175–186.
57. Ratan, Z. A.; Mashrur, F. R.; Chhoan, A. P.; Shahriar, S. M.; Haidere, M. F.; Runa, N. J.; Kim, S.; Kweon, D.-H.; Hosseinzadeh, H.; Cho, J. Y. Silver nanoparticles as potential antiviral agents. *Pharmaceutics* **2021**, 13, 2034. https://doi.org/10.3390/pharmaceutics13122034.

58. Haggag, E. G.; Elshamy, A. M.; Rabeh, M. A.; Gabr, N. M.; Salem, M.; Youssif, K. A.; Abdelmohsen, U. R. Antiviral potential of green synthesized silver nanoparticles of *Lampranthus coccineus* and *Malephora lutea*. *International Journal of Nanomedicine* **2019**, 14, 6217.
59. Mehmood, Y.; Farooq, U.; Yousaf, H.; Riaz, H.; Mahmood, R. K.; Nawaz, A.; Khalid, I. Antiviral activity of green silver nanoparticles produced using aqueous buds extract of *Syzygium aromaticum*. *Pakistan Journal of Pharmaceutical Sciences* **2020**, 2, 839–845.
60. Sharma, V.; Kaushik, S.; Pandit, P.; Dhull, D.; Yadav, J. P.; Kaushik, S. Green synthesis of silver nanoparticles from medicinal plants and evaluation of their antiviral potential against chikungunya virus. *Applied Microbiology and Biotechnology* **2019**, 103, 881–891.
61. American Cancer Society. Cancer facts and figures. *Genes & Development* **2017**, 21, 2525–2538.
62. Srivastava, N.; Saxena, S. K. Opportunities in clinical translation and commercialization of nanomedicine. *Nano BioMedicine* **2020**, 501–517, Springer, Singapore.
63. Zhang, W.; Zhang, Z.; Zhang, Y. The application of carbon nanotubes in target drug delivery systems for cancer therapies. *Nanoscale Research Letters* **2011**, 6, 555.
64. Ahmad, S. S.; Reinius, M. A.; Hatcher, H. M.; Ajithkumar, T. V. Anticancer chemotherapy in teenagers and young adults: Managing long term side effects. *BMJ* **2016**, 354, i4567.
65. Hossen, S.; Hossain, M. K.; Basher, M. K.; Mia, M. N. H.; Rahman, M. T.; Uddin, M. J. Smart nanocarrier-based drug delivery systems for cancer therapy and toxicity studies: A review. *Journal of Advanced Research* **2019**, 15, 1–18.
66. Patil, S.; Chandrasekaran, R. Biogenic nanoparticles: A comprehensive perspective in synthesis, characterization, application and its challenges. *Journal of Genetic Engineering and Biotechnology* **2020**, 18(1), 1–23.
67. Lim, Z. J.; Li, J. J.; Ng, C.; Yung, L. L.; Bay, B. Gold nanoparticles in cancer therapy. *National Public Growth* **2011**, 32, 983–990. https://doi.org/https://doi.org/10.1038/aps.
68. Patil, S.; Sivaraj, R.; Vanathi, R. Anticancer potential of green synthesized silver nanoparticles: A review. *International Journal of Current Research* **2015**, 7, 21539–21544.
69. Kim, C. G.; Castro-Aceituno, V.; Abbai, R.; Lee, H. A.; Simu, S. Y.; Han, Y.; Yang, D. C. Caspase-3/MAPK pathways as main regulators of the apoptotic effect of the phyto-mediated synthesized silver nanoparticle from dried stem of *Eleutherococcus senticosus* in human cancer cells. *Biomedicine & Pharmacotherapy* **2018**, 99, 128–133.
70. Ansar, S.; Tabassum, H.; Aladwan, N. S. M.; Ali, M. N.; Almaarik, B.; AlMahrouqi, S.; Abudawood, M.; Banu, N.; Alsubki, R. Ecofriendly silver nanoparticles synthesis by *Brassica oleracea* and its antibacterial, anticancer and antioxidant properties. *Scientific Reports* **2020**, 10, 18564.
71. Sankar, R.; Maheswari, R.; Karthik, S.; Shivashangari, K. S.; Ravikumar, V. Anticancer activity of *Ficus religiosa* engineered copper oxide nanoparticles. *Materials Science and Engineering: C* **2014**, 44, 234–239. https://doi.org/10.1016/j.
72. Gurunathan, S.; Kim, E.; Han, J. W.; Park, J. H.; Kim, J. Green chemistry approach for synthesis of effective anticancer palladium nanoparticles. *Molecules* **2015**, 20, 22476–22498. https://doi.org/10.3390/molecules201219860.
73. Castro-Aceituno, V.; Ahn, S.; Simu, S. Y.; Singh, P.; Mathiyalagan, R.; Lee, H. A.; Yang, D. C. Anticancer activity of silver nanoparticles from Panax ginseng fresh leaves in human cancer cells. *Biomed Pharmacother* **2016**, 84, 158–165. https://doi.org/10.1016/j.biopha.(2016).09.016.
74. Wypij, M.; Jedrzejewski, T.; Trzcínska-Wencel, J.; Ostrowski, M.; Rai, M.; Golínska, P. Green synthesized silver nanoparticles: Antibacterial and anticancer activities, biocompatibility, and analyses of surface-attached proteins. *Frontiers in Microbiology* **2021**, 12, 632505.
75. El-Naggar, N. E. A.; Hussein, M. H.; El-Sawah, A. A. Bio-fabrication of silver nanoparticles by phycocyanin, characterization, *in vitro* anticancer activity against breast cancer cell line and *in vivo* cytotxicity. *Scientific Reports* **2017**, 7(1), 1–20.
76. He, Y.; Du, Z.; Ma, S.; Liu, Y.; Li, D.; Huang, H.; Zheng, X. Effects of green-synthesized silver nanoparticles on lung cancer cells *in vitro* and grown as xenograft tumors *in vivo*. *International Journal of Nanomedicine* **2016**, 11, 1879.
77. Khalifa, K. S.; Hamouda, R. A.; Hamza, H. A. Antitumor activity of silver nanoparticles biosynthesized by micro algae. *Journal of Chemical and Pharmaceutical Research* **2016**, 8(3), 1–6.
78. El Bialy, B. E.; Hamouda, R. A.; Khalifa, K. S.; Hamza, H. A. Cytotoxic effect of biosynthesized silver nanoparticles on Ehrlich Ascites tumor cells in mice. *International Journal of Pharmaceutics* **2017**, 13(2), 134–144.
79. Wang, L.; Xu, J.; Yan, Y.; Liu, H.; Karunakaran, T.; Li, F. Green synthesis of gold nanoparticles from *Scutellaria barbata* and its anticancer activity in pancreatic cancer cell (PANC-1) activity in pancreatic cancer cell (PANC-1). *Artificial Cells, Nanomedicine and Biotechnology* **2019**, 47, 1617–1627. https://doi.org/10.1080/21691401.(2019).1594862.

80. Fan, T. X.; Chow, S. K.; Zhang, D. Biomorphic mineralization: From biology to materials. *Progress in Materials Science* **2009**, 54(5), 542–659.
81. Hsu, P.-C.; Chen, C.-M.; Ou, H.-Y.; Chang, H.-T. Extremely high inhibition activity of photoluminescent carbon nanodots toward cancer cells. *Journal of Materials Chemistry B* **2013**, 1, 1774–1781.
82. Datkhile, K. D.; Durgawale, P. P.; Patil, M. N. Biogenic silver nanoparticles are equally cytotoxic as chemically synthesized silver nanoparticles. *Biomedical and Pharmacology Journal* **2017**, 10, 337–344.
83. Umar, H.; Kavaz, D.; Rizaner, N. Biosynthesis of zinc oxide nanoparticles using *Albizia lebbeck* stem bark, and evaluation of its antimicrobial, antioxidant, and cytotoxic activities on human breast cancer cell lines. *International Journal of Nanomedicine* **2019**, 14, 87–100.
84. Hamouda, R. A.; Abd El Maksoud, A. I.; Wageed, M.; Alotaibi, A. S.; Elebeedy, D.; Khalil, H.; Abdella, A. Characterization and anticancer activity of biosynthesized Au/Cellulose nanocomposite from *Chlorella vulgaris*. *Polymers* **2021**, 13(19), 3340.
85. Khan, M. S.; Alomari, A.; Tabrez, S.; Hassan, I.; Wahab, R.; Bhat, S. A.; Ismael, M. A. Anticancer potential of biogenic silver nanoparticles: A mechanistic study. *Pharmaceutics* **2021**, 13(5), 707.
86. Buttacavoli, M.; Albanese, N. N.; Di Cara, G.; Alduina, R.; Faleri, C.; Gallo, M.; Cancemi, P. Anticancer activity of biogenerated silver nanoparticles: An integrated proteomic investigation. *Oncotarget* **2018**, 9(11), 9685.
87. Dinparvar, S.; Bagirova, M.; Allahverdiyev, A. M.; Abamor, E. S.; Safarov, T.; Aydogdu, M.; Aktas, D. A nanotechnology-based new approach in the treatment of breast cancer: Biosynthesized silver nanoparticles using *Cuminum cyminum* L. seed extract. *Journal of Photochemistry and Photobiology B: Biology* **2020**, 208, 111902.
88. Chabattula, S. C.; Gupta, P. K.; Tripathi, S. K.; Gahtori, R.; Padhi, P.; Ahapatra, S.; Kesari, K. K. Anticancer therapeutic efficacy of biogenic Am-ZnO nanoparticles on 2D and 3D tumor models. *Materials Today Chemistry* **2021**, 22, 100618.
89. Gowdhami, B.; Jaabir, M.; Archunan, G.; Suganthy, N. Anticancer potential of zinc oxide nanoparticles against cervical carcinoma cells synthesized via biogenic route using aqueous extract of *Gracilaria edulis*. *Materials Science and Engineering: C* **2019**, 103, 109840.
90. Hussain, A.; Oves, M.; Alajmi, M. F.; Hussain, I.; Amir, S.; Ahmed, J.; Ali, I. Biogenesis of ZnO nanoparticles using Pandanus odorifer leaf extract: Anticancer and antimicrobial activities. *RSC Advances* **2019**, 9(27), 15357–15369.
91. Kumar, P. S.; Jeyalatha, M. V.; Malathi, J.; Ignacimuthu, S. Anticancer effects of one-pot synthesized biogenic gold nanoparticles (Mc-Au-NPs) against *laryngeal carcinoma*. *Journal of Drug Delivery Science and Technology* **2018**, 44, 118–128.
92. Botteon, C. E. A.; Silva, L. B.; Ccana-Ccapatinta, G. V.; Silva, T. S.; Ambrosio, S. R.; Veneziani, R. C. S.; Marcato, P. D. Biosynthesis and characterization of gold nanoparticles using Brazilian red propolis and evaluation of its antimicrobial and anticancer activities. *Scientific Reports* **2021**, 11(1), 1–16.
93. Gomez-Orellana, I. Strategies to improve oral drug bioavailability. *Expert Opinion on Drug Delivery* **2005**, 2, 419–433.
94. Hanafi-Bojd, M. Y.; Jaafari, M. R.; Ramezanian, N.; Xue, M.; Amin, M.; Shahtahmassebi, N.; Malaekeh-Nikouei, B. Surface functionalized mesoporous silica nanoparticles as an effective carrier for epirubicin delivery to cancer cells. *European Journal of Pharmaceutics and Biopharmaceutics* **2015**, 89, 248–258.
95. Fang, J.; Nakamura, H.; Maeda, H. The EPR effect: Unique features of tumor blood vessels for drug delivery, factors involved, and limitations and augmentation of the effect. *Advanced Drug Delivery Reviews* **2011**, 63(3), 136–151.
96. Cheng, R.; Meng, F.; Deng, C.; Klok, H. A.; Zhong, Z. Dual and multi-stimuli responsive polymeric nanoparticles for programmed site-specific drug delivery. *Biomaterials* **2013**, 34(14), 3647–3657.
97. Seo, J. M.; Kim, E. B.; Hyun, M. S.; Kim, B. B.; Park, T. J. Self-assembly of biogenic gold nanoparticles and their use to enhance drug delivery into cells. *Colloids and Surfaces B: Biointerfaces* **2015**, 135, 27–34.
98. Yang, Q.; Jones, S. W.; Parker, C. L.; Zamboni, W. C.; Bear, J. E.; Lai, S. K. Evading immune cell uptake and clearance requires PEG grafting at densities substantially exceeding the minimum for brush conformation. *Molecular Pharmaceutics* **2014**, 11, 1250–1258. https://doi.org/10.1021/mp400703d.
99. Zylberberg, C.; Matosevic, S. Pharmaceutical liposomal drug delivery: A review of new delivery systems and a look at the regulatory landscape. *Drug Delivery* **2016**, 23, 3319–3329. https://doi.org/10.1080/10717544.2016.1177136.
100. Chen, Y.; Zhu, X.; Zhang, X.; Liu, B.; Huang, L. Nanoparticles modified with tumor-targeting scFv deliver siRNA and miRNA for cancer therapy. *Molecular Therapy* **2010**, 18, 1650–1656. https://doi.org/10.1038/mt.2010.136.
101. Amreddy, N.; Babu, A.; Muralidharan, R.; Panneerselvam, J.; Srivastava, A.; Ahmed, R., et al. Recent advances in nanoparticle-based cancer drug and gene delivery. *Advances in Cancer Research* **2018**, 137, 115–170. https://doi.org/10.1016/bs.acr.2017.11.003.

102. Han, G.; Ghosh, P.; Rotello, V. M. Functionalized gold nanoparticles for drug delivery. *Nanomedicine* **2007**, 2, 113–123. https://doi.org/10.2217/17435889.2.1.113.
103. Cheng, J.; Gu, Y. J.; Cheng, S. H.; Wong, W. T. Surface functionalized gold nanoparticles for drug delivery. *Journal of Biomedical Nanotechnology* **2013**, 9, 1362–1369. https://doi.org/10.1166/jbn.2013.1536.
104. Basoglu, H.; Goncu, B.; Akbas, F. Magnetic nanoparticle-mediated gene therapy to induce Fas apoptosis pathway in breast cancer. *Cancer Gene Therapy* **2018**, 25, 141–147. https://doi.org/10.1038/s41417-018-0017-2.
105. Zhao, X., Li, F.; Li, Y.; Wang, H.; Ren, H.; Chen, J., et al. Co-delivery of HIF1α siRNA and gemcitabine via biocompatible lipid-polymer hybrid nanoparticles for effective treatment of pancreatic cancer. *Biomaterials* **2015**, 46, 13–25. https://doi.org/10.1016/j.biomaterials 2015,.2014.12.028.
106. Gao, F.; Zhang, J.; Fu, C.; Xie, X.; Peng, F.; You, J., et al. iRGD-modified lipid-polymer hybrid nanoparticles loaded with isoliquiritigenin to enhance anti-breast cancer effect and tumor-targeting ability. *International Journal of Nanomedicine* **2017**, 12, 4147–4162. https://doi.org/10.2147/ijn.s134148.
107. Wang, X.; Liu, X.; Li, Y.; Wang, P.; Feng, X.; Liu, Q., et al. Sensitivity to antitubulin chemotherapeutics is potentiated by a photoactivable nanoliposome. *Biomaterials* **2017**, 141, 50–62. https://doi.org/10.1016/j.biomaterials.2017.06.034.
108. Colapicchioni, V.; Palchetti, S.; Pozzi, D.; Marini, E. S.; Riccioli, A.; Ziparo, E., et al. Killing cancer cells using nanotechnology: Novel poly(I:C) loaded liposome-silica hybrid nanoparticles. *Journal of Materials Chemistry B* **2015**, 3, 7408–7416. https://doi.org/10.1039/c5tb01383.
109. Karunakaran, S.; Ramanujam, S.; Gurunathan, B. Green synthesised iron and iron-based nanoparticle in environmental and biomedical application – A review. *IET Nanobiotechnology* **2018**, 12(8), 1003–1008.
110. Singh, R.; Hano, C.; Tavanti, F.; Sharma, B. Biogenic synthesis and characterization of antioxidant and antimicrobial silver nanoparticles using flower extract of *Couroupita guianensis* Aubl. *Materials* **2021**, 14(22), 6854.
111. Stozhko, N. Y.; Bukharinova, M. A.; Khamzina, E. I.; Tarasov, A. V.; Vidrevich, M. B.; & Brainina, K. Z. The effect of the antioxidant activity of plant extracts on the properties of gold nanoparticles. *Nanomaterials* **2019**, 9(12), 1655. https://doi.org/10.3390/nano9121655.
112. Bedlovičová, Z.; Strapáč, I.; Baláž, M.; Salayová, A. A brief overview on antioxidant activity determination of silver nanoparticles. *Molecules* **2020**, 25(14), 3191. https://doi.org/10.3390/molecules25143191.
113. Paul, R. K.; Kesharwani, P.; Raza, K. Recent update on nano-phytopharmaceuticals in the management of diabetes. *Journal of Biomaterials Science, Polymer Edition* **2021**, 32(15), 2046–2068.
114. Aynalem, S. B.; Zeleke A. J. Prevalence of diabetes mellitus and its risk factors among individuals aged 15 years and above in Mizan-Aman town, Southwest Ethiopia, 2016: A cross sectional study. *International Journal of Endocrinology* **2018**, 2018. https://doi.org/10.1155/2018/9317987.
115. Reinehr, T. Type 2 diabetes mellitus in children and adolescents. *World Journal of Diabetes* **2013**, 4(6), 270.
116. Samadder, A.; Khuda-Bukhsh, A. R. Nanotechnological approaches in diabetes treatment: A new horizon. *World Journal of Translational Medicine* **2014**, 3(2), 84–95.
117. Bahman, F.; Greish, K.; Taurin, S. Nanotechnology in insulin delivery for management of diabetes. *Pharmaceutical Nanotechnology* **2019**, 7(2), 113–128.
118. Gupta, R. Diabetes treatment by nanotechnology. *Journal of Biotechnology & Biomaterials* **2017**, 7(3), 268.
119. Erzengin, S.; Guler, E.; Eser, E.; Polat, E. B.; Gunduz, O.; Cam, M. E. *In vitro* and *in vivo* evaluation of 3D printed sodium alginate/polyethylene glycol scaffolds for sublingual delivery of insulin: Preparation, characterization, and pharmacokinetics. *International Journal of Biological Macromolecules* **2022**, 204, 429–440.
120. Hamouda, R. A.; Salman, A. S.; Alharbi, A. A.; Alhasani, R. H.; Elshamy, M. M. Assessment of the antigenotoxic effects of alginate and ZnO/alginate–nanocomposites extracted from brown alga fucus vesiculosus in mice. *Polymers* **2021**, 13(21), 3839.
121. Raus, R. A.; Nawawi, W. M. F. W.; Nasaruddin, R. R. Alginate and alginate composites for biomedical applications. *Asian Journal of Pharmaceutical Sciences* **2021**, 16(3), 280–306.
122. Bagheri, M.; Validi, M.; Gholipour, A.; Makvandi, P.; Sharifi, E. Chitosan nanofiber biocomposites for potential wound healing applications: Antioxidant activity with synergic antibacterial effect. *Bioengineering & Translational Medicine* **2022**, 7(1), e10254.
123. Azmana, M.; Mahmood, S.; Hilles, A. R.; Rahman, A.; Arifin, M. A. B.; Ahmed, S. A review on chitosan and chitosan-based bionanocomposites: Promising material for combatting global issues and its applications. *International Journal of Biological Macromolecules* **2021**, 185, 832–848.

124. Li, J.; Wu, H.; Jiang, K.; Liu, Y.; Yang, L.; Park, H. J. Alginate calcium microbeads containing chitosan nanoparticles for controlled insulin release. *Applied Biochemistry and Biotechnology* **2021**, 193(2), 463–478.
125. Shanker, K.; Naradala, J.; Mohan, G. K.; Kumar, G. S.; Pravallika, P. L. A sub-acute oral toxicity analysis and comparative *in vivo* anti-diabetic activity of zinc oxide, cerium oxide, silver nanoparticles, and Momordica charantia in streptozotocin-induced diabetic Wistar rats. *RSC Advances* **2017**, 7(59), 37158–37167.
126. Tiwari, M.; Narayanan, K.; Thakar, M. B.; Jagani, H. V.; Rao, J. V. Biosynthesis and wound healing activity of copper nanoparticles. *IET Nanobiotechnology* **2014**, 8(4), 230–237.
127. Naraginti, S.; Kumari, P. L.; Das, R. K.; Sivakumar, A.; Patil, S. H.; Andhalkar, V. V. Amelioration of excision wounds by topical application of green synthesized, formulated silver and gold nanoparticles in albino Wistar rats. *Materials Science and Engineering C: Materials for Biological Applications* **2016**, 62, 293–300.
128. Boomi, P.; Ganesan, R.; Poorani, G. P.; Jegatheeswaran, S.; Balakumar, C.; Prabu, H. G.; Saravanan, M. Phyto-engineered gold nanoparticles (AuNPs) with potential antibacterial, antioxidant, and wound healing activities under *in vitro* and *in vivo* conditions. *International Journal of Nanomedicine* **2020**, 15, 7553.
129. Chinnasamy, G.; Chandrasekharan, S.; Bhatnagar, S. Biosynthesis of silver nanoparticles from *Melia azedarach*: Enhancement of antibacterial, wound healing, antidiabetic and antioxidant activities. *International Journal of Nanomedicine* **2019**, 11(14), 9823–9836.
130. Zhang, Y.; Zhang, C.; Liu, K.; Zhu, X.; Liu, F.; Ge, X. Biologically synthesized titanium oxide nanostructures combined with morphogenetic protein as wound healing agent in the femoral fracture after surgery. *Journal of Photochemistry and Photobiology B: Biology* **2018**, 182, 35–41.
131. Vijayakumar, V.; Samal, S. K.; Mohanty, S.; Nayak, S. K. Recent advancements in biopolymer and metal nanoparticle-based materials in diabetic wound healing management. *International Journal of Biological Macromolecules* **2019**, 1(122), 137–148.
132. Shao, F.; Yang, A.; Yu, D. M.; Wang, J.; Gong, X.; Tian, H. X. Bio-synthesis of Barleria gibsoni leaf extract mediated zinc oxide nanoparticles and their formulation gel for wound therapy in nursing care of infants and children. *Journal of Photochemistry and Photobiology B* **2018**, 189, 267–273.
133. Liyaskina, E.; Revin, V.; Paramonova, E.; Nazarkina, M.; Pestov, N.; Revina, N.; Kolesnikova, S. Nanomaterials from bacterial cellulose for antimicrobial wound dressing. *Journal of Physics: Conference Series* **2017**, 784, 012034, IOP Publishing.
134. Trujillo-Nolasco, R. M.; Morales-Avila, E.; Ocampo-Garcia, B. E.; Ferro-Flores, G.; Gibbens-Bandala, B. V.; Escudero-Castellanos, A.; Isaac-Olive, K. Preparation and *in vitro* evaluation of radiolabeled HA-PLGA nanoparticles as novel MTX delivery system for local treatment of rheumatoid arthritis. *Materials Science and Engineering*: *C* **2019**, 103, 109766.
135. Gaikwad, G.; Rohra, N.; Kumar, C.; Jadhav, S.; Sarma, H. D.; Borade, L.; Chakravarty, R. A facile strategy for synthesis of a broad palette of intrinsically radiolabeled chitosan nanoparticles for potential use in cancer theranostics. *Journal of Drug Delivery Science and Technology* **2021**, 63, 102485.
136. Makadia, H. K.; Siegel, S. J. Poly lactic-co-glycolic acid (PLGA) as biodegradable controlled drug delivery carrier. *Polymers* **2011**, 3(3), 1377–1397.
137. Pandey, R. K.; Prajapati, V. K. Molecular and immunological toxic effects of nanoparticles. *International Journal of Biological Macromolecules* **2018**, 107, 1278–1293.
138. El Bialy, B. E.; Hamouda, R. A.; Abd Eldaim, M. A.; El Ballal, S. S.; Heikal, H. S.; Khalifa, H. K.; Hozzein, W. N. Comparative toxicological effects of biologically and chemically synthesized copper oxide nanoparticles on mice. *International Journal of Nanomedicine* **2020**, 15, 3827–3842.
139. Hamouda, R. A.; Abd El-Mongy, M.; Eid, K. F. Comparative study between two red algae for biosynthesis silver nanoparticles capping by SDS: Insights of characterization and antibacterial activity. *Microbial pathogenesis* **2019**, 129, 224–232.
140. Hamouda, R. A., Yousuf, W. E., Mohammed, A. A., Mohammed, R. S., Darwish, D. B., Abdeen, E. E. Comparative study between zinc oxide nanoparticles synthesis by biogenic and wet chemical methods in vivo and in vitro against *Staphylococcus aureus*. *Microbial pathogenesis* **2020**, 147, 104–384.

4 Biogenic Metallic Nanoparticles as Eco-Friendly Biocidal Agents

Marwa M. Eltarahony and Ayman Kamal

4.1 INTRODUCTION

Nanotechnology, which is deemed as "the sixth revolutionary technology", following the industrial, nuclear energy, green, information technology, and biotechnology revolutions, opens up new avenues for multiple significant applications in various sectors, including the agricultural, industrial, medical, environmental, and therapeutic sectors. Via manipulating the molecular level of materials, the intrinsic traits of them are altered, and other new properties are gained. Particles in sizes smaller than 100 nm are considered as nanoparticles [1]. Owing to this size, which mimics the dimension of biological entities, nanoparticles (NPs) exhibit vital characteristics, generating superior functionalities in several applications, in particular a biological one, which allow easier penetration of biomolecules and effective interaction with complex biological systems [2].

Metals and their oxide NPs (MNPs/MONPs) are categorized among inorganic NPs that received great appeal as a materials science discipline and in technological applications owing to their size/shape-dependent structural, physicochemical, optical, biological, and functional characteristics. Generally, silver (Ag) NPs are commonly utilized in medical, textile, water treatment, and food packaging applications, while gold (Au) and palladium (Pd) NPs are widely employed in biomedical and pharmaceutical research. However, for industrial applications, platinum (Pt) NPs are mainly recruited. Remarkably, other metals, such as iron oxide, titanium oxide, and copper oxide, can serve as catalysts for the removal of common environmental contaminants, sensing poisonous gases, and as antifouling agents in paints and fabrics. It is important to report that MNPs/MONPs are considered multifunctional technical tools [3].

Despite the physical and chemical or even hybrid synthesis approaches that provide considerable quantities with excellent performance, green synthesis means are the most preferable. That is attributed to sustainability, economical feasibility, low toxicity, and being eco-friendly, with high biocompatibility features of biologically synthesized NPs. The biological means of synthesis can be implemented either by plants/extracts or microbes (bacteria, yeast, mold, viruses, and algae). Interestingly, all such biogenic methods furnish biomolecules that serve not only as reducing agents but also as functionalizing and stabilizing agents simultaneously, which are generated either intracellularly or extracellularly [1]. Metals (e.g., Au, Bi, Pt, Ag, Se, Ti, Cu, Fe, Co, Ni, Zn, etc.) and their oxides (MNPs/MONPs) are the most common NPs that are extensively fabricated via such bottom-up biological methods [4].

4.2 METALS/OXIDES NANOPARTICLES (MNPS/MONPS) EXAMPLES

4.2.1 Silver (Ag)

From the ancient era, a colloidal solution of silver was utilized in China and Egypt for combating pathogenic infections, which proved its antimicrobial potential [5]. So AgNPs in recent products exhibit venomous features against a broad spectrum of pathogens. The AgNPs are also reported in medical and pharmaceutical applications directed in human consumption products, such as

DOI: 10.1201/9781003430087-4

shampoos, soaps, detergents, toothpaste, and cosmetics [6]. The biomedical use of AgNPs comprises their employment as antibacterial [7], antifungal [8], anti-inflammatory, antiviral [9], and antidiabetic agents [10] and as drug carriers either by active or passive strategy [11]. Recently, the prevalence of antibiotic-resistant microbes has become alarming and disturbing due to the concomitance of certain pathogens with epidemic diseases that elevate the mortality rate in humans, thus an urgent requirement for developing biocompatible, cost-effective, and biosafe AgNPs as antimicrobial agents with good potency, especially against multidrug-resistant strains (MDR).

4.2.1.1 Antibacterial Activity of AgNPs

As AgNPs are the most commonly examined metal among noble metals, the following table (Table 4.1) denotes the antibacterial potency of biogenic AgNPs against different bacterial species, demonstrating their size and effective concentration. As illustrated, there is a variation in the effective doses between gram-negative and gram-positive species, implying that the antagonistic proficiency varies according to the differences in cell wall composition, structure, and polarity.

4.2.1.2 Antifungal Activity of AgNPs

Table 4.2 summarizes the antifungal potency of biosynthesized AgNPs obtained by different biological reducing/capping agents.

4.2.1.3 Antibiofilm Activity of AgNPs

Many studies reported positive effect of AgNPs against microbial biofilm *S. aureus* [19], *P. aeruginosa* [20], *Acinetobacter baumannii, P. aeruginosa, S. aureus, Streptococcus mutans, C. albicans* [21], and *S. epidermidis* [22]. [23] studied the antibiofilm potentiality of AgNPs against antibiotic-susceptible and multidrug-resistant *P. aeruginosa* strains. The authors attributed the antibiofilm activity to the devastation of biofilm exopolysaccharide (EPS) structure and oxidative stress at doses ranging from 1 to 600 μg/mL of AgNPs. However, AgNPs produced by endophytic fungus *A. tubingensis* were impregnated on polyester fabric and examined against *B. subtilis* biofilm. The results indicated powerful antibiofilm efficacy at 8 μM. Scanning electron microscopy (SEM) micrographs depicted the inhibition of biofilm cells and their EPS matrix by the dint of small-sized AgNPs, which recorded 35 nm [24].

TABLE 4.1
Antibacterial Activity of AgNPs Synthesized Biologically against Different Bacterial Genera with Their Effective Doses

Source of AgNPs	Target	Effective Concentration	References
Aspergillus tubingiensis and *Bionectria ochroleuca*	*Pseudomonas aeruginosa*	0.28 μg/mL	[12]
Proteus mirabilis 10B	*E. coli, Staphylococcus aureus, Salmonella typhi, P. aeruginosa, Clostridium perfringens, Enterococcus faecalis, Bacillus cereus*	100 μg/mL	[1]
Streptomyces fulvissimus EM1 and *Streptomyces mediolani* EM2	*E. coli, S. typhi, S. aureus, P. aeruginosa, P. vulgaris, K. pneumoniae, E. faecalis, B. cereus*	50 μg/mL	[13]
Penicillium aculeatum	*E. coli, B. subtilis, S. aureus, P. aeruginosa*	50, 100, and 200 μg/mL	[14]
Stenotrophomonas rhizophila	*P. aeruginosa, S. typhi, E. coli, Klebsiella sp., Microbacterium* sp., *Bacillus* and *Lactobacillus* sp.	100 μg/mL	[15]

TABLE 4.2
Fungicidal Activity of Biogenic AgNPs versus Different Fungal Genera with Their Effective Doses

Source of AgNPs	Target Fungal Pathogen	Effective Concentration	References
Ligustrum lucidum leaf extract	*Setosphaeria turcica*	170.20 μg/mL	[16]
Bacillus sp.	*A. niger, A. spergillus fumigatus, F. soleni*	10–50 μg/mL	[17]
Trichoderma harzianum	*Botrytis cinerea, F. oxysporum, A. parasiticus Alternaria alternata, A. niger*, and *P. chrysogenum*	1–10 mg/ml	[18]

TABLE 4.3
Virucidal Potency of Biosynthesized AgNPs against Diverse Types of Viruses with Their Effective Doses and Lethal Effects

Target	Characteristic Features of AgNPs	Effective Concentration and Lethal Effects	References
Tacaribe virus	10 nm	50, 25, and 10 μg/mL; significant reduction in the progeny virus titer	[26]
Hepatitis B virus	10 nm and 50 nm	10 nm (5 μM) inhibited 38%; 10 nm (50 μM) inhibited 80%; 50 nm (5 μM) inhibited 53%; 50 nm (50 μM) inhibited 92%	[27]
T-tropic, M-tropic, and dual-tropic strains of HIV-1	30–50 nm	Inhibited viral entry and replication at IC_{50} in the range of 0.44–0.91 mg/mL	[28]

4.2.1.4 Antialgal Activity of AgNPs

The algicidal potency of AgNPs on photochemical reactions of photosynthesis was scrutinized in *Chlamydomonas reinhardtii*. Algal cells were treated with 1, 5, and 10 μmol L^{-1} of AgNPs. By the aid of the fluorescence imaging system, the structural disruption of photosystem II (PSII) photochemical reaction center, the suppression of electron transport activity, and the changes in oxygen-evolving complex were detected and compared in control and treated algal cultures. This study gave an insinuation about how to use photosynthetic-based fluorescence parameters, which are valuable and informative, in evaluating the aquatic toxicological risk of AgNPs [25]. However, about 300 μg/mL of bacterially prepared AgNPs that caused inhibition in the photosynthetic activity of *Chlorella vulgaris* reached 96% compared to silver nitrate, which inhibited only 65%, implying the effectiveness of small size with larger surface area for Ag in its nanoform.

4.2.1.5 Antiviral Activity of AgNPs

The virucidal potentiality of AgNPs versus different viral types and also at different stages is demonstrated in Table (4.3).

4.2.1.6 Antiparasitic Activity of AgNPs

In a recent study, AgNPs (9 nm) was tested versus two of the most omnipresent parasites in Egypt (*Cryptosporidium parvum* and *Entamoeba histolytica*). A noticeable reduction for cysts viability was detected significantly in their oocysts at lethal concentration 50 (LC_{50}) of AgNPs for *E. histolytica* and *C. parvum*, reaching 0.34 and 0.54 mg/l, respectively. Hereby, AgNPs could be proposed as a new, safe, and effective nanoform treatment against *E. histolytica* and *C. parvumparasites* [29]. On the other hand, [30] found that infection index values for amastigotes of *Leishmania tropica* exposed to AgNPs were appraised as 35, 23, and 15 when amastigotes were treated with 1, 5, and 10 μg/mL of AgNPs, respectively, revealing the superior antagonistic potency via several mechanisms, such as reactive oxygen species (ROS). Infection index values for amastigotes were evaluated as 35, 23, and 15 when amastigotes were exposed to 1, 5, and 10 μg/mL of Ag-NPs, respectively, revealing antagonistic potency via several mechanisms.

4.2.2 Iron (Fe)

Despite the powerful antimicrobial activity of AgNPs in several sectors, their cytotoxicity remains the main concern that restrict their application [31]. However, several investigations have listed their toxicity in low doses (10–50 μg/mL). Therefore, all attentions are directed to naturally originating elements that are essential for physiological processes of living creatures, hence symbolized as biosafe, biocompatible, and low-toxic impact in returning to human being and surrounding milieu. On the other hand, [32] found that bacterially fabricated FeNPs kept normal human cell viability at safe dose (concentration that achieves 100% cell viability) of EC_{100} up to 1999.7 μg/mL. Meanwhile, [33] illustrated that FeNPs showed low or no cytotoxicity at 100 μg/mL as biocidal concentration. Interestingly, the Food and Drug Administration (FDA) gave approval for some FeNP formulations to be used in iron-deficiency anemia in adults and as magnetic resonance imaging agents of liver pathologies. Herein, the antimicrobial potentiality of biologically synthesized Fe-based nanoformulation would be displayed.

4.2.2.1 Antibacterial Activity of FeNPs

The phyto-extract of *Phoenix dactylifera* was utilized as a reducing agent in the preparation of FeNPs in a study implemented by [34]. The results revealed the potency of as-prepared FeNPs against *K. pneumonia* and *E. coli* more than that observed against *B. subtilis*. However, [35] prepared both zerovalent iron (ZVINPs) and magnetite NPs (MTNPs) under aerobic and anaerobic growth of *P. mirabilis* and examined their antagonistic activity against a broad spectrum of pathogens. Moreover, in this study, the authors examined the disinfection potency of both types of NPs to eradicate the microbial loads in eight water and wastewater samples collected from different environmental localities. The results unveiled the superior effectiveness of MTNPs over ZVINPs, assigning these differences to the smallest particle size of MTNPs, which provided higher-surface-area-to-volume ratio and, eventually, more lethal effect. Additionally, [36] visualized the potentiality of hematite NPs (HNPs) in deformation and deterioration of the propagation of *E. coli*, *C. perfringens*, and *S. aureus* using both electron microscopy techniques, which illustrated the presence of pits and pores in the examined pathogens upon HNPs treatment; also, the formation of ghost cells of *S. aureus* was recorded, while *C. perfringens* cells exposed to HNPs manifested an irregular cell wall and deformation of peripheral spore, which showed vanishing of the peptidoglycan layer with an irregular and thin spore coat, implying the release of immature spore.

4.2.2.2 Antifungal Activity of FeNPs

Regarding fungal growth frustration, Table 4.4 illustrates the different types of FeNPs with different properties fabricated from different biomolecules, either microbial or plant.

4.2.2.3 Antibiofilm Activity of FeNPs

As a biofilm-eradicating agent, FeNPs were utilized to prevent the formation of biofilm in its multistage process, as represented in Table 4.5.

TABLE 4.4
Antifungal Potency of Biofabricated FeNPs versus Different Fungal Species, with Their Properties and Effective Doses

Target	Preparation Method	NPs Characterization and Effective Concentration	References
A. fumigatus and *C. albicans*	Extracellular extract of *A. flavus*	Spherical IONPs (28.6–33.8 nm); increasing the permeability of membranes and ROS at 10 mg/ml	[37]
A. Brasiliensis and *C. albicans*	*Proteus mirabilis 10B*	MNPs (1.44–1.92 nm and −31.8 mV ξ potential) and ZVINPs (11.7–60.8 nm and −66.4 mV ξ potential); effective inhibition at 250 μg/mL	[35]
A. mali, Botryosphaeria dothidea, and *Diplodia seriata*	Leaf extract of *Azadirachta indica*	20–80 nm, SPR at 272 nm; anti-phytopathogenic activity in dose-dependent manner (10, 25, 50, and 100 ppm)	[38]

TABLE 4.5
Antibiofilm Activity of Biogenic FeNPs against Different Biofilm-Forming Species, with Their Types, Properties, and Effective Doses

Target	Preparation Method	Characterization and Effective Concentration	References
Methicillin-resistant *Staphylococcus aureus* (MRSA), *C. albicans*, and *P. aeruginosa*	Extract of polyherbal drug Liv52 (L52E)	1.25 mg/ml of maghemite (γ-Fe_2O_3) NPs; suppression percentage range 55.5%–82.3%	[39]
S. aureus and *E. coli*	*Lactobacillus sp.*	Catheters coated by IONPs (68.28 nm); inhibition by 33.97% (*S. aureus*) and 16.92% (*E. coli*)	[40]
Microfouling bacteria in seawater	*Sargassum vvulgare*, *Ulva fasciata*, and *Jania rubens*	As biofouling paint, IONPs (Fe_3O_4NPs), with 17–34 nm size, blocked biofilm growth on coated wood and steel	[41]

4.2.2.4 Antialgal Activity of FeNPs

IONPs were utilized as antialgal agent in phytoremediation as a modern tool in environmental biotechnology. [42] prepared IONPs (Fe_3O_4) using brown seaweeds, including *Petalonia fascia*, *Colpomenia sinuosa*, and *Padina pavonica*. The as-fabricated IONPs were characterized and examined in the remediation of marine water from the blooming of harmful algae. The results of this investigation showed the diminution of more than 90, 89.9, and 93.9% of total chlorophyll related to phytoplankton, total nitrogen, and total phosphorous, respectively. Considering the same target, 150 and 300 μg/mL of HNPs led to the reduction of *C. vulgaris* viability by 54.5 and 86.3%. Upon visualization by electron microscopy, HNPs triggered drastic ultrastructure changes in algal cells via cell wall decomposition, protoplast retraction, cytoplasmic organelle degradation, and thylakoid dilatation, compared to control cells without treatment, which remained healthy with normal cell morphology and cellular organelles. Generally, FeNP-mediated chloroplasts dysfunction, corruption of photosynthesis process, and finally cell death [36].

4.2.2.5 Antiviral Activity of FeNPs

As a general observation, the virucidal and antiparasitic activity of biogenic MNPs/MONPs is scarce, except from AgNPs; this entails more studies regarding this goal, in particular with continuous publications reporting day-to-day in MNPs/MONPs synthesis biogenically [43]. Nonetheless, it is vital to shed light on the studies focused on virucidal capacity of MNPs/MONPs. In this concern, [44] investigated the antiviral potency of IONPs on anti-influenza activity and H1N1 cell viability, which was assessed by MTT assay, quantitative real-time PCR, and plaque inhibition. On the other hand, [45] examined the impact of ZVINPs on human norovirus surrogates, Tulane virus, and murine norovirus. He and coauthors found that ZVINPs enhanced their removal from water columns.

4.2.2.6 Antiparasitic Activity of FeNPs

Virtually, [46] performed surface modification for artesunate (antimalaria drug) with IONPs and examined them versus *Plasmodium falciparum* and in an experimental malaria mouse model. The results unveiled a retardation of *P. falciparum* growth and malfunction of macromolecules induced by the enhanced ROS production. Additionally, preclinical *in vivo* studies indicated a radical reduction in parasitemia with eight- to tenfold decrease in drug dosage upon conjugation with IONPs; by such way, this nanocombination formula represents an efficient solution against artemisinin-resistant parasites. Meanwhile, [47] studied the conjugation of IONPs with isoniazid and amphotericin B to detect antiamoebic impact in response to *Acanthamoeba castellanii*. The results showed that IONPs enhanced trophicidal and cysticidal influences with limited cytotoxicity versus rat and human cells.

4.2.3 Copper (Cu)

Copper is a transition metal grouped among semiconducting compounds that has different oxidation states (Cu0, Cu^{+}, Cu^{2+}, and Cu^{3+}) and a narrow bandgap (Eg), which all such peculiar features collectively encourage its wide applications [48], let alone its natural origin and distinguished role in physiological processes. Recently, by the dint of availability with low cost and relative physicochemical stability, it replaced other noble metals, such as Ag, Pt, Au, etc., in several nanotechnological applications. In recent decades, the US Environmental Protection Agency (EPA) approved copper and its nanostructures as an antimicrobial agent.

4.2.3.1 Antibacterial Activity of CuNPs

Table 4.6 summarizes some of the studies focused on the synthesis, characterization, and environmental application of biogenic CuNPs.

TABLE 4.6
Bactericidal Potentiality of CuNPs Biosynthesized versus Bacterial Genera, with Their Properties and Effective Doses

Target	Preparation Method	NPs Type, Characterization, and Effective Concentration	References
B. cereus, E. faecalis, S. aureus, E. coli, P. aeruginosa, S. typhi	*Streptomyces cyaneofuscatus*	12–36 nm size of Cu^0, CuO, and Cu_2O nanocomposites; 750 µg/mL inhibited bacterial growth in the following order: *B. cereus* > *S. aureus* > *E. faecalis* > *E. coli* > *S. typhi* > *P. aeruginosa*	[32]
S. aureus, E. coli, S. typhi, P. aeruginosa, C. perfringens, E. faecalis, Bacillus cereus	*Proteus mirabilis* 10B	Intracellularly zero-dimensional CuONPs (10 nm); higher toxicity in gram-negative than gram-positive Extracellularly one-dimensional CuONPs (17–37.5 nm)	[48]

4.2.3.2 Antifungal Activity of CuNPs

Via an eco-friendly synthesis of CuNPs, *Celastrus paniculatus* leaves extract was employed as a reducing and stabilizing agent in this green method. The phyto-synthesized NPs with 2–10 nm size and negative zeta potential (-22.2 mV) exhibited mycocide efficiency versus *F. oxysporum*. By determining the radial growth of mycelia, the maximum mycelial inhibition reached 76.29 ± 1.52 at a concentration of 0.12% of CuNPs; relative to $CuSO_4$ (1.0%), the inhibition of fungal mycelial growth showed 20.74 ± 1.52 [49]. However, [50] produced CuONPs by *Azadirachta indica* and characterized them, which proved their antifungal capacity against plant pathogenic fungi, viz. *A. mali*, *Diplodia seriata*, and *B. dothidea*, with 40% of mycelial growth inhibition at a concentration of 1 g/ml.

4.2.3.3 Antibiofilm Activity of CuNPs

As nanotherapy, [51] employed marine entophytic actinomycetes in CuONPs production and examined its antibiofilm potency against *E. coli* and *P. mirabilis*, where 1,000 µg/mL implemented 92% and 94% biofilm eradication efficiency for *E. coli* and *P. mirabilis*, respectively, ascribing such result to the capability of CuONPs to disrupt the structure of the exopolysaccharide layer, the cytoplasmic membrane via continuous release of ROS, and also arresting of the quorum-sensing system. More so, [52] utilized *E. faecalis* to prepare CuONPs extracellularly by enzymatic method. The antibiofilm activity recorded was 36.54% and 28.74% for *P. aeruginosa* and *S. aureus*, respectively.

4.2.3.4 Antialgal Activity of CuNPs

Notably, [53] studied the impact of CuNPs exposure to the green alga *Chlamydomonas reinhardtii* in comparison to the essential micronutrient copper. By applying a high-throughput ATP-based fluorescence assay, the toxicity to algae was determined, where IC_{50} recorded 14–16 mg/L. Similarly, [54] prepared CuNPs form red seaweed extracts (e.g., *Corallina officinalis Linnaeus* and *Corallina mediterranea*). Their algicide potency against harmful cyanobacterium *Lyngbya majuscule* recorded was 85 ± 4% growth reduction at a concentration of 2 g/mL. Notably, with increasing the exposed dose to 4 g/mL, further reduction was noticed by 91 ± 6%. In the same perspective, 300 µg/mL of spherical shape of CuONPs caused inhibition of *C. vulgaris* growth by 92.3%, while only 20% growth suppression was noticed at exact concentration of CuO nanowires, reflecting the impact of size and morphology in antagonistic efficiency.

4.2.3.5 Antiviral Activity of CuNPs

Remarkably, [55] evaluated the antiviral potency of CuONPs against hepatitis C virus (HCV), wherein CuONPs didn't exert any influence on HCV replication; nonetheless, the main mechanisms summarized in blocking the entry and attachment of HCV pseudoparticle. Additionally, [56] declared the antiviral traits of CuNPs in response to the non-enveloped virus feline calicivirus (FCV) as an alternative for human norovirus. Upon treatment with 1,000 µg/ml of CuNPs, the infectivity of FCV to CRFK cells was significantly diminished by 10^7 time via hydroxyl radical production, which was proven through electron spin resonance (ESR) analysis. A recent study addressed CuONPs in the therapy of SARS-COV-2, in comparison to AgNPs and ZnNPs [57]. The CuONP film showed a significant reduction (54% reduction) in the viral load after 30 min and reached more than 76% reduction after 120 min.

4.2.3.6 Antiparasitic Activity of CuNPs

The powerful capability of CuNPs as an antiparasitic agent was evaluated by [29] against two parasites, *E. histolytica* and *C. parvum*. Via small sizes of synthesized CuONPs (29 nm), LC_{50}-3h of CuONPs recorded 0.13 and 0.72 mg/l for *E. histolytica* and *C. parvum*, in respective order. Similarly, [58] examined the effectiveness of CuNPs (10–200 µg/mL) as *in vitro* antileishmanial agent, in comparison to meglumine antimoniate with exact concentration. The results referred to the growth suppression of *Leishmania major amastigotes* by the action of CuNPs, which enhanced the production of NO ($p < 0.05$) in a concentration-dependent way. On the other hand, [59] stated that CuNPs exerted an equivalent effect as metronidazole at a concentration of 0.6 mg/ml in killing of *Giardia* cysts.

4.2.4 Zinc (Zn)

ZnO is an n-type semiconducting metal oxide that is distinguished by a large bandgap (3.37 eV) and considerable exciton binding energy (60 meV). Recently, **the** US-FDA has enrolled ZnO as a safe metal oxide [60]. In addition, Zn plays pivotal regulatory, catalytic, and structural roles in more than 300 proteins. It exerts prominent functions in amino acid biosynthesis, protein stabilization/folding, gene expression management, DNA replication/repair, cofactor of virulence-related proteins, and buffering intracellular redox potential [61, 62].

4.2.4.1 Antibacterial Activity of ZnNPs

Table 4.7 sums up some investigation concerning the biological synthesis of ZnONPs and their effectiveness as antimicrobial agents.

4.2.4.2 Antifungal Activity of ZnNPs

Commonly, phytopathogens devastate a vast array of crop yields not only in the fields but also post-harvest. Therefore, commercial fungicides have been extensively applied to control fungal spread; however, the generation of resistance against them generates worrisome problems. Subsequently, the emergence of a new class of anti-phytopathogens such as ZnONPs is deemed as promising solution. As indicated in Table 4.8, different fungal responses were observed upon treatment with biogenic ZnONPs.

4.2.4.3 Antibiofilm Activity of ZnNPs

To combat biofilm formation and overcome its adverse effect, [65] investigated the antibiofilm of ZnNPs prepared by leaf extract of *Ochradenus baccatus*. The spherical/oval shape of the ZnNPs displayed eminent antibiofilm potency against *S. marcescens*, *K. pneumoniae*, *E. coli*, *Listeria monocytogenes*, and *P. aeruginosa*. In a dose-dependent behavior, the inhibition of biofilm formation happened in the range of 64–84%; however, the range of 59–81% reduction was determined in EPS production in the biofilms of *E. coli*, *P. aeruginosa* PAO1, *L. monocytogenes*, *S. marcescens*, and *K. pneumoniae*. On the other hand, [66] examined the antibiofilm potentiality of ZnONPs, synthesized by leaf extract of *Plectranthus amboinicus*, against the growth of MRSA biofilms.

TABLE 4.7
Bactericidal Capacity of Biogenic ZnNPs against Different Bacterial Species, with Their Properties and Effective Doses

Target	Preparation Method	Characterization and Effective Csoncentration	References
S. aureus, *L. monocytogenes*, *E. faecalis*, *K. pneumoniae E. coli*, *E. faecium*, *S. pneumoniae*, *B. cereus*, *S. typhi*, and *M. catarrhalis*	Pomegranate, *Punica granatum* (flower and leaf extract)	Spherical ZnONPs (50.95–54.84 nm); MIC at 1,250 μg/mL as for *S. aureus*, *E. coli*, *L. monocytogenes*, *E. faecium*, *E. faecalis*, *K. pneumoniae*, *S. typhi*, *S. pneumoniae*, and 0.61 ± 0.54 μg/mL for *M. catarrhalis* and *B. cereus*	[61]
E. coli and *S. aureus*	*Cassia fistula* and *Melia azedarach*	ZnONPs spheres (68 and 3.6 nm); effective potency at a concentration range of 50–1,000 μg/mL	[62]

TABLE 4.8
Mycocid Potentiality of ZnNPs Fabricated Biologically versus Different Bacterial Genera, with Their Properties and Effective Doses

Target	Preparation Method	Characterization and Effective Concentration	References
A. alternata, *B. cinerea*, *A. niger*, *P. expansum*, and *F. oxysporum*	Flower extract of *Nyctanthes arbor-tristis*	12–32 nm ZnONPs exhibited the lowest minimum inhibitory concentration (MIC) at 16 μg/mL	[63]
A. mali, *B. dothidea*, and *Diplodia seriata*	*Eucalyptus globules*	ZnNPs (20 nm); inhibition range 55.2%–76.7% at 100 ppm	[64]
C. albicans	*Proteus mirabilis* 10B	ZnONPs nanospheres (19.1 nm, −53 mV); 200 μg/ml	[1]

Confocal laser scanning microscope (CLSM) imaged the damage of biofilm architecture upon treatment with 20–50 nm ZnONPs.

4.2.4.4 Antialgal Activity of ZnNPs

The algal blooms dispersion caused serious environmental issues, such as eutrophication and biofouling. As a model of green algae, *C. vulgaris* was used to determine the performance of bacterially synthesized ZnONPs at 150 and 300 μg/mL. The results signalized the potency of both doses in suppression of 44.5 and 73% of algal growth. The blocking in photosynthesis and damage of chloroplasts were also noticed through yellowish to faint green color of algal cells relative to the dark green of fresh, healthy untreated cells [1]. However, the toxic effect of ZnONPs on *Spirulina platensis* showed time-dependent cytotoxicity, where the maximum cytotoxicity was achieved at 96 h of treatment with ZnO NPs (200 mg/L) by 87.3 ± 1% viability reduction and 92.5 ± 2.2%, 76.2 ± 2.2%, and 74.1 ± 3.4% lowering in the contents of chlorophyll-a, carotenoids, and phycocyanin, respectively [67].

4.2.4.5 Antiviral Activity of ZnNPs

Regarding the virucidal capacity of ZnNPs, different studies detected its considerable activity. In a study conducted by [68], ZnONPs led to 2.8 log^{10} median tissue culture infectious dose ($TCID_{50}$) reduction in H1N1 virus. At the highest non-toxic doses, the post-exposure of influenza virus to ZnONPs offered 94.6% viral inhibition. However, [69] studied the recruitment of ZnONPs synthesized phytologically in virucidal potency. Upon treatment of herpetic HSV-1 with ZnONPs, a promising virucidal activity was noticed by 50% cytotoxicity concentration (CC_{50}) and IC_{50} of 43.96 ± 1.39 and 23.17 ± 2.29 μg/mL, respectively. The authors proposed the strategy of antiviral activity through hindering the virus's attachment/entry and also changing the virostatic potential. Interestingly, [70] tested the influence of ZnONPs on hepatitis A virus (HAV virus HM175 strain, ATCC VR-1402), and they found that ZnONPs offered advantageous antiviral activity by 58.83% at maximum noncytotoxic concentration and EC_{50} assessed by 176.3 μg/mL.

4.2.4.6 Antiparasitic Activity of ZnNPs

In this concern, [71] employed biocompatible ZnONPs as an anthelmintic agent to overcome *Gigantocotyle explanatum* infection. By exposure of *G. explanatum* to 0.012% ZnONPs, significant abatement of some enzyme activities, like superoxide dismutase and glutathione S- transferase, which serve as antioxidant enzymes, was observed. Likewise, in another study performed by [72], the ascaridida nematode (*Parascaris equorum*) exposed to ZnONPs (100, 200, 300, 400 mg/L) had a 50% mortality rate detected after treatment with 100 mg/l ZnONPs. In addition, as anti-leishmaniasis agent, ZnONPs (5 μg/ml) was applied against *Leishmania tropica* in both phases (promastigote and amastigote) *in vitro*.

4.2.5 Titanium (Ti)

TiO_2 is the oxide form of titanium metal and naturally occupying the Earth's crust in different polyforms, such as anatase, rutile, or brookite. The unique physicochemical properties of TiO_2, like higher refractive index, absorption of ultraviolet (UV) light, photocatalytic and antimicrobial activities, empower its application in food preparation, sunscreens, cosmetics, and drug delivery systems [73]. Recently, it was approved by FDA to be applied as food coloring additive.

4.2.5.1 Antibacterial Activity of TiO2NPs

In the study achieved by [74], Propolis extract was used in the reduction and stabilization of TiO_2NPs (57.3 ± 4nm), which exhibited not only considerable antibacterial activity against MRSA, *S. epidermidis*, *K. pneumoniae*, *P. aeruginosa*, *L. monocytogenes*, *P. vulgaris*, and *Acinetobacter baumannii* at 16, 8, 32, 64, 64, and 128 µg/mL as MIC but also exerted significant antioxidant and anticancer activity both *in vitro* and *in vivo*. On the other hand, [75] produced TiO_2NPs bacterially with spherical shape (10–20 nm) and offered their potent biocidal activity against *B. subtilis* and *E. coli* either in their free-living or biofilm lifestyles. Recently, [76] phyco-synthesized TiO_2NPs, which showed a crystalline structure and negative zeta potential with cubic-, square-, and spherical-shaped particles (50–90 nm). The powerful antibacterial impact was verified against *S. aureus*, *P. vulgaris*, *S. epidermidis*, *K. pneumoniae*, *E. coli*, and *P. aeruginosa*, also in both planktonic and biofilm states.

4.2.5.2 Antifungal Activity of TiO2NPs

In applications in the field, TiO_2NPs succeeded in preventing fungal infections. Table 4.9 displays some examples on the fungicidal potency of TiO_2NPs.

4.2.5.3 Antibiofilm Activity of TiO2NPs

Interestingly, [80] produced mixed-sphere and square-shaped TiO_2NPs (50–90 nm, –24mV zeta potential) using root extract of *Withania somnifera*, which offered potent suppression of biofilm formation of *L. monocytogenes* and *S. marcescens* at 64 and 8 µg/ml as MIC. Regarding *C. albicans* biofilm, 64 µg/ml TiO_2NPs blocked biofilm synthesis via decreasing the number of microbial cells that were scattered within less-dense polymeric matrix, as visualized by confocal laser scanning microscopy. Furthermore, significant eradication and disintegration were noticed for mature pre-established biofilms of *E. coli*, *P. aeruginosa*, *L. monocytogenes*, *MRSA*, *S. marcescens*, and *C. albicans* by 45%, 65%, 49%, 64%, 24%, and 46%, respectively.

4.2.5.4 Antialgae Activity of TiO2NPs

As mentioned previously, due to scarcity and fewer availability of studies regarding the biocidal activity of biogenic TiO_2NPs, other than antibacterial and antifungal potency, it is important to

TABLE 4.9
Antifungal Activity of TiO_2NPs Prepared Biologically versus Different Fungal Genera, with Their Characteristic Features and Effective Concentrations

Target	Preparation Method	Characterization and Effective Concentration	References
Ustilago tritici (wheat rust)	*Trianthema portulacastrum* and *Chenopodium quinoa*	15 nm; inhibition at 2.5 µg/mL at range of 36%–54%	[77]
A. flavus, *A. niger*, *Rhizopus oryzae*, and *Sclerotium rolfsii*	*Luffa acutangular* leaf extract	10–49 nm, SPR at 322 nm; 40 mg/ml suppressed the growth (27 ± 0.36–42 ± 0.25 mm)	[78]
C. albicans	*Andrographis paniculate* (king of bitter herbal plant)	22 nm; full eradication at 50 µL	[79]

address the other biocidal impacts of TiO_2NPs, whatever its synthesis route. In a study achieved by [81], 10, 50, and 200 nm of TiO_2NPs (10 mg/L) were examined to determine the toxicity of TiO_2NPs on phytoplankton, such as *Scenedesmus obliquus*. Overall, TiO_2NPs induced vacuoles and cell membrane disruption, implying stimulation of oxidative stress that influences severely on the entire photosynthesis process. In the same extent, [82] compared the toxic effect of three types of TiO_2NPs, which were significantly differed in their physicochemical properties, in opposes to freshwater algae *Chlorella pyrenoidosa* in short- and long-time exposure. The biological responses of green algae in terms of esterase activity, growth rate, and pigment autofluorescence were defined.

4.2.5.5 Antiviral Activity of TiO2NPs

In facing the Newcastle virus (NDV), [83] assessed the antiviral activity of TiO_2 nano-colloids against NDV, which displayed remarkable virucidal potency at 6.25 ug/ml through lipid damage. In addition, TiO_2NPs exhibited considerable virucidal activity against the influenza virus (H3N2) through a direct interaction pathway, as explained by [84]. Meanwhile, [85] declared that the multiplication of H5N1 and H1N1 viruses was blocked via utilizing DNA-tagged TiO_2NPs. However, the antagonistic potentiality of TiO_2NPs against broad bean strain virus in faba beans was performed in a study by [86].

4.2.5.6 Antiparasitic Activity of TiO2NPs

The as-synthesized TiO_2NPs produced by green route of *Mangifera indica* extract showed a remarkable larvicidal potency in facing blood-feeding larvae, such as *Rhipicephalus microplus*, *Haemaphysalis bispinosa*, *Hyalomma anatolicum anatolicum*, *Culex quinquefasciatus*, and *Anopheles subpictus*, by 28.56, 33.17, 23.81, 4.34, and 5.84 mg/L as LC_{50} values, respectively [87]. On the other hand, [88] examined the acaricidal potentiality of TiO_2NPs, fabricated by *Calotropis gigantea* flower extract, toward *Rhipicephalus microplus* and *Haemaphysalis bispinosa*. He and coauthors found that LC_{50} recorded 24.63 and 35.22 mg/L for *R. microplus* and *H. bispinosa*, respectively.

4.3 BIOCIDAL SCENARIOS FOR INHIBITING MICROBIAL GROWTH

In brief, MNPs/MONPs serve merely by contacting with microbes' cell walls via several means, such as van der Waals forces, receptor/ligand, electrostatic attraction, and hydrophobic interactions. Once successful interaction happens, MNPs/MONPs invade through the inner membranes, then interfere with metabolic paths, causing lethal adverse effects. However, oxidative stress, non-oxidative stress, and dissolved metal ions are the common antimicrobial tactics. Whatever the followed mechanism, which is considered as unclear and undefined, there are general features accompanying MNPs/MONPs treatment. Broadly, MNPs/MONPs deteriorate cellular machinery and block physiological developments through modifying the membrane penetrability of microbial cells, mutilating cell membranes, disrupting electrolyte balance, interacting with functional groups, disabling proteins functions, inactivating enzymes, binding with nucleic acids, malfunctioning of cell replication, altering gene expression scale, and inducing oxidative stress. Remarkably, MNPs/MONPs stimulate oxygen and form reactive oxygen species (ROS), like single oxygen, hydrogen peroxide (H_2O_2), superoxide radical (O^-), and hydroxyl radical (OH), which change the intracellular redox state of the cells and subsequently induce oxidative stress instantly. The detailed descriptions concerning mechanisms are elaborated extensively in various studies by several research groups, such as [89, 90].

4.4 FUTURE PERSPECTIVES OF MPS/MONPS

The prevalence of multidrug-resistance (MDR) phenomena among pathogens, due to misuse and over usage of antibiotics, leads to the development of new classes of antimicrobial therapeutics. Among such promising medications are nanoparticles. However, different microbial species produce excessive types of barriers that tolerate such antimicrobial mediators. Nevertheless, based on the nano concept, the quick progression in nanotechnology promotes fabrication of various nanoformulations that focus their effectiveness on arresting microbial resistance. The fabrications of doped metal/metal oxide/metal nanomaterials, metal organic frameworks, bimetallic NPs, enzyme-based metal nanocomposites, polymer-based metal nanocomposites, and carbon-based metal nanocomposites are considered as novel crystalline nanomaterials. All such nanocomposites exhibit eminent structural characteristics with informal functionalization, unusual biodegradability, effective biological activity, and good biocompatibility. The decoration of several materials as listed previously with different types of metals in their nanoscale would exert outstanding synergism in their antimicrobial properties, which represent as shock for microbes to develop sudden resistance mechanism, therefore encouraging their potential applications.

4.5 CONCLUSION

In the post-antibiotic era, metal nanomaterials offer a promising alternative means to counter MDR infections. In order to substitute antibiotic usage, numerous and diverse metallic nanoparticles were developed in considerable amounts and tunable properties by chemical and physical means. However, there were a number of challenges that obstructed the utilization of chemically/physically prepared nanomaterials in biological applications, regarding safety, biocompatibility, and effectiveness. Therefore, the biological means of synthesis presented an easy and better solution for such issues. In this chapter, we shed light on MNPs/MONPs and their biocidal activity against a wide spectrum of unicellular, multicellular, free-living, sessile encapsulated biofilm, prokaryotic, and eukaryotic pathogens, considering the biological tools in the fabrication process, which make the generated nanomaterials more safe, cost-effective, stabilized, and efficacious.

4.6 CONFLICT OF INTEREST

The authors declare that they have no conflicts of interest.

REFERENCES

1. Eltarahony, M.; Zaki, S.; ElKady, M.; Abd-El-Haleem, D. **2018**. Biosynthesis, characterization of some combined nanoparticles, and its biocide potency against a broad spectrum of pathogens. *Journal of Nanomaterials* 2018.
2. Abdulla, N.; Siddiqui, S.; Fatima, B. et al. **2021**. Silver based hybrid nanocomposite: A novel antibacterial material for water cleansing. *Journal of Clean Production* 284: 124746.
3. Nguyen, N.; Padil, V.; Slaveykova, V.; Černík, M.; Ševců, A. **2018**. Green synthesis of metal and metal oxide nanoparticles and their effect on the unicellular alga *Chlamydomonas Reinhardtii: Nanoscale Research Letters* 13: 1–3.
4. Lombardo, D.; Calandra, P.; Pasqua, L.; Magazù, S. **2020**. Self-assembly of organic nanomaterials and biomaterials: The bottom-up approach for functional nanostructures formation and advanced applications. *Materials* 13: 1048.
5. Chugh, H.; Sood, D.; Chandra, I.; Tomar, V.; Dhawan, G.; Chandra, R. **2018**. Role of gold and silver nanoparticles in cancer nano-medicine. *Artificial Cells, Nanomedicine, and Biotechnology* 46: 1210–1220.
6. Narasaiah, B.; Mandal, B.; Chakravarthula, S. **2018**. Mitigation of textile industries generated pollution by agro-waste cotton peels mediated synthesized silver nanoparticles. *Biointerface Research in Applied Chemistry* 8: 3602–3610.
7. Amini, E.; Azadfallah, M. **2018**. In situ synthesis of silver nanoparticles on fiber matrix for preparing antibacterial paper. *Biointerface Research in Applied Chemistry* 8: 3449–3456.
8. Kumar, S.; Shukla, A.; Baul, P.; Mitra, A.; Halder, D. **2018**. Biodegradable hybrid nanocomposites of chitosan/gelatin and silver nanoparticles for active food packaging applications. *Food Package and Shelf Life* 16: 178–184.
9. Lazíc, V.; Vivod, V.; Peršin, Z. et al. **2020**. Dextran-coated silver nanoparticles for improved barrier and controlled antimicrobial properties of nanocellulose films used in food packaging. *Food Package and Shelf Life* 26: 100575.

10. Liao, S.; Zhang, Y.; Pan, X. et al. **2019**. Antibacterial activity and mechanism of silver nanoparticles against multidrug-resistant Pseudomonas aeruginosa. *International Journal of Nanomedicine* 14: 1469–1487.
11. Samoilova, N.; Krayukhina, M.; Popov, D.; Anuchina, N.; Piskarev, V. **2018**. 30-sialyllactose-decorated silver nanoparticles: Lectin binding and bactericidal properties. *Biointerface Research in Applied Chemistry* 8: 3095–3099.
12. Rodrigues, A.; Ping, L.; Marcato, P. **2013**. Biogenic antimicrobial silver nanoparticles produced by fungi. *Applied Microbiology and Biotechnology* 97: 775–782.
13. Eltarahony, M.; Ibrahim, A.; El-Shall, H.; Ibrahim, E.; Althobaiti, F.; Fayad, E. **2021**. Antibacterial, antifungal and antibiofilm activities of silver nanoparticles supported by crude bioactive metabolites of bionanofactories isolated from Lake Mariout. *Molecules* 26: 3027.
14. Ma, L.; Su, W.; Liu, J. et al. **2017**. Optimization for extracellular biosynthesis of silver nanoparticles by Penicillium aculeatum Su1 and their antimicrobial activity and cytotoxic effect compared with silver ions. *Materials Science and Engineering: C* 77: 963–971.
15. Zaki, S.; Kamal, A.; Elkady, M.; Abu-Elreesh, G.; Abd-El-Haleem, D. **2014**. Biosynthesis of silver nanoparticles using *Stenotrophomonas rhizophila* and its application as a disinfectant agent of water. *European Journal of Experimental Biology* 4: 662–669.
16. Huang, W.; Yan, M.; Duan, H.; Bi, Y.; Cheng, X.; Yu, H. **2020**. Synergistic antifungal activity of green synthesized silver nanoparticles and epoxiconazole against setosphaeria turcica. *Journal of Nanomaterials* 2020: 1–7.
17. Khan, T.; Yasmin, A.; Townley, H. **2020**. An evaluation of the activity of biologically synthesized silver nanoparticles against bacteria, fungi and mammalian cell lines. *Colloids and Surfaces B: Biointerfaces* 194: 111156.
18. Yilmaz, M.; İspirli, H.; Taylan, O.; Dertli, E. **2020**. Synthesis and characterisation of alternan-stabilised silver nanoparticles and determination of their antibacterial and antifungal activities against foodborne pathogens and fungi. *Lebensmittel-Wissenschaft & Technologie* 128: 109497.
19. Namasivayam, S.; Christo, B.; Arasu, S.; Kumar, K.; Deepak, K. **2013**. Anti-biofilm effect of biogenic silver nanoparticles coated medical devices against biofilm of clinical isolate of Staphylococcus aureus. *Global Journal of Medical Research* 13: 1–7.
20. Park, H.; Park, S.; Roh, J. et al. **2013**. Biofilm-inactivating activity of silver nanoparticles: A comparison with silver ions. *Journal of Industrial and Engineering Chemistry* 19: 614–619.
21. Martinez-Gutierrez, F.; Boegli, L.; Agostinho, A. et al. **2013**. Anti-biofilm activity of silver nanoparticles against different microorganisms. *Biofouling* 29: 651–660.
22. Qin, H.; Cao, H.; Zhao, Y. et al. **2014**. *In vitro* and *in vivo* anti-biofilm effects of silver nanoparticles immobilized on titanium. *Biomaterials* 35: 9114–9125.
23. de Lacerda Coriolano, D.; de Souza, J. B.; Bueno, E.; Medeiros, S.; Cavalcanti, I.; Cavalcanti, I. **2020**. Antibacterial and antibiofilm potential of silver nanoparticles against antibiotic-sensitive and multidrug-resistant *Pseudomonas aeruginosa* strains. *Brazilian Journal of Microbiology* 52: 267–278.
24. Rodrigues, A.; Ruiz, R.; Selari, P.; Araújo, W.; Souza, A. **2021**. Anti-biofilm action of biological silver nanoparticles produced by *Aspergillus tubingensis* and antimicrobial activity of fabrics carrying it. *Biointerface Research in Applied Chemistry* 11: 14764–14774.
25. Dewez, D.; Oukarroum, A. **2012**. Silver nanoparticles toxicity effect on photosystem II photochemistry of the green alga *Chlamydomonas reinhardtii* treated in light and dark conditions. *Toxicological & Environmental Chemistry* 94: 1536–1546.
26. Speshock, J.; Murdock, R.; Braydich-Stolle, L.; Schrand, A.; Hussain, S. **2010**. Interaction of silver nanoparticles with Tacaribe virus. *Journal of Nanobiotechnology* 8: 1–9.
27. Lu, L.; Sun, R.; Chen, R. et al. **2008**. Silver nanoparticles inhibit hepatitis B virus replication. *Antiviral Therapy* 13: 253–262.
28. Lara, H.; Ayala-Nuñez, N.; Ixtepan-Turrent, L.; Rodriguez-Padilla, C. **2010**. Mode of antiviral action of silver nanoparticles against HIV-1. *Journal of Nanobiotechnology* 8: 1–10.
29. Saad, A.; Soliman, M.; Azzam, A. **2015**. Antiparasitic activity of silver and copper oxide nanoparticles against entamoeba histolytica and cryptosporidium parvum cysts. *Journal of the Egyptian Society of Parasitology* 45: 593–602.
30. Allahverdiyev, A.; Abamor, E.; Ustundag, C. B.; Kaya, R. **2011**. Antileishmanial effect of silver nanoparticles and their enhanced antiparasitic activity under ultraviolet light. *International Journal of Nanomedicine* 2705.
31. Keat, C.; Aziz, A.; Eid, A.; Elmarzugi, N. **2015**. Biosynthesis of nanoparticles and silver nanoparticles. *Bioresources and Bioprocessing* 2: 1–10.

32. Eltarahony, M.; Abu-Serie, M.; Hamad, H.; Zaki, S.; Abd-El-Haleem, D. **2021**. Unveiling the role of novel biogenic functionalized CuFe hybrid nanocomposites in boosting anticancer, antimicrobial and biosorption activities. *Scientific Reports* 11: 1–22.
33. Prodan, A.; Iconaru, S.; Chifiriuc, C. **2013**. Magnetic properties and biological activity evaluation of iron oxide nanoparticles. *Journal of Nanomaterials* 2013.
34. Batool, F.; Iqbal, M.; Khan, S.; Khan, J.; Ahmed, B.; Qadir, M. **2021**. Biologically synthesized iron nanoparticles (FeNPs) from Phoenix dactylifera have anti-bacterial activities. *Scientific Reports* 11: 1–9.
35. Zaki, S.; Eltarahony, M.; Abd-El-Haleem, D. **2019**. Disinfection of water and wastewater by biosynthesized magnetite and zerovalent iron nanoparticles via NAP-NAR enzymes of Proteus mirabilis 10B. *Environmental Science and Pollution Research* 26: 23661–23678.
36. Eltarahony, M.; Zaki, S.; Kheiralla, Z.; Abd-El-Haleem, D. **2019**. Study on the antagonistic potential of biosynthesized hematite nanoparticles during water and wastewater treatment. *Clean – Soil, Air, Water* 47: 1800418.
37. Sidkey, N. **2020**. Biosynthesis, characterization and antimicrobial activity of iron oxide nanoparticles synthesized by fungi. *Al-Azhar Journal of Pharmaceutical Sciences* 62: 164–179.
38. Ahmad, H.; Rajagopal, K.; Shah, A.; Bhat, A.; Venugopal, K. **2017**. Study of bio-fabrication of iron nanoparticles and their fungicidal property against phytopathogens of apple orchards. *IET Nanobiotechnology* 11: 230–235.
39. Ansari, M.; Asiri, S. **2021**. Green synthesis, antimicrobial, antibiofilm and antitumor activities of superparamagnetic γ-Fe_2O_3 NPs and their molecular docking study with cell wall mannoproteins and peptidoglycan. *International Journal of Biological Macromolecules* 171: 44–58.
40. Salman, J.; Marjani, M.; Abdulrazaq, R.; Kamil, I.; Kamil, N. **2015**. Antibiofilm effect of iron oxide nanoparticles synthesized by lactobacillus fermentum on catheter. *World Journal of Pharmaceutical Research* 4: 317–328.
41. Salem, D.; Ismail, M.; Tadros, H. **2020**. Evaluation of the antibiofilm activity of three seaweed species and their biosynthesized iron oxide nanoparticles (Fe_3O_4-NPs). *The Egyptian Journal of Aquatic Research* 46: 333–339.
42. El-Sheekh, M.; El-Kassas, H.; Shams El-Din, N.; Eissa, D.; El-Sherbiny, B. **2021**. Green synthesis, characterization applications of iron oxide nanoparticles for antialgal and wastewater bioremediation using three brown algae. *International Journal of Phytoremediation* 23: 1538–1552.
43. Tortella, G.; Rubilar, O.; Fincheira, P. **2021**. Bactericidal and virucidal activities of biogenic metal-based nanoparticles: Advances and perspectives. *Antibiotics* 10: 783.
44. Kumar, R.; Nayak, M.; Sahoo, G. **2019**. Iron oxide nanoparticles based antiviral activity of H1N1 influenza A virus. *Journal of Infection and Chemotherapy* 25: 325–329.
45. Shearer, A.; Kniel, K. **2018**. Enhanced removal of norovirus surrogates, murine norovirus and Tulane virus, from aqueous systems by zero-valent iron. *Journal of Food Protection* 81: 1432–1438.
46. Kannan, D.; Yadav, N.; Ahmad, S. et al. **2019**. Pre-clinical study of iron oxide nanoparticles fortified artesunate for efficient targeting of malarial parasite. *EBioMedicine* 45: 261–277.
47. Iqbal, K.; Abdalla, S.; Anwar, A. et al. **2020**. Isoniazid conjugated magnetic nanoparticles loaded with amphotericin B as a potent antiamoebic agent against acanthamoeba Castellanii. *Antibiotics* 9: 276.
48. Eltarahony, M.; Zaki, S.; Abd-El-Haleem, D. **2018**. Concurrent synthesis of zero-and one-dimensional, spherical, rod-, needle-, and wire-shaped CuO nanoparticles by proteus mirabilis 10B. *Journal of Nanomaterials* 2018.
49. Mali, S.; Dhaka, A.; Githala, C.; Trivedi, R. **2020**. Green synthesis of copper nanoparticles using Celastrus paniculatus Willd leaf extract and their photocatalytic and antifungal properties. *Biotechnology Reports* 27: e00518.
50. Pariona, N.; Mtz-Enriquez, A. I.; Sánchez-Rangel, D. **2019**. Green-synthesized copper nanoparticles as a potential antifungal against plant pathogens. *RSC Advances* 9: 18835–18843.
51. Zhao, H.; Maruthupandy, M.; Al-mekhlafi, F.; Chackaravarthi, G.; Ramachandran, G.; Chelliah, C. **2022**. Biological synthesis of copper oxide nanoparticles using marine endophytic actinomycetes and evaluation of biofilm producing bacteria and A549 lung cancer cells. *Journal of King Saud University-Science* 34: 101866.
52. Ashajyothi, C.; Harish, K.; Dubey, N.; Chandrakanth, R. **2016**. Antibiofilm activity of biogenic copper and zinc oxide nanoparticles-antimicrobials collegiate against multiple drug resistant bacteria: A nanoscale approach. *Journal of Nanostructure in Chemistry* 6: 329–941.
53. Reyes, V.; Spitzmiller, M.; Hong-Hermesdorf, A. **2016**. Copper status of exposed microorganisms influences susceptibility to metallic nanoparticles. *Environmental Toxicology and Chemistry* 35: 1148–1158.

54. El-Kassas, H.; Okbah, A. **2017**. Phytotoxic effects of seaweed mediated copper nanoparticles against the harmful alga: Lyngbya majuscula. *Journal of Genetic Engineering and Biotechnology* 15: 41–48.
55. Hang, X.; Peng, H.; Song, H.; Qi, Z.; Miao, X.; Xu, W. **2015**. Antiviral activity of cuprous oxide nanoparticles against Hepatitis C Virus *in vitro*. *Journal of Virological Method* 222: 150–157.
56. Shionoiri, N.; Sato, T.; Fujimori, Y. **2012**. Investigation of the antiviral properties of copper iodide nanoparticles against feline calicivirus. *Journal of Bioscience and Bioengineering* 113: 580–586.
57. Merkl, P.; Long, S.; McInerney, G.; Sotiriou, G. **2021**. Antiviral activity of silver, copper oxide and zinc oxide nanoparticle coatings against SARS-CoV-2. *Nanomaterials* 11: 1312.
58. Albalawi, A.; Abdel-Shafy, S.; Khudair, A. **2021**. Therapeutic potential of green synthesized copper nanoparticles alone or combined with meglumine antimoniate (glucantime®) in cutaneous leishmaniasis. *Nanomaterials* 11: 891.
59. Malekifard, F.; Tavassoli, M.; Vaziri, K. **2020**. *In vitro* assessment antiparasitic effect of selenium and copper nanoparticles on *Giardia deodenalis* cyst. *Iranian Journal of Parasitology* 15: 411–417.
60. Pulit-Prociak, J.; Chwastowski, J.; Kucharski, A.; Banach, M. **2016**. Functionalization of textiles with silver and zinc oxide nanoparticles. *Applied Surface Science* 385: 543–553.
61. Ifeanyichukwu, U.; Fayemi, O.; Ateba, C. **2020**. Green synthesis of zinc oxide nanoparticles from pomegranate (Punica granatum) extracts and characterization of their antibacterial activity. *Molecules* 25: 4521.
62. Naseer, M.; Aslam, U.; Khalid, B.; Chen, B. **2020**. Green route to synthesize zinc oxide nanoparticles using leaf extracts of cassia fistula and Melia azadarach and their antibacterial potential. *Scientific Reports* 10: 1–10.
63. Jamdagni, P.; Khatri, P.; Rana, J. S. **2018**. Green synthesis of zinc oxide nanoparticles using flower extract of Nyctanthes arbor-tristis and their antifungal activity. *Journal of King Saud University-Science* 1: 168–175.
64. Ahmad, H.; Venugopal, K.; Rajagopal, K. et al. **2020**. Green synthesis and characterization of zinc oxide nanoparticles using eucalyptus globules and their fungicidal ability against pathogenic fungi of apple orchards. *Biomolecules* 10: 425.
65. Al-Shabib, N.; Husain, F. A.; Hassan, I. et al. **2018**. Biofabrication of zinc oxide nanoparticle from Ochradenus baccatus leaves: Broad-spectrum antibiofilm activity, protein binding studies, and *in vivo* toxicity and stress studies. *Journal of Nanomaterials* 1–14.
66. Vijayakumar, S.; Vinoj, G.; Malaikozhundan, B.; Shanthi, S.; Vaseeharan, B. **2015**. Plectranthus amboinicus leaf extract mediated synthesis of zinc oxide nanoparticles and its control of methicillin resistant Staphylococcus aureus biofilm and blood sucking mosquito larvae. *Spectrochimica Acta Part A: Molecular and Biomolecular Spectroscopy* 25: 886–891.
67. Djearamane, S.; Lim, Y.; Wong, L.; Lee, P. **2018**. Cytotoxic effects of zinc oxide nanoparticles on cyanobacterium Spirulina (Arthrospira) platensis. *Peer Journal* 1: e4682.
68. Ghaffari, H.; Tavakoli, A.; Moradi, A. et al. **2019**. Inhibition of H1N1 influenza virus infection by zinc oxide nanoparticles: Another emerging application of nanomedicine. *Journal of Biomedical Science* 26: 1–10.
69. Melk, M.; El-Hawary, S.; Melek, F. et al. **2021**. Antiviral activity of zinc oxide nanoparticles mediated by plumbago indica L. extract against herpes simplex virus type 1 (HSV-1). *International Journal of Nanomedicine* 16: 8221–8233.
70. Attia, G.; Moemen, Y.; Youns, M.; Ibrahim, A.; Abdou, R.; El Raey, M. **2021**. Antiviral zinc oxide nanoparticles mediated by hesperidin and in silico comparison study between antiviral phenolics as anti-SARS-CoV-2. *Colloids and Surfaces B: Biointerfaces* 203: 111724.
71. Khan, Y.; Singh, B.; Ullah, R.; Shoeb, M.; Naqvi, A.; Abidi, S. **2015**. Anthelmintic effect of biocompatible zinc oxide nanoparticles (ZnO NPs) on Gigantocotyle explanatum, a neglected parasite of Indian water buffalo. *PLoS One* 15: e0133086.
72. Morsy, K.; Fahmy, S.; Mohamed, A.; Ali, S.; El–Garhy, M.; Shazly, M. **2019**. Optimizing and evaluating the antihelminthic activity of the biocompatible zinc oxide nanoparticles against the ascaridid nematode, parascaris equorum *in vitro*. *Acta Parasitologica* 64: 873–886.
73. Grande, F.; Tucci, P. **2016**. Titanium dioxide nanoparticles: A risk for human health? *Mini Reviews in Journal of Medicinal Chemistry* 16: 762–769.
74. Hamed, M.; Bakr, B.; Shahin, Y. et al. **2021**. Novel synthesis of titanium oxide nanoparticles: Biological activity and acute toxicity study. *Bioinorganic Chemistry and Applications* 11: 2021.
75. Landage, K.; Arbade, G.; Khanna, P.; Bhongale, C. **2020**. Biological approach to synthesize TiO_2 nanoparticles using staphylococcus aureus for antibacterial and anti-biofilm applications. *Journal of Microbiology & Experimentation* 8: 36–43.

76. Shabir Mahr, M.; Hübert, T.; Stephan, I.; Militz, H. **2013**. Decay protection of wood against brown-rot fungi by titanium alkoxide impregnations. *International Biodeterioration & Biodegradation* 77: 56–62.
77. Irshad, M. A.; Nawaz, R.; Ur Rehman, M. Z. et al. **2020**. Synthesis and characterization of titanium dioxide nanoparticles by chemical and green methods and their antifungal activities against wheat rust. *Chemosphere* 258: 127352.
78. Anbumani, D.; Dhandapani, K. V.; Manoharan, J. et al. **2022**. Green synthesis and antimicrobial efficacy of titanium dioxide nanoparticles using Luffa acutangula leaf extract. *Journal of King Saud University-Science* 34: 101896.
79. Rajeshkumar, S.; Santhoshkumar, J.; Jule, L.; Ramaswamy, K. **2021**. Phytosynthesis of titanium dioxide nanoparticles using king of bitter Andrographis paniculata and its embryonic toxicology evaluation and biomedical potential. *Bioinorganic Chemistry and Applications* 7: 2021.
80. Al-Shabib, N.; Husain, F.; Qais, F. et al. **2020**. Phyto-mediated synthesis of porous titanium dioxide nanoparticles from Withania somnifera root extract: Broad-spectrum attenuation of biofilm and cytotoxic properties against HepG2 cell lines. *Frontiers in Microbiology* 11: 1680.
81. Li, Z.; Juneau, P.; Lian, Y. et al. **2020**. Effects of titanium dioxide nanoparticles on photosynthetic and antioxidative processes of Scenedesmus obliquus. *Plants* 9: 1748.
82. Abdulrahman, A.; Zhang, L.; Yang, J.; Wei, F.; Chen, C.; Sun, D. **2021**. Toxicity assessment of synthesized titanium dioxide nanoparticles in fresh water algae *Chlorella pyrenoidosa* and a zebrafish liver cell line. *Ecotoxicology and Environmental Safety* 15: 111948.
83. Akhtar, S.; Shahzad, K.; Mushtaq, S.; Ali, I.; Rafe, M.; Fazal-ul-Karim, S. **2019**. Antibacterial and antiviral potential of colloidal Titanium dioxide (TiO_2) nanoparticles suitable for biological applications. *Materials Research Express* 6: 105409.
84. Mazurkova, N. A.; Spitsyna, Y. E.; Shikina, N. V. et al. **2010**. Interaction of titanium dioxide nanoparticles with influenza virus. *Nanotechnology Russia* 5: 417–420.
85. Fouad, G. **2021**. A proposed insight into the anti-viral potential of metallic nanoparticles against novel coronavirus disease-19 (COVID-19). *Bulletin of the National Research Centre* 45: 1–22.
86. Elsharkawy, M.; Derbalah, A. **2019**. Antiviral activity of titanium dioxide nanostructures as a control strategy for broad bean strain virus in Faba bean. *Pest Management Science* 75: 828–834.
87. Rajakumar, G.; Rahuman, A.; Roopan, S.; Chung, I.; Anbarasan, K.; Karthikeyan, V. **2014**. Efficacy of larvicidal activity of green synthesized titanium dioxide nanoparticles using Mangifera indica extract against blood-feeding parasites. *Parasitology Research* 114: 571–581.
88. Marimuthu, S.; Rahuman, A.; Jayaseelan, C. et al. **2013**. Acaricidal activity of synthesized titanium dioxide nanoparticles using Calotropis gigantea against Rhipicephalus microplus and Haemaphysalis bispinosa. *Asian Pacific Journal of Tropical Medicine* 6: 682–688.
89. Wang, L.; Hu, C.; Shao, L. **2017**. The antimicrobial activity of nanoparticles: Present situation and prospects for the future. *International Journal of Nanomedicine* 12: 1227.
90. Slavin, Y.; Asnis, J.; Häfeli, U.; Bach, H. **2017**. Metal nanoparticles: Understanding the mechanisms behind antibacterial activity. *Journal of Nanobiotechnology* 15: 1–20.

5 Role of Nanoparticles in Precision Medicine and Targeted Drug Delivery

Meliha Koldemir-Gündüz

5.1 INTRODUCTION

Focusing on diseases in conventional medicine, it was thought that the same treatment methods would be appropriate for individuals with the same disease. Today, however, it has become clear that the same disease progresses differently in different people, and therefore, the same treatment methods cannot be used. Individuals, rather than diseases, are the focus of today's medicine, and the concept of precision medicine has made its way into our daily lives [1]. The concept of precision medicine was initially introduced in Clayton Christensen's book published in 2009 [1, 2]. However, the acceptance of this concept was mainly after the publication of a report related to precision medicine by the US National Research Council (NRC) in 2011. In this report, it is recommended to classify diseases according to genetic or genomic basis, not symptoms [1, 3].

Precision medicine is described as medical care that targets the needs of individual patients based on genetic, phenotypic, biomarker, or psychological variables to improve diagnostic efficiency or therapeutic effect [4]. The goal of precision medicine is to minimize unnecessary side effects for individuals who do not benefit or are unlikely to benefit from a particular treatment. The focus of precision medicine is on the genetic and genomic basis of disease. The Institute of Precision Medicine defines *precision medicine* as "personalized service tailored to each patient according to their genetic profile and medical history" [5]. Unlike conventional and all-physician drugs, precision medicine examines a patient's entire genome, revealing the cause of the disease and detecting specific genetic changes that guide it [1].

Conventional drug use consists of frequent and repeated doses. It's possible that the dose employed to determine the concentration of the active drug released into the system is either insufficient or hazardous. Drug delivery systems that minimize the active substance's dose, extend the dosing interval, guard against side and harmful effects, and transport the active material to the target area help avoid these circumstances [6]. *Drug targeting*, which aims for the drug to act only in the pathological region, is defined as the ability of selective and quantitative aggregation in the target tissue or organ, independent of the chemical structure and uptake of the active drug substance. Drug targeting allows traditional, biotechnology, and gene-derived medications to be delivered to specific organs, tissues, and cells in the body. The use of various targeted nanocarriers has become one of the most important areas in nanomedicine. These carriers are nanosized materials that can carry a variety of drugs and/or imaging agents.

The goal of drug targeting is to transport, absorb, and distribute the pharmacological substance selectively to the site of action [7]. Unwanted side effects of medications are decreased as a result of this selective targeting, the most appropriate therapeutic response is attained, and a safer use is guaranteed by lowering toxic consequences that occur with high dose use [8]. Thanks to nanocarriers, when drugs reach the target area, uptake in organs, tissues, and cells increases. These structures use active and passive targeting strategies to deliver the active substance to the targeted cells [9]. The active targeting ligands used are due to peptides or monoclonal antibodies that increase cellular uptake. The use of active targeting increases transport to cellular levels. The release movements

DOI: 10.1201/9781003430087-5

are optimized during the circulation of drugs or their concentration in other biological fluids. The use of low or high doses provides effective and safe treatment and increases the stability of drugs. Since there is no undesirable effect in other parts of the body, the desired level of pharmacological response is obtained in the target area [7]. The use of nanocarriers in drug delivery systems provides many advantages. In order to develop different nanocarrier systems, lipid-, polymer-, and inorganic-based structures are used. Today, the most studied and developed nanocarrier systems are polymeric conjugates, polymeric micelles, polymeric nanoparticles, dendrimers, polymersomes, liposomes, carbon nanotubes, and gold nanoparticles [10].

Nanotechnology is defined as a theoretical and experimental field of applied science and technology that deals with the control and use of substances at the atomic and molecular level. Medical applications of nanotechnology cover application areas such as diagnosis and imaging, drug delivery, diagnostic applications, and reporting the efficacy of combined treatments and therapeutic agents [11]. Nanoparticles are solid particles ranging in size from 10 to 1000 nm [12] that can allow for the transport of dissolved, entrapped, encapsulated, or surfaced drugs as well as for their controlled release [12], and medicine, biotechnology, sensors, and environmental technology are gaining importance [13, 14]. Polymeric nanoparticles are promising drug carriers used to deliver drugs to the targeted site in a specific way in the treatment of targeted metabolic diseases and cancer [15]. Nanoparticles effectively transport DNA, drug substance, and proteins to target cells and organs. With its nano dimensions, it increases effective permeability across cell membranes and stability in blood circulation [11].

Obesity, a chronic condition, has become a major public health issue worldwide in the previous decade. Obesity increases the risk of numerous major diseases, including type 2 diabetes, coronary heart disease, cancer, and respiratory ailments. Pharmacotherapy has become an important technique for weight loss, alongside lifestyle changes. Most anti-obesity medicines, however, do not have the desired impact and have substantial adverse effects [16]. In this section, the fundamentals of nanoparticles and nanocarriers, with an emphasis on enhanced therapeutic efficacy, are described, as well as the latest developments in the use of nanoparticles in precision medicine, an emerging new therapeutic strategy for the treatment of obesity.

5.2 THERAPEUTIC POTENTIAL OF NANOPARTICLES

Nanotechnology is a multidisciplinary discipline that combines physics, chemistry, biology, and materials science to provide nanoparticle-based medication delivery, diagnostics, gene therapy, biosensing, imaging, and targeted thermal ablation. Many clinical trials examining the efficiency of nanotechnology-based drug carriers targeting oncology or metabolic illnesses have been conducted in the field of nanotechnology (ClinicalTrial.Gov). For effective therapeutics, it is important to understand the mechanism of disease and the complex interplay between fabricated nanoparticle biology. Nanobiosensor research has advanced to the point where it can now detect multiple important pathways, regulatory events, and disease biomarkers at an early stage [17–19].

Nanoparticles are microscopic crystals with a diameter of less than 100 nm that are made up of ions and inorganic or organic compounds. Nanoparticles are employed in nanomedicine to improve drug delivery and effect. Gold, silica, fullerene, quantum dots, carbon nanotubes, and magnetic nanoparticles are only a few of the nanoparticles that have been developed [20]. Bioimaging and biosensor studies for early diagnosis of diseases and development of personalized treatments, therapeutic drug delivery, and gene delivery are important biomedical application areas of nanoparticles [21]. Nanoparticles have recently been investigated quite frequently in drug delivery systems. Its nano size gives it the ability to pass through cell and tissue barriers. It provides a better distribution of drugs with low solubility in water [22]. Nanoparticles come to the forefront with advantages, such as the controlled release of drugs in the desired amount specific to a certain region, and the improvement of the half-life of drugs by encapsulation [23]. The advantages of nanoparticle-based delivery methods include improved biocompatibility, multifunctional encapsulation of active compounds, decreased degradation during blood circulation, passive or active targeting, effective administration, and reduced side effects [24].

By their design, nanoparticle carriers are administered for therapeutic purposes [25]. When nanocarriers are coated with polymers such as polyethylene glycol (PEG), they spend more time in the circulation, making them invisible to the immune system. Nanocarriers are imbued with ligands that bind to receptors that are overexpressed in tumor cells. Nanocarriers can be made sensitive to stimuli and then triggered to release their biological charge at the exact location where it is needed [26]. The creation of multifunctional medicinal nanocarriers has the potential to improve diagnostic and therapeutic outcomes. Pharmaceutical nanocarriers, such as nanoparticles, micelles, and liposomes, are currently used because of their accumulation in pathological regions; specific targeting to specific disease or organ locations, thanks to bound open ligands on their surfaces; better penetration into the cell, thanks to surface particles; and responsiveness to stimuli that allow the drug to be released from the carriers under special physiological conditions [25].

Clinical use of effective nanotechnology-based inventions is fast expanding, and it has completely transformed the field of disease management [27]. Because of the rapid advancement of nanotechnology, pharmacological applications in the diagnosis and treatment of various diseases, particularly oncology, have increased, allowing for better diagnostics, effective drug administration, and focused therapeutic options. Because of the physio-biochemical features of nanoparticles for clinical applications, nanotherapy deals with tailored nanoparticle drug delivery that decreases drug side effects while also promising a reduction in traditional treatment costs, which is a key hurdle in therapy [28].

5.3 TARGETED DELIVERY OF NANOPARTICLES IN PRECISION MEDICINE

Nanotechnology's application in health, medicine, and therapy has opened up the possibility of using nanoparticle platforms in the development of novel therapeutic agents [29, 30]. The effective dose in certain areas is very important for the treatment of the disease. The physiological milieu can destroy therapeutic reagents, such as small molecule medicines, nucleic acids, peptides, and proteins, but nanomedicine boosts drug delivery effectiveness due to its superior targeting capabilities [31]. From chemistry to biology to medicine, the use of nanoparticles in precision medicine has an impact on the drug discovery cycle. Various types of nanoparticles have been created and proposed for therapeutic uses using various materials [32]. Nanoparticle characteristics can be tailored for size, shape, surface chemistry, and site-specific medication delivery in the organ [31]. The identification of unique platforms for efficient transport of bioactive compounds and their particular release in the microenvironment of sick tissues is possible thanks to the development of targeted drug delivery nanosystems [33].

In order to achieve successful delivery of the targeted drug in an effective treatment, it is possible with an understanding of passive and active targeting systems and knowledge of the role of various physiological barriers and their pathophysiological status [34, 35]. As is known, most of the available drugs do not show great differences between diseased and healthy cells. This causes systemic toxicity and side effects. Side effects of drugs against healthy cells or tissues can be eliminated by applying passive or active targeting approaches [11].

Passive targeting based on modifying the physicochemical features of nanoparticles or surface modification with specific ligands that can recognize receptors in targeted cells improves targeting efficiency. For nanoparticle-mediated drug delivery, proteins, peptides, polysaccharides, aptamers, and small molecules are commonly used ligands [31].

5.3.1 Passive Targeting Approach

Due to the rapid and uncontrolled proliferation of cells in tumor tissues, cells form their own vascular networks to meet their nutritional needs and accelerate their growth. Tumor neo-vessels, unlike normal blood vessels, increase permeability, which allows the flow of oxygen and nutrients necessary for the growth of tumor cells through the interstitial space. The presence of vasodilation disorder in the tumor region facilitates the extravasation of nanosized carriers. The absence of the lymphatic system in tumors ensures that the intravenously transmitted nanocarriers are mostly trapped in the tumor [36]. The increased permeability and retention (EPR) effect enables nanosized carriers smaller than 200 nm, with high stability in the blood circulation, to accumulate in the tumor region [36, 37]. Polymeric micelles, which are used as a carrier system in cancer treatment, are considered as one of the most promising nanocarriers that can carry especially hydrophobic anticancer drugs by the effect of EPR in the tumor [36, 38].

5.3.2 Active Targeting Approach

In passive targeting, the nanosized carrier used can accumulate in tumor tissue in high amounts. As a result, the drug loaded on the nanoparticle may not enter the cells directly [36]. In addition, there are many difficulties and limitations in the passive targeting process, such as the inability to specifically transport and take nanoparticles into tumor cells, especially mucosal barriers. This situation causes the active targeting approach to be preferred [39]. In the treatment of receptor-mediated cancer, the method that has been applied many times for an effective treatment is to increase the cellular uptake capabilities of the drug carrier by binding many specific ligands, such as antibodies, growth factors, or folic acid, to the drug delivery system [40]. Active targeting approach, antibody–antigen, ligand–receptor, etc., is based on specific interactions. This approach requires designing the active agent or carrier system by conjugating it with a tissue- or cell-specific ligand. However, nano-sized materials used as carriers in cancer chemotherapy applied with active targeting can be directed to bind to specific sites using various affinity ligands such as folic acid [39]. Whether the targeted carrier–ligand conjugates will be assimilated after specific binding to the target cells is an important criterion in the selection of the appropriate targeting ligands to be used. It's vital to remember that a successful cancer treatment will require both passive and aggressive targeting. Active targeting, which is accepted as a complementary approach to the passive targeting approach, promotes the cellular uptake of the carrier and increases the specificity of the carrier and the effectiveness of the treatment [41]. To date, many polymeric nanoparticle carriers have been conjugated with ligands and used in *in vitro* and/or *in vivo* studies.

Fundamental concerns in nanomedicine include medication delivery to sick organs and tissues via blood circulation, differentiation of cellular absorption of nanoparticles, and preparation of drug delivery at the subcellular level [31]. Therapeutics are classified based on their *in vivo* phases; targeted nanoparticles must have certain properties. As a result of the improvement of drug pharmacokinetics, upon systemic administration, formulations with nanoparticles face adsorption from serum proteins in the blood, which can change the physicochemical properties of nanoparticles that affect their retention in the biological system and target activity to a specific region [42]. When the nanoparticle enters the body, it can be detected as a foreign stimulus and eliminated by the liver or spleen. As a result, the dose is reduced at the expected target [43]. By preventing this manipulation, non-specific distribution is prevented, and drug accumulation in the diseased area is facilitated. Nanoparticles distinguish healthy and diseased cells, avoid side effects, and show optimal therapeutic effect [44]. Surface ligand that targets the particular receptor on the plasma membrane of sick cells can be used to fix cargo-loaded nanoparticles. Because the disease is characterized by subcellular disruptions, targeting organelles with custom-designed nanoparticles could advance the therapy to a more precise stage [31].

5.4 DRUG DELIVERY STRATEGIES FOR THE TREATMENT OF DISEASES

Drug delivery systems consist of organic/inorganic nanoparticles, viral systems, and recombinant proteins. Intelligent drug delivery systems have been developed due to the disadvantages of drug delivery systems, such as inability to resist the immune system, toxic effects, and inability to release sufficient doses of drugs to the targeted location. Two different applications, passive targeting and active targeting, are used in these systems. Active targeting system was needed because passive targeting, where drug delivery is made by utilizing vascular permeability, is limited to some tumor types. The active targeting system is based on intermolecular interaction as well as antibody–antigen affinity. It is possible to modify the organic or inorganic nanocarriers used for active targeting. Inorganic nanoparticles have significant advantages over organic ones in terms of thermal, chemical, and mechanical stability and good biocompatibility under physiological conditions. With surface coatings, nanocarriers gain targeting ability. Antibodies, proteins, peptides, aptamers, carbohydrates, and small molecules are among the various ligands used for active targeting. Nanoparticles containing these ligands on the surface bind to their targets with high affinity [45].

5.4.1 Drug Delivery Systems

Nanoparticles are called nanocarriers when they are used as transport modules for other substances [46]. Surface charges and sizes are very important in taking nanoparticles into the cell [47]. Negatively charged cellular membranes and positively charged nanoparticles have an electrostatic interaction [48]. Endocytosis and pinocytosis are the most common ways for polycation DNA gene delivery systems to reach the cell. In order to perform endocytosis by cells under suitable conditions, the size of the nanocarrier system to be sent to the cell must be smaller than 100 nm [49].

Drug delivery systems with nanoparticles are used both in drug research and in the clinical environment to increase the *in vivo* efficiency of many drugs that have proven themselves in recent years [50]. For example, it is possible to control the normal biological properties of liposomes, micelles, nanocapsules, polymeric nanoparticles, solid lipid particles, niosomes, and many other pharmaceutical nanocarriers in a desired manner [45].

Nanoparticles have been developed in terms of drug delivery amounts and are optimized systems in terms of drug stability. Classical nanoparticles do not have the ability to carry drugs to the targeted cell or organ with external or internal stimulation and to release targeted drug at the right concentration [46, 51]. Modified and functionalized smart nanoparticles are able to release the drug at reduced dosages and in a spatially controlled manner after carrying the agent they transmit to the target areas [50, 52]. Unmodified drug carriers usually represent foreign particles for the body defense system and can be removed from the blood circulation before they can transport the drug they carry to the target cell or organ [53]. The positive results of various modifications used to make smart nanocarrier systems, such as increasing the stability and half-life of circulating nanocarriers, providing the required biodistribution, and passive or active targeting to the desired pathological site, have proven that the classical drug delivery systems and drugs are renewable [48].

Drug delivery systems can be divided into two groups [54]: (a) non-viral systems, such as organic cationic compounds, recombinant proteins, and inorganic nanoparticles [54]; and (b) viral carrier systems [54]. Although viral carriers are the most effective carrier systems, their side effects, such as immune response, high production, low loading capacity, and insertion mutagenesis, limit their successful application in cellular transmission [54]. Cationic transporters are generally toxic to cells [55] and are unstable in the long run [56]. Inorganic and organic nanoparticles are among the promising nanoparticles as a new alternative to viral carriers and cationic carriers with their low toxicity and controlled application properties [54]. The most commonly used non-viral inorganic nanocarrier systems are carbon nanotube (CNT), dendrimer, micelle, liposome, quantum dot (QD), superparamagnetic iron oxide nanoparticle (SPION), gold nanoparticle (GNP), and mesoporous silica nanoparticle (MSN) [52].

5.4.2 Nanoparticular Drug Delivery Systems

Loading drugs into nanoparticles improves biopharmaceutical properties, such as solubility, stability, overcoming barriers, long-term stay in the application area, and prolonged release [57].

Nanomaterials have a high ratio of inner and exterior surface area to volume, making them ideal candidates for transporting various medications and attaching the targeting group [58]. Furthermore, various nanoparticles can be created to have a wide range of sizes, shapes, and chemical characteristics on their surfaces [59]. In this way, nanoparticle drug delivery systems have been developed in recent years to be used in many areas, such as obesity, oncology, cardiology, infectious diseases, eye diseases, and genetic therapy applications.

The most commonly used methods in the treatment of obesity are lifestyle modification and bariatric surgery. Obesity drugs are not often preferred because they can also damage healthy cells. Although this undesirable effect reduces the success of classical drug use, it can cause it to fail. Combining anti-obesity therapeutics with different effects may help use lower doses and reduce intolerable side effects. Various strategies are being developed in order to obtain the most appropriate response and long-term effect in the treatment of obesity. Intelligent drug delivery systems have been developed so that the anti-obesity agent can be targeted at the required location and show its effectiveness. Smart drug delivery systems use physicochemical differences between fat cells and healthy cells to identify fat cells.

Nanocarriers encounter many biological barriers on their way to the targeted site, including clearance by the reticuloendothelial system (RES) [60]. The first feature expected from smart nanocarriers is that the body's immune system avoids the cleansing process [61, 62]. RES removes the nanocarrier from the circulation in a short time and deposits it in the liver, spleen, or bone marrow. Coating with polyethylene glycol (PEG) is a good solution to avoid this clearing of the immune system [61]. Second, the nanocarriers are modified to gain selectivity against fat cells. Some proteins on the surface of white adipose tissue cells are the main targets of smart nanocarriers. The surface of the nanocarriers is modified with ligands that have the ability to pair with these proteins, and they are transported to the white adipose tissue cells, where there are many receptor proteins on the surface. Transporting the drug to the target area may not be sufficient for the success of pharmacotherapy [61, 63]. It can be expected that a smart nanocarrier will release the drug it carries under external or internal stimulation at the target location at the right concentration [63]. Various chemical groups can be added to the surface of nanocarriers to make them sensitive to the environmental conditions in which they will be released [64]. An ideal smart nanocarrier system is to find the area to be treated by our body's immune system without being involved in the cleansing process, deliver the pharmacotherapeutic substance it carries, and release the pharmacotherapeutic substance at the right concentration.

5.5 FOR THE TREATMENT OF OBESITY: TARGETED DRUG DELIVERY STRATEGIES

The World Health Organization classifies obesity as a chronic condition [65]. A sedentary lifestyle and high calorie intake cause the obese population to increase dramatically worldwide [66]. Obesity is becoming one of the most common health concerns, impacting 15% of the global population [67].

Obesity is a disease caused by certain factors, including behavioral, environmental, and genetic interactions [68]. For the treatment of obesity, the management of diets and the regulation of lifestyle are important throughout life, starting from the prenatal and perinatal periods [67, 69, 70]. The use of pharmacotherapy to decrease appetite or block fat absorption is the mainstay of obesity treatment [71, 72]. Anti-obesity medicines, on the other hand, cause major side effects, such intestinal hemorrhage and even suicide [73, 74].

Thanks to innovations in nanotechnology [75–77], integrating anti-obesity medications into nanomedicine has proven to have a significant therapeutic effect in the treatment of obesity [78]. Thanks to the nanoparticles with the targeting moiety, it is possible to minimize the side effects in the treatment of obesity [79].

5.5.1 Obesity Treatment Using Nanoparticles

Obesity has been treated with a variety of treatments, including lifestyle changes, bariatric surgery, and medicine [80]. Lifestyle changes are a difficult process to achieve and maintain. Therefore, pharmacotherapy is an additional treatment option [81]. Conventional treatment options are insufficient and carry risks of side effects. To prevent this situation, it is necessary to develop treatments with high efficacy and specificity. Nanoparticles are used in drug delivery systems for therapeutic efficacy and specificity by stabilizing and spatiotemporally controlling the release of anti-obesity drugs [82].

In recent years, several materials and chemical techniques have been used to build drug delivery systems. Biocompatible polymers are attractive drug carrier matrices due to their diversity in polymer chemistry and ease of modifying their physicochemical and biological properties [83]. Anti-obesity medications have been delivered through a variety of carriers that target distinct metabolic pathways. When compared to conventional pharmacotherapeutics, these new drug delivery systems show improved therapeutic efficacy and reduced side effects, indicating clinical translation potential [82].

5.5.2 Obesity Treatments in the Conventional and Their Limitations

Conventional treatments for obesity are usually planned according to the severity of weight and coexisting chronic diseases. Because of their low cost and lack of adverse effects, healthy lifestyle therapies are the foundation of obesity treatment [80]. However, bariatric surgery and pharmacotherapy [82, 84] are used because lifestyle changes alone are difficult to maintain and cannot provide sufficient weight loss.

5.5.3 Developed Nanoparticle Drug Delivery Systems for Obesity Treatment

For the treatment of obesity, despite the research done in the last decade, traditional treatments are insufficient to maintain metabolic balance and prevent life-threatening complications. Therefore, there is a need for new strategies that increase treatment efficacy and reduce adverse effects. Carriers intended to release therapeutic substances for extended periods of time and target specific tissue or cell types in the body are making substantial breakthroughs in medicine thanks to the development of biomaterials for drug delivery [85]. Polymer conjugates, hydrogels, microneedles, micro- and nanoparticles, and liposomes are examples of these carriers [82].

Nanoparticles have a high surface-to-volume ratio, intracellular drug release, and variable surface chemistry. Nanoparticles have these qualities, making them intriguing delivery vehicles for a variety of illnesses, including obesity [86]. Its structural characteristics can be tweaked during the production process to improve drug loading and specificity in reaching target cells and tissues. The treatment efficacy is raised while side effects are reduced in this manner [87]. The researchers minimized nanoparticle aggregation and protein modification and opsonization in the blood after intravascular injection by using rosiglitazone-encapsulated PLGA nanoparticles enclosed by a polyvinyl alcohol film [88]. They discovered a decrease in inflammatory macrophages in adipose tissue and liver in mice treated with nanoparticles, with no deleterious effects on lipid metabolism or cardiac tissues, as shown in animals treated with standard rosiglitazone [88]. Effective drug delivery to specific targets in obesity can be challenging. Targeting adipose tissue to the vasculature as a result of modifications of nanoparticles may be a potential therapeutic intervention for obesity. Xue et al. encapsulated the obesity drugs using PLGA-b-poly (ethylene glycol) and modified the surface of the resulting nanoparticles with adipose vasculature-targeted peptides that bind to antigens specifically expressed in the endothelium of the angiogenic vasculature [89]. As a result, white adipose tissue browning was increased and body weight gain of obese mice was prevented. Mitochondrial dysfunction in adipose tissue is associated with obesity [82]. Researchers have developed mitochondria-targeted polymeric nanoparticles, achieving high endolysosomal escape and mitochondrial uptake efficiency [90]. Inhibition of differentiation in 3T3-L1 preadipocytes was achieved at a lower dose using these nanoparticles compared to a traditional medication [90]. According to Jiang et al., local medication delivery in the treatment of obesity was examined due to its capacity to retain bioavailability and allow dosage reduction. Obese mice injected nanoparticles into adipose tissue to increase browning of white adipocytes, improve glucose homeostasis, and reduce body weight by inhibiting the Notch signaling system [91]. The nanoparticles were endocytosed by adipocytes after delivery, restricting the therapeutic agent's distribution in the body and decreasing negative effects in other cells [91]. Nanoparticles also inhibited Notch signaling at a much lower dosage than using intraperitoneal injection or microparticles [91]. Orlistat, an obesity drug, has a low water solubility. Therefore, oral administration reduces bioavailability [92]. Orlistat nanoemulsion improved its *in vivo* dissolution [93]. Chen et al. developed an orlistat-conjugated polymer nanocarrier [94]. Orlistat was encapsulated in nanocarriers in this method. Thus, bioavailability was improved, and a mechanism for delivering drugs on demand was provided through the negative feedback loop that minimized its side effects. Lipid deposition vascularization is required for the growth of adipocytes [95–97]. Antiangiogenic treatment decreases the course of adipocyte hyperplasia and reduces weight gain by inducing death in endothelial cells. The apoptosis-inducing [D(KLAKLAK)2] peptide has been shown to increase endothelial cell death in obese animals' adipocyte tissue [98, 99]. Researchers developed a targeting nanoparticle that delivers its cargo to vascular endothelial cells to increase the accumulation of antiangiogenic drug in white adipose tissue [100]. They demonstrated that inhibitory targeting peptide nanoparticles, which can bind specifically to the inhibitory protein on the endothelial cell surface, are specifically taken up by adipose vascular endothelial cells via inhibition-mediated endocytosis, resulting in apoptosis [97, 101, 102].

5.6 CONCLUSION

Precision medicine applications offer fewer side effects and more precise solutions than traditional treatment methods. By reducing digestibility and raising energy expenditure, nanoparticle-based therapies have demonstrated potential preclinical activity for weight reduction. Nanocarriers demonstrate an effective weight control approach. The efficacy of therapy may also be affected by different drug delivery carriers/materials and administration routes. To achieve definitive therapy, adipose tissue targeting groups can be included into nanoparticle-based drug delivery systems [103, 104]. However, because weight loss is a long-term treatment, a thorough assessment of biocompatibility is required [105]. Nanotechnology-based strategies will be a critical pillar for future discoveries and improvements in precision medicine, paving the door for further collaboration among experts in pharmacokinetics, pharmacotherapy, immunology, toxicology, and nanotechnology.

5.7 CONFLICT OF INTEREST

The authors declare that they have no conflict of interest

REFERENCES

1. Tekpınar, L.; Erdem, R. **2019**. Evaluation of personalized type and genome studies in the context of health outcomes. *Hacettepe Sağlık İdaresi Dergisi* 22: 843–862.
2. Kleinke, J.; Christensen, C. M.; Grossman, J. H.; Hwang, J. **2009**. The innovator's prescription: A disruptive solution for health care. Project Hope 7500 Old Georgetown RD, STE 600, Bethesda, MD 20814–6133 USA,
3. Özer, M.; Bostancı, İ. **2021**. Precision medicine concept and its role in the treatment of immunological diseases. *J Curr Pediatr* 19: 259–270.
4. Jameson, J. L.; Longo, D. L. **2015**. Precision medicine – personalized, problematic, and promising. *N Engl J Med* 372: 2229–2234.

5. https://ipm.pitt.edu/(accessed 05.09.2020).
6. Danckwerts, M.; Fassihi, A. **1991**. Implantable controlled release drug delivery systems: A review. *Drug Dev Ind Pharm* 17: 1465–1502.
7. Soloviev, M. **2007**. Nanobiotechnology today: Focus on nanoparticles. *Journal of Nanobiotechnology* 5: 1–3.
8. Canefe, K.; Duman, G. **1994**. Selective transport and targeting of drugs. *Ankara Ecz Fak Derg* 23: 1–2.
9. Değim, İ. **2011**. Pharmaceutical nanotechnology. *Mised* 8–13: 198–205.
10. Saltzman, W. M. **2001**. *Drug Delivery: Engineering Principles for Drug Therapy*. New York: Oxford University Press.
11. Tietilu, S. D. **2020**. *Synthesis and characterization of targeted drug delivery systems and investigation of kinetic drug release*, İstanbul University, Cerrahpaşa Lisansüstü Eğitim Enstitüsü, Master Thesis, Turkey.
12. Mohanraj, V. J.; Chen, Y. **2006**. Nanoparticles – a review. *Trop J Pharm Res* 5: 561–573.
13. Shon, Y. S. **2010**. Nanoparticle-cored dendrimers and hyperbranched polymers: Synthesis, properties, and applications. *Adv Nanomat* 743–766.
14. Wang, X.; Summers, C. J.; Wang, Z. L. **2004**. Large-scale hexagonal-patterned growth of aligned ZnO nanorods for nano-optoelectronics and nanosensor arrays. *Nano Lett* 4: 423–426.
15. Shokri, N.; Javar, H. A.; Fouladdel, S. H.; et al. **2011**. Preparation and evaluation of poly (caprolactone fumarate) nanoparticles containing doxorubicin HCI. *DARU J Fac Pharm Tehran Univ Med Sci* 19: 12.
16. Zhang, Y.; Yu, J.; Qiang, L.; Gu, Z. **2018**. Nanomedicine for obesity treatment. *Sci China Life Sci* 61: 373–379.
17. Shandilya, R.; Bhargava, A.; Bunkar, N.; Tiwari, R.; Goryacheva, I. Y.; Mishra, P. K. **2019**. Nanobiosensors: Point-of-care approaches for cancer diagnostics. *Biosens Bioelectron* 130: 147–165.
18. Kwon, S. J.; Bard, A. J. **2012**. DNA analysis by application of Pt nanoparticle electrochemical amplification with single label response. *J Am Chem Soc* 134: 10777–10779.
19. Kydd, J.; Jadia, R.; Velpurisiva, P.; Gad, A.; Paliwal, S.; Rai, P. **2017**. Targeting strategies for the combination treatment of cancer using drug delivery systems. *Pharmaceutics* 9: 46.
20. Tang, X.; Loc, W. S.; Dong, C.; et al. **2017**. The use of nanoparticulates to treat breast cancer. *Nanomedicine* 12: 2367–2388.
21. Faraji, A. H.; Wipf, P. **2009**. Nanoparticles in cellular drug delivery. *Bioorg Med Chem* 17: 2950–2962.
22. Sivasankar, M.; Kumar, B. P. **2010**. Role of nanoparticles in drug delivery system. *Int J Res Pharm Biomed Sci* 1: 41–66.
23. Sun, M.; Lee, J.; Chen, Y.; Hoshino, K. **2020**. Studies of nanoparticle delivery with *in vitro* bioengineered microtissues. *Bioact Mater* 5: 924–937.
24. Adair, J. H.; Parette, M. P.; Altınoglu, E. I.; Kester, M. **2010**. Nanoparticulate alternatives for drug delivery. *ACS Nano* 4: 4967–4970.
25. Bolhassani, A.; Saleh, T. **2013.** Challenges in advancing the field of cancer gene therapy: An overview of the multi-functional nanocarriers. In *Novel Gene Therapy Approaches*. Rijeka, Croatia: IntechOpen, 197–259.
26. Viseu, T.; Lopes, C. M.; Fernandes, E.; Real Oliveira, M. E. C.; Lúcio, M. **2018**. A systematic review and critical analysis of the role of graphene-based nanomaterials in cancer theranostics. *Pharmaceutics* 10: 282.
27. Ferrari, M. **2005**. Cancer nanotechnology: Opportunities and challenges. *Nat Rev Cancer* 5: 161–171.
28. Mirza, Z.; Karim, S. **2021**. Nanoparticles-based drug delivery and gene therapy for breast cancer: Recent advancements and future challenges. *Semin Cancer Biol* 69: 226–237.
29. Peer, D.; Karo, J. M.; Seungpyo, H.; Farokhzad, O. C.; Margalit, S.; Langer, R. **2007**. Nanocarriers as an emerging platform for cancer therapy. *Nat Nanotechnol* 2: 751–760.
30. Kim, B. Y.; Rutka, J. T.; Chan, W. C. **2010**. Nanomedicine. *N Engl J Med* 363: 2434–2443.
31. Liang, T.; Xing, Z.; Jiang, L.; Zhu, J. J. **2021**. Tailoring nanoparticles for targeted drug delivery: From organ to subcellular level. *View* 2: 20200131.
32. Min, Y.; Caster, J. M.; Eblan, M. J.; Wang, A. Z. **2015**. Clinical translation of nanomedicine. *Chem Rev* 115: 11147–11190.
33. Sanna, V.; Sechi, M. **2020**. Therapeutic potential of targeted nanoparticles and perspective on nanotherapies. *ACS Med Chem Lett* 11: 1069–1073.
34. Ahmad, A.; Khan, F.; Mishra, R. K.; Khan, R. **2019**. Precision cancer nanotherapy: Evolving role of multifunctional nanoparticles for cancer active targeting. *J Med Chem* 62: 10475–10496.
35. Blanco, E.; Shen, H.; Ferrari, M. **2015**. Principles of nanoparticle design for overcoming biological barriers to drug delivery. *Nat Biotechnol* 33: 941–951.

36. Mahmud, A.; Xiong, X. B.; Aliabadi, H. M.; Lavasanifar, A. **2007**. Polymeric micelles for drug targeting. *J Drug Target* 15: 553–584.
37. Maeda, H.; Wu, J.; Sawa, T.; Matsumura, Y.; Hori, K. **2000**. Tumor vascular permeability and the EPR effect in macromolecular therapeutics: A review. *J Control Release* 65: 271–284.
38. Aliabadi, H. M.; Lavasanifar, A. **2006**. Polymeric micelles for drug delivery. *Expert Opin Drug Deliv* 3: 139–162.
39. Yadav, H. K.; Almokdad, A. A.; Sumia, I. M.; Debe, M. S. **2019**. Polymer-based nanomaterials for drug-delivery carriers. In *Nanocarriers for Drug Delivery*. Amsterdam, The Netherlands: Elsevier, 531–556.
40. Dubé, D.; Francis, M.; Leroux, J. C.; Winnik, F. M. **2002**. Preparation and tumor cell uptake of poly (N-isopropylacrylamide) folate conjugates. *Bioconjug Chem* 13: 685–692.
41. Torchilin, V. **2011**. Tumor delivery of macromolecular drugs based on the EPR effect. *Adv Drug Deliv Rev* 63: 131–135.
42. Tenzer, S.; Docter, D.; Kuharev, J.; et al. **2013**. Rapid formation of plasma protein corona critically affects nanoparticle pathophysiology. *Nat Nanotechnol* 8: 772–781.
43. Wu, T.; Tang, M. **2018**. Review of the effects of manufactured nanoparticles on mammalian target organs. *J Appl Toxicol* 38: 25–40.
44. Bazak, R.; Houri, M.; El Achy, S.; Kamel, S.; Refaat, T. **2015**. Cancer active targeting by nanoparticles: A comprehensive review of literature. *J Cancer Res Clin Oncol* 141: 769–784.
45. Gizem, K. **2020**. *Targeting of magnetic nanoparticles to the cell by modification of pancreatic cancer cell line specific DNA nanoparticles*, Istanbul Medipol Üniversitesi Sağlik Bilimleri Enstitüsü, Master Thesis, Turkey.
46. Din, F. U.; Aman, W.; Ullah, I.; et al. **2017**. Effectiveuse of nanocarriers as drug delivery systems for the treatment of selected tumors. *Int J Nanomedicin* 12: 7291–7309.
47. Mansouri, S.; Cuie, Y.; Winnik, F.; et al. **2006**. Characterization of folate-chitosan-DNA nanoparticles for gene therapy. *Biomaterials* 27: 2060–2065.
48. Mao, H. Q.; Roy, K.; Troung-Le, V. L.; et al. **2001**. Chitosan-DNA nanoparticles as gene carriers: Synthesis, characterization and transfection efficiency. *J Control Release* 70: 399–421.
49. Erbacher, P.; Zou, S.; Bettinger, T.; Steffan, A. M.; Remy, J. S. **1998**. Chitosan-based vector/DNA complexes for gene delivery: Biophysical characteristics and transfection ability. *Pharm Res* 15: 1332–1339.
50. Torchilin, V. P. **2014**. Multifunctional, stimuli-sensitive nanoparticulate systems for drug delivery. *Nat Rev Drug Discov* 13: 813–827.
51. Khanmohammadi, M.; Elmizadeh, H.; Ghasemi, K. **2015**. Investigation of size and morphology of chitosan nanoparticles used in drug delivery system employing chemometric technique. *Iran J Pharm Res* 14: 665–675.
52. Hossen, S.; Hossain, M. K.; Basher, M. K.; Mia, M. N. H.; Rahman, M. T.; Uddin, M. J. **2019**. Smart nanocarrier-based drug delivery systems for cancer therapy and toxicity studies: A review. *J Adv Res* 15: 1–18.
53. Torchilin, V. P. **2012**. Multifunctional nanocarriers. *Adv Drug Deliv Rev* 64: 302–315.
54. Xu, Z. P.; Zeng, Q. H.; Lu, G. Q.; Yu, A. B. **2006**. Inorganic nanoparticles as carriers for efficient cellular delivery. *Chem Eng Sci* 61: 1027–1040.
55. Brown, M. D.; Schätzlein, A. G.; Uchegbu, I. F. **2001**. Gene delivery with synthetic (non viral) carriers. *Int J Pharm* 1: 1–21.
56. Bai, Z.; Wei, J.; Yu, C.; et al. **2019**. Non-viral nanocarriers for intracellular delivery of microRNA therapeutics. *J Mater Chem B* 7: 1209–1225.
57. Kaur, I. P.; Kakkar, S. **2014**. Nanotherapy for posterior eye diseases. *J Control Release* 193: 100–112.
58. Labhasetwar, V. **2005**. Nanotechnology for drug and gene therapy: The importance of understanding molecular mechanisms of delivery. *Curr Opin Biotechnol* 16: 674–680.
59. Selvaraj, K.; Gowthamarajan, K.; Karri, V. V.; Barauah, U. K.; Ravisankar, V.; Jojo, G. M. **2017**. Current treatment strategies and nanocarrier based approaches for the treatment and management of diabetic retinopathy. *J Drug Target* 25: 386–405.
60. Uğurlu, N. **2021**. *Evaluation of nanoparticular carbonic anhydrase inhibitor effectiveness in the treatment of diabetic macular edema*. Hacettepe University Graduate School of Health Sciences PhD. Thesis in Biopharmaceutics and Pharmacokinetics, Ankara.
61. Wisniewska-Kruk, J.; Hoeben, K. A.; Vogels, I. M.; et al. **2012**. A novel co-culture model of the blood retinal barrier based on primary retinal endothelial cells, pericytes and Astrocytes. *Exp Eye Res* 96: 181–190.
62. Taddei, A.; Giampietro, C.; Conti, A.; et al. **2008**. Endothelial adherens junctions control tightjunctions by VE-cadherin mediated upregulation of claudin-5. *Nat Cell Biol* 10: 923–934.

63. Sagaties, M. J.; Raviola, G.; Schaeffer, S.; Miller, C. **1987**. The structural basis of the inner blood retina barrier in the eye of Macaca mulatta. *Invest Ophthalmol Vis Sci* 28: 2000–2014.
64. Toda, R.; Kawazu, K.; Oyabu, M.; Miyazaki, T.; Kiuchi, Y. **2011**. Comparison of drug permeabilities across the bloode retinal barrier, bloode aqueous humor barrier, and bloode brain barrier. *J Pharm Sci* 100: 3904–3911.
65. World Health Organization. **2014**. Obesity and overweight. *J Physiother* 60: 114.
66. Malik, V. S.; Willett, W. C.; Hu, F. B. **2013**. Global obesity: Trends, risk factors and policy implications. *Nat Rev Endocrinol* 9: 13–27.
67. Friedman, J. M. **2009**. Causes and control of excess body fat. *Nature* 459: 340–342.
68. Jackson, V. M.; Breen, D. M.; Fortin, J. P.; et al. **2015**. Latest approaches for the treatment of obesity. *Expert Opin Drug Discov* 10: 825–839.
69. Kushner, R. F.; Ryan, D. H. **2014**. Assessment and lifestyle management of patients with obesity. *JAMA* 312: 943–952.
70. Wadden, T. A.; Butryn, M. L.; Wilson, C. **2007**. Lifestyle modification for the management of obesity. *Gastroenterology* 132: 2226–2238.
71. Mun, E. C.; Blackburn, G. L.; Matthews, J. B. **2001**. Current status of medical and surgical therapy for obesity. *Gastroenterology* 120: 669–681.
72. Yanovski, S.; Dietz, W.; Goodwin, N.; et al. **1996**. Long-term pharmacotherapy in the management of obesity. *JAMA* 276: 1907–1915.
73. George, M.; Rajaram, M.; Shanmugam, E. **2014**. New and emerging drug molecules against obesity. *J Cardiovasc Pharmacol Ther* 19: 65–76.
74. Kakkar, A. K.; Dahiya, N. **2015**. Drug treatment of obesity: Current status and future prospects. *Eur J Intern Med* 26: 89–94.
75. Mura, S.; Couvreur, P. **2012**. Nanotheranostics for personalized medicine. *Adv Drug Deliver Rev* 64: 1394–1416.
76. Parveen, S.; Misra, R.; Sahoo, S. K. **2012**. Nanoparticles: A boon to drug delivery, therapeutics, diagnostics and imaging. *Nanomedicine* 8: 147–166.
77. Sun, W.; Hu, Q.; Ji, W.; Wright, G.; Gu, Z. **2017**. Leveraging physiology for precision drug delivery. *Physiol Rev* 97: 189–225.
78. Hossen, M. N.; Kajimoto, K.; Akita, H.; Hyodo, M.; Harashima, H. **2013**. A comparative study between nanoparticle-targeted therapeutics and bioconjugates as obesity medication. *J Control Release* 171: 104–112.
79. Yameen, B.; Choi, W. I.; Vilos, C.; Swami, A.; Shi, J.; Farokhzad, O. C. **2014**. Insight into nanoparticle cellular uptake and intracellular targeting. *J Control Release* 190: 485–499.
80. Heymsfield, S. B.; Wadden, T. A. **2017**. Mechanisms, pathophysiology, and management of obesity. *N Engl J Med* 376: 254–266.
81. Apovian, C. M.; Aronne, L. J.; Bessesen, D. H.; et al. **2015**. Pharmacological management of obesity: An endocrine society clinical practice guideline. *J Clin Endocrinol Metab* 100: 342–362.
82. Huang, D.; Deng, M.; Kuang, S. **2019**. Polymeric carriers for controlled drug delivery in obesity treatment. *Trends Endocrinol Metab* 30: 974–989.
83. Kumbar, S.; Laurencin, C.; Deng, M. (Eds.). **2014**. *Natural and Synthetic Biomedical Polymers*. USA: Newnes. San Diego, USA.
84. Angrisani, L.; Santonicola, A.; Iovino, P.; Formisano, G.; Buchwald, H.; Scopinaro, N. **2015**. Bariatric surgery worldwide 2013. *Obes Surg* 25: 1822–1832.
85. Fenton, O. S.; Olafson, K. N.; Pillai, P. S.; Mitchell, M. J.; Langer, R. **2018**. Advances in biomaterials for drug delivery. *Adv Mater* 30: 1705328.
86. Tsou, Y. H.; Wang, B.; Ho, W.; et al. **2019**. Nanotechnology-mediated drug delivery for the treatment of obesity and its related comorbidities. *Adv Healthc Mater* 8: 1801184.
87. Ash, G. I.; Kim, D.; Choudhury, M. **2019**. Promises of nanotherapeutics in obesity. *Trends Endocrinol Metab* 30: 369–383.
88. Di Mascolo, D.; Lyon, C. J.; Aryal, S.; et al. **2013**. Rosiglitazone-loaded nanospheres for modulating macrophage-specific inflammation in obesity. *J Control Release* 170: 460–468.
89. Xue, Y.; Xu, X.; Zhang, X. Q.; Farokhzad, O. C.; Langer, R. **2016**. Preventing diet-induced obesity in mice by adipose tissue transformation and angiogenesis using targeted nanoparticles. *Proc Natl Acad Sci* 113: 5552–5557.
90. Marrache, S.; Dhar, S. **2012**. Engineering of blended nanoparticle platform for delivery of mitochondria-acting therapeutics. *Proc Natl Acad Sci* 109: 16288–16293.

91. Jiang, C.; Cano-Vega, M. A.; Yue, F.; et al. **2017**. Dibenzazepine-loaded nanoparticles induce local browning of white adipose tissue to counteract obesity. *Mol Ther* 25: 1718–1729.
92. Ballinger, A.; Peikin, S. R. **2002**. Orlistat: Its current status as an antiobesity drug. *Eur J Pharmacol* 440: 109–117.
93. Sangwai, M.; Sardar, S.; Vavia, P. **2014**. Nano emulsified orlistat embedded multi-unit pellet system (MUPS) with improved dissolution and pancreatic lipase inhibition. *Pharmaceut Dev Tech* 19: 31–41.
94. Chen, Y. L.; Zhu, S.; Zhang, L.; et al. **2016**. Smart conjugated polymer nanocarrier for healthy weight loss by negative feedback regulation of lipase activity. *Nanoscale* 8: 3368–3375.
95. Cao, Y. **2007**. Angiogenesis modulates adipogenesis and obesity. *J Clin Invest* 117: 2362–2368.
96. Carmeliet, P.; Jain, R. K. **2000**. Angiogenesis in cancer and other diseases. *Nature* 407: 249–257.
97. Voros, G.; Maquoi, E.; Demeulemeester, D.; Clerx, N.; Collen, D.; Lijnen, H. R. **2005**. Modulation of angiogenesis during adipose tissue development in murine models of obesity. *Endocrinology* 146: 4545–4554.
98. Barnhart, K. F.; Christianson, D. R.; Hanley, P. W.; et al. **2011**. A peptidomimetic targeting white fat causes weight loss and improved insulin resistance in obese monkeys. *Sci Transl Med* 3: 108ra112.
99. Kolonin, M. G.; Saha, P. K.; Chan, L.; Pasqualini, R.; Arap, W. **2004**. Reversal of obesity by targeted ablation of adipose tissue. *Nat Med* 10: 625–632.
100. Kajimura, S.; Spiegelman, B. M.; Seale, P. **2015**. Brown and beige fat: Physiological roles beyond heat generation. *Cell Metab* 22: 546–559.
101. Cao, Y. **2013**. Angiogenesis and vascular functions in modulation of obesity, adipose metabolism, and insulin sensitivity. *Cell Metab* 18: 478–489.
102. Harms, M.; Seale, P. **2013**. Brown and beige fat: Development, function and therapeutic potential. *Nat Med* 19: 1252–1263.
103. Ma, L.; Liu, T. W.; Wallig, M. A.; et al. **2016**. Efficient targeting of adipose tissue macrophages in obesity with polysaccharide nanocarriers. *ACS Nano* 10: 6952–6962.
104. Thovhogi, N.; Sibuyi, N.; Meyer, M.; Onani, M.; Madiehe, A. **2015**. Targeted delivery using peptide-functionalised gold nanoparticles to white adipose tissues of obese rats. *J Nanopart Res* 17: 112.
105. Lu, Y.; Aimetti, A. A.; Langer, R.; Gu, Z. **2016**. Bioresponsive materials. *Nat Rev Mater* 2: 16075.

6 Application of Bioactive Gold Nanoparticles in Lung Cancer Diagnosis

Md Arifuzzaman, Shakil Ahmed Polash,
Md Monir Hossain, and Satya Ranjan Sarker

6.1 INTRODUCTION

Among various types of cancer, lung cancer (LC) is a prevalent, fast-spreading, as well as invasive type of tumor and is considered the leading reason for cancer-associated deaths. Every year, approximately 1.76 million deaths and 2 million new cases are discovered worldwide in this regard [1, 2]. The use of tobacco and smoking are mainly responsible for around 90% of lung cancer cases globally. Additional factors that contribute to the development of lung cancer include exposure to excessive radon gas, asbestos, extreme air pollution, as well as enduring infections [3]. Besides, the genetic predisposition as well as the acquired pathways of susceptibility could also be the cause of lung cancer [4]. As an extremely heterogeneous disease, lung cancer can occur at various sites in the bronchial tree, resulting in diverse indications and symptoms that largely rely on the specific location of lung anatomy [5]. The possibility of survival of the patients depends on the time/stage of diagnosis of lung cancer [6]. The most common method for diagnosing lung cancer is X-ray, although alternative methods, including bronchoscopy, biopsy, and ultrasound, are also available [7]. Early diagnosis of lung cancer is critical for the effective treatment and speedy recovery of the patients. Detecting lung cancer early is a highly sought-after topic for healthcare professionals as the tumor often grows without showing any symptoms for years, making it challenging to identify in most cases [8].

The management of LC holds promise in nanotechnology, which recommends innovative treatment approaches that can selectively target cancerous cells for drug delivery while also aiding in early detection [9]. It allows scientists to identify any molecular alterations in a small fraction of cells responsible for the initiation of lung cancer. Nanotechnology-based molecular diagnostics, such as the detection of biomarkers, can reliably and swiftly diagnose malignancies [x]. The unique physicochemical properties of nanoscale materials make them an attractive molecular toolkit for respiratory oncology, as they can be easily modified for various applications. Various factors, including gases, proteins, as well as DNA, contribute to disease development and can be detected by integrating nanomaterials with current biosensing platforms [11, 12]. Recently, nanotechnology has been utilized to develop nanomaterials, including gold nanoparticles and quantum dots, which are used to diagnose lung cancer at the molecular level [13].

Among several bioactive nanomaterials [14–18], bioactive gold nanoparticles (AuNPs) are widely used as preferred material for a variety of applications due to their distinct shape, size, and architectures [19–22]. For instance, bioactive Au nanomaterials have shown enormous potential in identifying and imaging disease biomarkers, particularly in the development of different types of biosensors, sensors for electrochemical assay, as well as various chromogenic assays [23]. Bioactive AuNPs, in particular, have been widely used for this purpose due to their unique properties, such as surface-enhanced Raman scattering, LSPR (localized surface plasmon resonance), redox activity, FRET (fluorescence resonance energy transfer), conductivity, and quantized charging effect [24]. Moreover, these properties make them ideal tools for imaging and detecting target molecules,

 DOI: 10.1201/9781003430087-6

as they can scatter visible and near-infrared light and interact with various technologies, including microscopic techniques, such as confocal microscopy, electron microscopy, as well as dark-field light-scattering microscopy [25], and also other techniques involving UV-vis spectroscopy, Raman spectroscopy, different types of PT heterodyne imaging modalities, as well as CT (computed tomography) [26]. Additionally, bioactive AuNPs, in particular, have outstanding properties that make them promising tools for the diagnosis and treatment of lung cancer. Due to their stability and capability to amplify signals, Au nanomaterials are highly effective transmitters for diagnostic purposes. Furthermore, their surface functionalization and size modification can be manipulated to enhance better therapeutic results and outcomes. Therefore, the use of AuNPs in lung cancer diagnosis holds great potential for providing accurate results at low concentrations and improving therapeutic interventions [27]. This book chapter aims to shed some light on how bioactive AuNPs have evolved into a powerful tool to diagnose lung cancer as well as novel platform to overcome the diagnostic challenges of lung cancer [9].

6.2 THE BIOLOGY AND DIFFERENT TYPES OF LUNG CANCER

LC (lung cancer) can develop in any part of the lungs or any airways of the lungs, and smoking is one of the primary causes of its development, accounting for more than 70% of lung cancer cases [2, 9]. However, a current research paper has shown that non-smokers exposed to airborne PM2.5 (particulate matter with an aerodynamic diameter of 2.5 mm or less) are also at risk of developing lung cancer (LC) [3, 28]. Early detection of lung cancer is difficult as symptoms usually develop as the disease progresses. These symptoms consist of continuing coughs, persistent chest infections, coughing up blood (hemoptysis), unexplained weight loss, tiredness, as well as breathlessness [9].

Furthermore, the molecular basis of LC is quite diverse and results from numerous genetic and epigenetic transformations that occur over a series of steps. The initiation of these processes can lead to the triggering of signaling pathways of growth-promoting factors as well as the inactivation of signaling pathways of tumor suppressor molecules [29, 30]. Like other malignancies, the development of tumors in LC is allied with the initiation of certain growth-promoting proteins, such as EGFR (epidermal growth factor receptor) (Figure 6.1) [31], KRAS (v-Ki-ras2 Kirsten rat sarcoma viral oncogene homolog), ALK, MET, HER2, MEK-1, BRAF, as well as a reorganization for the period of transfection, such as RET [31]. Furthermore, the deactivation of certain tumor suppressor genes, such as P53, PTEN (phosphatase with tensin homology), and LKB-1, is also commonly associated with LC [29, 30].

The activation of growth-promoting oncogenes through gene amplification, point mutations, or structural reorganizations can lead to uncontrolled signaling via oncogenic pathways. This was reported in [29]. When cellular survival is dependent on the continual stimulation of aberrant signaling pathways [32, 33], they become "oncogene addicted", which makes them perfect targets for targeted therapy. More than half of lung malignancies have been found to have oncogenic driver mutations, which not only initiate but also progress lung cancer (LC). These mutations are often exclusive of other driver mutations which are responsible for lung cancer [34, 35]. Additionally, tumor suppressor genes as well as oncogenes play a role in the control of signaling pathways that are occasionally connected with cross talk between pathways implicated in various carcinogenesis. Next-generation sequencing technologies, such as ERBB4, JAK2 [36], RET [37–39], FGFR1 (fibroblast growth factor receptor 1) [40], as well as DDR2 (discoidin domain receptor 2), have recently identified numerous new somatic alterations in entire cancer genomes or exosomes. Nonetheless, many oncogenic driver mutations have yet to be identified, posing a significant challenge to the development of effective personalized medicine for the treatment of lung cancer.

Early diagnosis of lung cancer is uncommon as it is typically detected at a later stage. Lung cancer can be classified into two types: NSCLC (non–small cell lung cancer) and SCLC (small cell

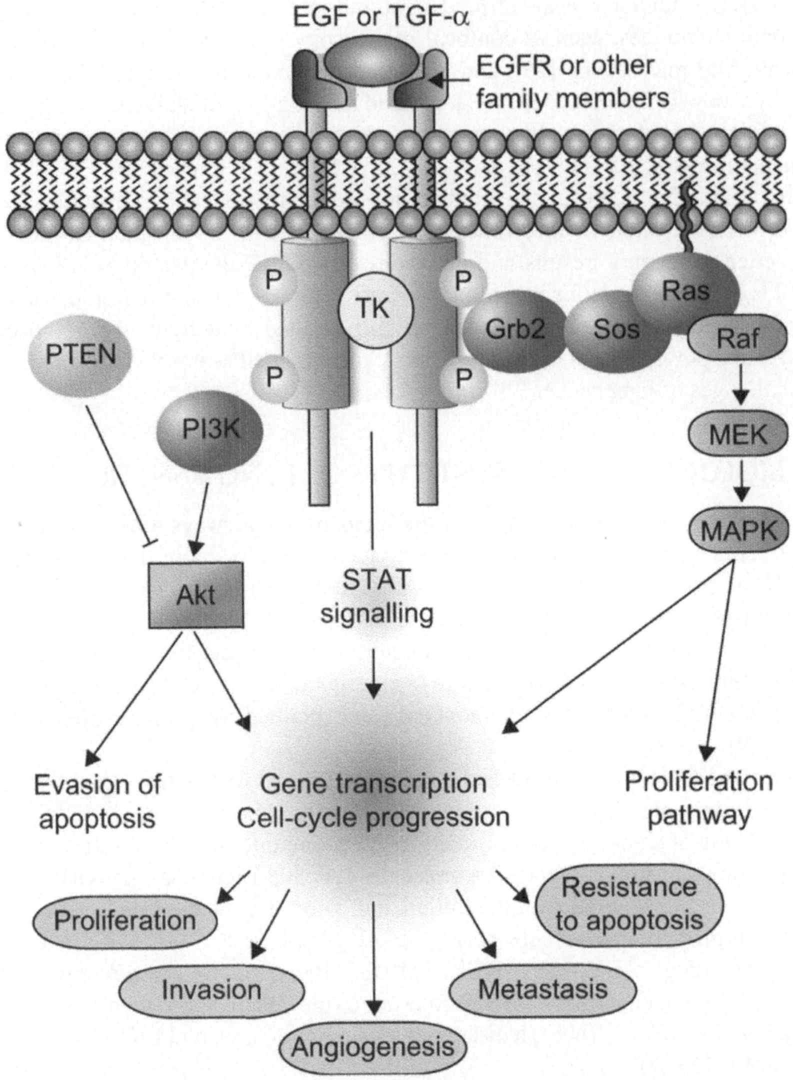

FIGURE 6.1 The pathway of the epidermal growth factor receptor (EGFR) during the formation of lung cancer. The activation of EGFR occurs when ligands, such as EGF (epidermal growth factor) or TGF-α (transforming growth factor-α), bind to the TK (kinase domain) of homo- and heterodimers, ensuing in transphosphorylation and activation of the receptor. As a result, this leads to the recruitment of Grb2 as well as Sos (adaptor proteins), which in turn employs PI3K (phosphatidylinositol 3-kinase) and Ras, ultimately resulting in the formation of two main signaling pathway branches: the first one is Ras/MAPK (mitogen-activated kinase-like protein), and the second one is PI3K/Akt. However, these pathways contribute to several processes, including but not limited to the evasion of apoptosis, proliferation, and different types of angiogenesis.

Source: Adapted from Brambilla, 2009, European Respiratory Society, Copyright © 2009. Reprinted with permission from European Respiratory Society Journals Ltd.

lung cancer). NSCLC is the utmost prevalent type, accounting for approximately 85% of all cases, and encompasses any kind of epithelial cell lung cancer. The three primary subtypes of NSCLC are squamous cell carcinoma, adenocarcinoma, as well as large cell carcinoma. However, SCLC accounts for around 10% of all lung cancer cases and exhibits a shorter cell doubling time, leading to earlier metastasis development than NSCLC [41].

6.3 CURRENT METHODS TO DIAGNOSE LUNG CANCER

6.3.1 Tests for Lung Cancer

Doctors usually ask about the patient's medical history and learn more about symptoms and potential risk factors. Doctors also look for any possible physical signs and symptoms of lung cancer and other health-related issues. Further diagnostic tests are performed if the history and results of the physical examination suggest that the person may be suffering from lung cancer. This includes imaging tests, biopsies, etc. of the lungs to look into the suspicious areas of the lungs to understand how far cancer has spread in the nearby tissues.

6.3.2 X-ray Scan

Chest X-ray is the very first diagnostic test to perform when lung cancer is suspected. Tumors in the lungs appear as whitish-gray masses. However, it is always difficult to distinguish tumors from other lung problems, such as lung abscesses. Because of their low resolution, chest X-rays are advantageous for detecting lung cancer in the advanced stage. They are, however, not suitable for diagnosing early-stage cancer [42]. The small tumors are usually masked by surrounding structures. The sensitivity of chest radiographs in symptomatic lung cancer is less than 80%, and further investigation by physicians is strongly recommended for patients whose chest radiographs are still negative [43].

6.3.3 Computed Tomography Scan

Computed tomography (CT) is a medical imaging technique where X-rays and computer processing are used simultaneously to create images of internal body parts from a different angle. It is usually used to visualize and diagnose diseased and/or damaged tissue for 30 min. It can detect any change in the lung that may require treatment. Recently, low-dose CT scans (about one-fifth of the radiation dose) are being used to diagnose people with the risk of developing lung cancer [44].

6.3.4 Positron Emission Tomography Scan

Positron emission tomography (PET) is an imaging tool that uses radioactive materials to construct detailed images of the body's internal tissues. Radioactive materials emit positrons, colloidal particles that produce gamma rays that can be detected by scanners. The patient is injected with the radiotracer, which then travels to the area of the body being examined. The PET scanner detects the gamma rays emitted by the radiotracer and uses them to create images of the area. These images show how the tissues and organs are functioning, providing information about the metabolic activity, blood flow, and other physiological processes. It is more effective and accurate than a chest CT scan. Diagnostic accuracy for lung cancer is about 90% [45]. While PET scans do involve exposure to radiation, the amount is generally considered safe and is similar to that of a CT scan. These days, doctors prefer to combine PET and CT scans to get more information about the cells and determine the stage of cancer [46].

6.3.5 Bone Scan

A bone scan is another medical imaging tool that is used to detect abnormalities in bones. It involves injecting a radioactive material into the patient's arteries. The tracer accumulates in areas of the bone that have high metabolic activity, such as those affected by cancer. A bone scan helps detect if cancer cells have already spread to the bone [47]. It can detect any changes in the bone much earlier than regular X-rays. This test is sensitive to differences in bone metabolism as highlighted *in vivo* by radiotracers [47]. The amount of radiation is considered safe for most patients, but pregnant women should not undergo bone scans unless necessary.

6.3.6 Sputum Cytology

Sputum cytology is a medical test that examines cells found in a patient's sputum, which is the mucus coughed up from the respiratory tract. The test is used to detect abnormal cells that may indicate lung cancer or other respiratory diseases. During the test, the patient is asked to cough up sputum into a sterile container that is examined by a cytotechnologist or pathologist under a microscope. Sputum cytology is a non-invasive test that can be useful in diagnosing lung cancer, particularly in patients with a high risk of developing the disease, such as smokers or those with a family history of lung cancer [48]. It can also be used to monitor the progress of lung cancer or other respiratory diseases and to evaluate the effectiveness of treatment. However, the test has limitations, as it may not detect non-cancerous lung diseases, such as pneumonia, inflammatory diseases, and tuberculosis.

6.3.7 Needle Biopsy

Needle biopsy for lung cancer is a medical procedure used to diagnose lung cancer by obtaining a small tissue sample from the lung using a needle. There are two main types of needle biopsies used for lung cancer: fine-needle aspiration biopsy (FNAB) and core-needle biopsy (CNB) [49]. During an FNAB, a thin needle is inserted through the chest wall and into the lung to obtain a small sample of tissue. A CNB is similar, but a slightly larger needle is used to obtain a larger tissue sample, which can be more helpful in making a definitive diagnosis. The sample is then examined under a microscope for signs of cancer. It is performed whenever there is an abnormality near the surface of the lung, the lung itself, or the chest wall. Needle biopsy for lung cancer is often used when X-rays, CT scans, or sputum cytology have shown abnormalities in the lungs.

6.3.8 Transthoracic Needle Biopsy

Transthoracic needle biopsy (TTNB) is an essential step in diagnosing and treating thoracic lung cancer. First, the lung is imaged using both CT scans and lung ultrasound techniques [50]. A local anesthetic is usually used to numb the area where the needle is placed. The doctor then inserts the needle into the desired location while looking at the lungs under fluoroscopy (X-ray) or CT scan. It is safer and faster than other surgical methods.

6.3.9 Molecular Tests

Molecular testing is the use of special tools to investigate the genome profile of an individual. Real-time PCR, Sanger sequencing, and next-generation sequencing (NGS) are common tools used for lung cancer diagnosis [51, 52]. Molecularly targeted therapy has greatly advanced the field of lung cancer treatment. Generally, pathogenesis involves the accumulation of various molecular anomalies over time [53]. Mutations within cells stimulate them to proliferate abnormally and thus initiate tumorigenesis. Molecular testing can contribute to the development of personalized medicine to correct these genetic abnormalities [54]. Furthermore, it helps identify lung cancer patients who may be benefited from targeted therapy or immunotherapy. Most mutated genes in lung cancer include *EGFR*, *KRAS*, *ALK*, and *EML4* [55, 56].

6.4 GOLD (AU) NANOMATERIALS

Gold (Au) is a chemically unreactive metal that exhibits remarkable biocompatibility with the human body. One of the most fascinating characteristics of Au is its ability to be engineered into nanoscale particles, which increases its surface area per unit mass, enabling more efficient fabrication. Various research groups have reported a diverse array of morphologies, including spherical particles, for these nanoparticles [9, 57] (as depicted in Figure 6.2), and they can be customized for a wide range of uses. These materials are collectively referred to as gold nanomaterials, indicating that they possess at least one dimension that is less than 100 nm.

Because of their small size, Au nanomaterials interact with light and cause a phenomenon called SPR (surface plasmon resonance). This happens when transmission electrons in the metal collectively oscillate due to excitation through incident light. Surface plasmon resonance is remarkably useful for medical applications because it allows for precise control of electromagnetic waves in a localized area around the material (as shown in Figure 6.3**A**). The SPR (surface plasmon resonance) wavelength is determined by the size and shape of the nanostructure, and it's important to find one that falls within the "biological window", where the light can infiltrate deep into various tissue without being absorbed by other biological macromolecules [9, 58]. The surface charge and coating material of nanoscale Au nanomaterials are also critical factors in their interaction with different types of biological macromolecules. However, coating Au nanomaterials with polymers can help achieve controlled hydrophobicity, reduce interaction with plasma proteins, decrease aggregation, and increase vascular circulation time. PEG (polyethylene glycol) is a commonly used surface coating for gold nanomaterials due to its ability to prevent protein corona formation and enhance the EPR effect within tumors. Furthermore, Au nanomaterials can be synthesized through various routes, including conjugating the thiol head group of polymers with the metal surface, binding the specific

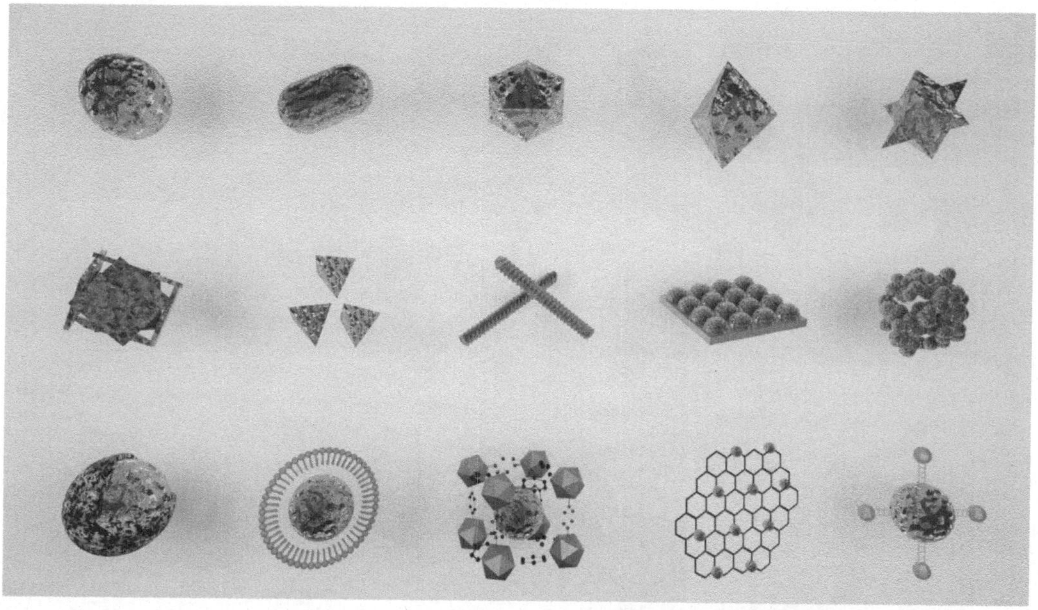

FIGURE 6.2 This figure illustrates a range of gold (Au) nanomaterials with different structures and configurations. The upper panel depicts basic gold nanostructures, including spherical in shape, rod-shaped, polygonal-shaped, and bipyramids, as well as nanostars from left to right. The middle panel displays nanoscale representatives based on Au, which include nanocages, diverged gold nanoparticles, gold nanohelices, self-accumulated layers of Au nanoparticles (AuNPs) on the surfaces, as well as gold nanoclusters from left to right. The bottom panel showcases hybrid nanoscale arrangements that utilize Au, such as silica nanoparticles with a nanocore of Au, AuNPs inside a vesicle of polymer, AuNPs encapsulated with metalorganic structures, AuNPs conjugated with graphene, as well as AuNPs with quantum dots from left to right.

Source: Reproduced with permission from Guinart (2020), Copyright © 2020, published by Portland Press Limited on behalf of the Biochemical Society and the Royal Society of Biology.

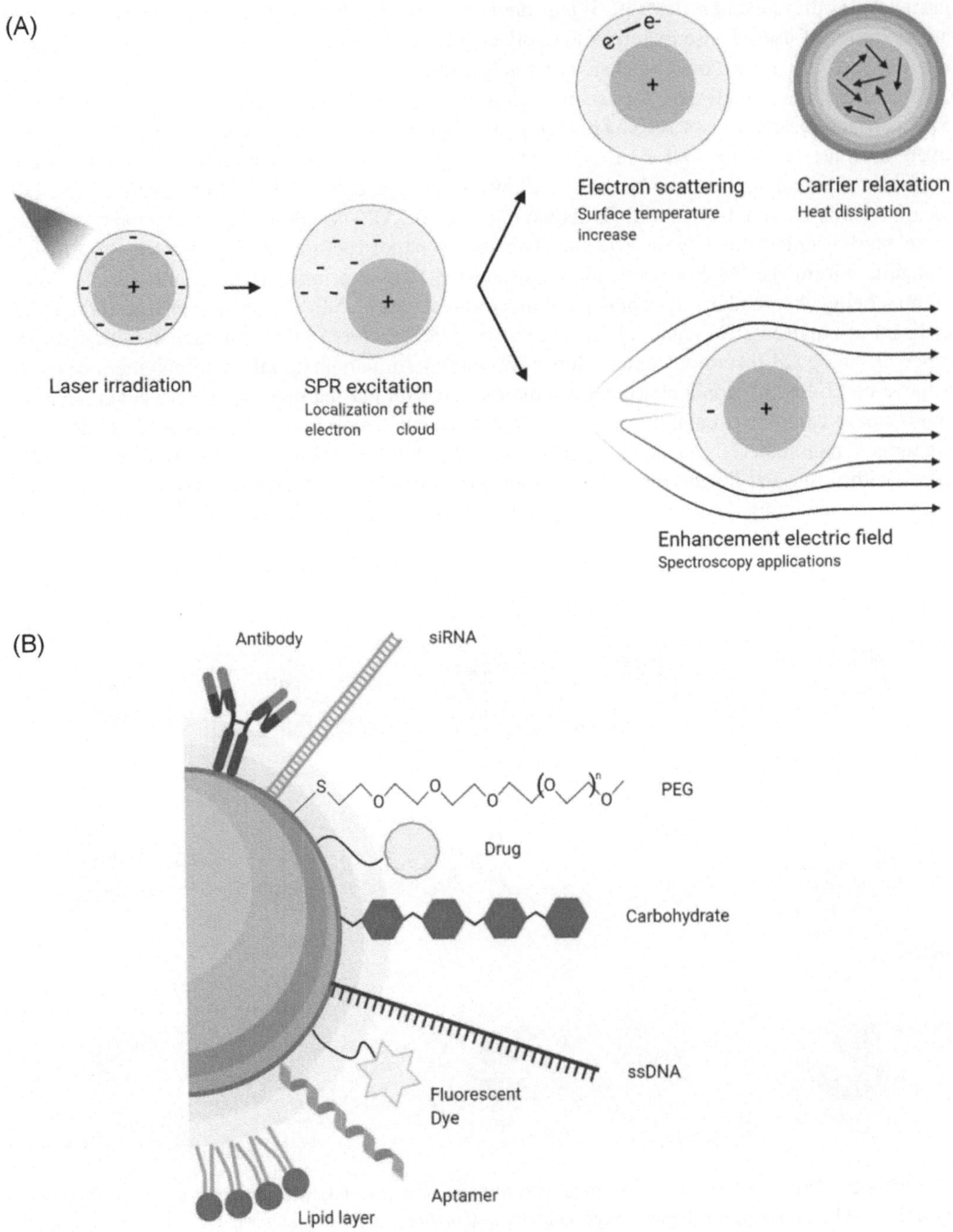

FIGURE 6.3 The physicochemical properties of Au nanomaterials. Panel (A) shows the phenomenon of SPR (surface plasmon resonance) of plasmonic Au nanoparticles and its potential applications in therapeutic purposes, imaging modalities, as well as diagnostics. Panel (B) depicts the different strategies and coatings for functionalizing Au nanoparticles.

Source: Adapted with permission from Guinart, 2020, Copyright © 2020, published by Portland Press Limited on behalf of the Biochemical Society and the Royal Society of Biology.

amino group of various biomolecules to terminal carboxylic acid groups through particular peptide coupling chemistry, or using the "click" chemistry (alkyne-azide cycloaddition reaction) to fabricate metal surfaces. This versatility allows Au nanoparticles to be coated with multiple polymers (such as glutathione, lipoic acid, sugars, antibodies, proteins, peptides, nucleic acids, and more), enabling controlled conjugation of many ligands on the Au surface (as depicted in Figure 6.3**B**). Due to their modular architecture and properties, like the lack of an oxide layer and prolonged intravascular circulation time, gold nanoparticles are excellent candidates for LC management [9, 59].

When developing diagnostic systems based on gold, scientists modify the surface of nanomaterials by attaching biomolecules that can identify disease biomarkers [12, 24]. One widely used diagnostic tool that utilizes nanotechnology is the LFA (lateral flow assays). These assays typically use Au nanomaterials that are around 30–40 nm in size, because particles that are too small or too large are not ideal for use in these types of assays [60]. In addition to the gold nanomaterials, other molecules and enzymes that can trigger changes in properties, like surface plasmon resonance (SPR), conductivity, as well as redox potentials, are also involved in these assays. When analytes interact with the gold nanomaterial conjugates, these changes produce detectable signals [24]. Whether a signal is present or not reveals whether the disease-causing target molecules/analytes are present or not. The specific signal created by Au nanomaterials is stable chemically, lasts long, as well as is consistent across various test systems, such as test tube experiments, strips, *in vitro*, as well as *in vivo* experiment [24]. Hence, using Au nanomaterials has significantly boosted the speed and success of diagnostic assays.

6.5 GOLD NANOMATERIAL–BASED DIAGNOSTIC SYSTEMS

In the clinical setting, various imaging techniques are used for lung cancer diagnosis. Currently, available diagnostic tools are invasive, time-consuming, and very expensive. Therefore, successful cancer treatment requires cost-effective and accurate detection as well as diagnosis of tumors in patients. Au nanoparticles play an important role in lung cancer treatment, especially in the case of photothermal therapy as well as photodynamic therapy. This is because they are inert and stable, are non-immunogenic, and have low toxicity to cells [61]. Moreover, Au nanoparticles and their functionalized derivatives, as well as Au nanoparticles loaded drugs, can pass through the tumor vessel wall, and the enhanced permeability and retention (EPR) effects facilitate the accumulation at tumor cells [62]. Hence, a significant effort has been made by scientists to improve disease diagnosis approaches by exploiting the physicochemical properties of Au nanoparticles.

6.5.1 Immunosensors and Serum Tumor Markers

Immunosensors and serum tumor markers are two types of medical tests used to help diagnose and monitor lung cancer. Immunosensors are a type of diagnostic test that use antibodies to detect specific substances, such as proteins or other molecules, in a patient's blood or other bodily fluids. On the other hand, serum tumor markers are proteins that are produced by cancer cells and can be revealed in a patient's fluid. Radioimmunoassay (RIA) and enzyme-linked immunosorbent assay (ELISA) are complex multi-step procedures that are used commonly for diagnosis. However, both are time-consuming, expensive, and difficult to use. Therefore, it is expected that a simple, precise, selective, and cost-effective immunosensor should be developed for detecting biomarkers. Cancer immunosensors are rapid and sensitive tools that are used to detect cancer biomarkers in patients. On the other hand, electrochemical immunosensors may be a good alternative to optical immunosensors. Recently, Au nanoparticles have attracted attention for the development of immunosensors to diagnose early lung cancer [63–68].

Kim et al. developed an amperometric immunosensor for diagnosing lung cancer by detecting annexin II and MUC5AC [69]. The Au nanoparticles–based technique offers easy and rapid detection of targets by mass spectrometry [70, 71]. Au nanoparticles are widely used as signal enhancers due to their high electrical conductivity, high surface energy, high adsorption capacity, and excellent biocompatibility [72, 73]. Additionally, the use of hydrophilic polymers such as polyethylene glycol (PEG) to fabricate Au nanoparticles helps immobilize the antibody on the electrode surface and stabilize its binding site. Ho et al. developed a simple electrochemical immunodetection system to detect lung cancer biomarkers [74]. A polyclonal secondary antibody was immobilized onto the PEG-functionalized Au nanoparticles. This Au nanoparticles–based sensor required only a few microliters of sample and exhibited a shorter response time and higher sensitivity and did not require any sophisticated logistical support for signal acquisition (Figure 6.4**a**).

Recently, the development of label-free electrochemical sensors has gained immense attention to eliminating the non-specific interactions of bound antibodies [75]. A label-free electrochemical immunosensor was developed by Kuntamung et al. for the detection of GM2 activator protein (GM2AP), a lung cancer biomarker [76]. Polyethylenimine-coated (PEI) Au nanoparticles were prepared and modified with phosphomolybdic acid to use as a redox platform. The limit of detection (LOD) of the biosensor (0.51 pg/mL) is lower than the clinically approved levels in human urine and serum (Figure 6.4**b**). The proposed biosensor offers ease of fabrication, cost-effectiveness, rapid analysis, and excellent reproducibility of the results over six weeks. On the other hand, Wang et al. developed a point-of-care testing system suitable for measuring tumor biomarkers [77]. Au nanoparticles–based sandwich immunoassay was performed for fast, sensitive, and discriminatory detection of common lung cancer biomarkers (Figure 6.4**c**). Moreover, Gao et al. reported a rapid and sensitive co-detection method using Au nanoparticles and microarray-based serum biomarkers [78]. Au nanoparticles antibody conjugate was synthesized as a detection probe for the diagnosis of early lung cancer. This antibody biochip can detect multiple biomarkers in serum samples within an hour (Figure 6.4**d**).

(a)

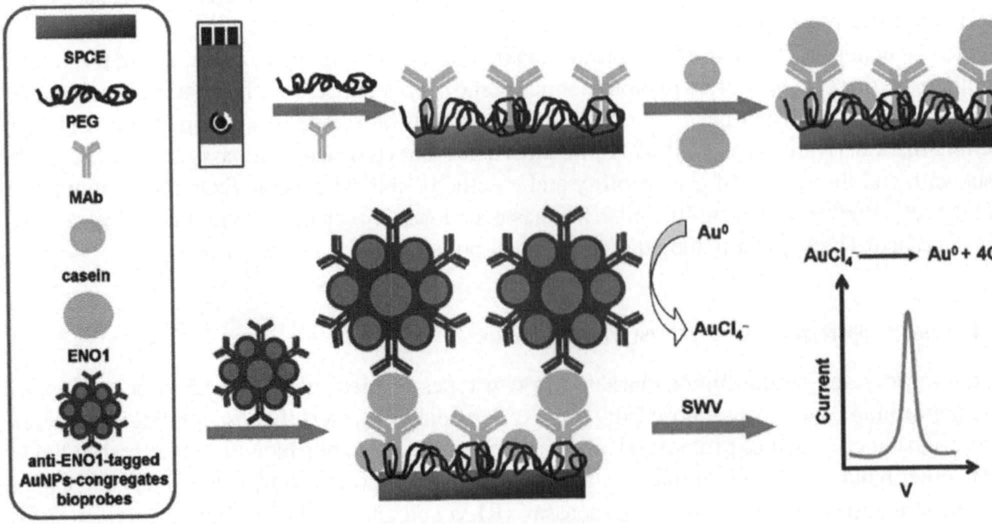

FIGURE 6.4 Au nanoparticles–based immunosensor to detect lung cancer biomarkers at the early stage: (a) electrochemical immunosensor for ENO1 detection; (b) tumor marker detection based on Au nanoparticles–based sensor; (c) point-of-care (POC) sensor development for early detection of common lung cancer biomarkers; and (d) construction procedure of electrochemical immunosensor for GM2AP detection in lung cancer sample.

Source: (a) Adapted from Ho et al. 2010 [74]. Copyright © 2010, American Chemical Society. (b) Adapted from Kuntamung et al. 2021 [76]. Copyright © 2021, Royal Society of Chemistry. (c) Adapted from Zou et al. 2014 [77]. Copyright © 2014, Springer Science Business Media New York. (d) Adapted from Gao et al. 2017 [78]. Copyright © 2016, Elsevier B.V.

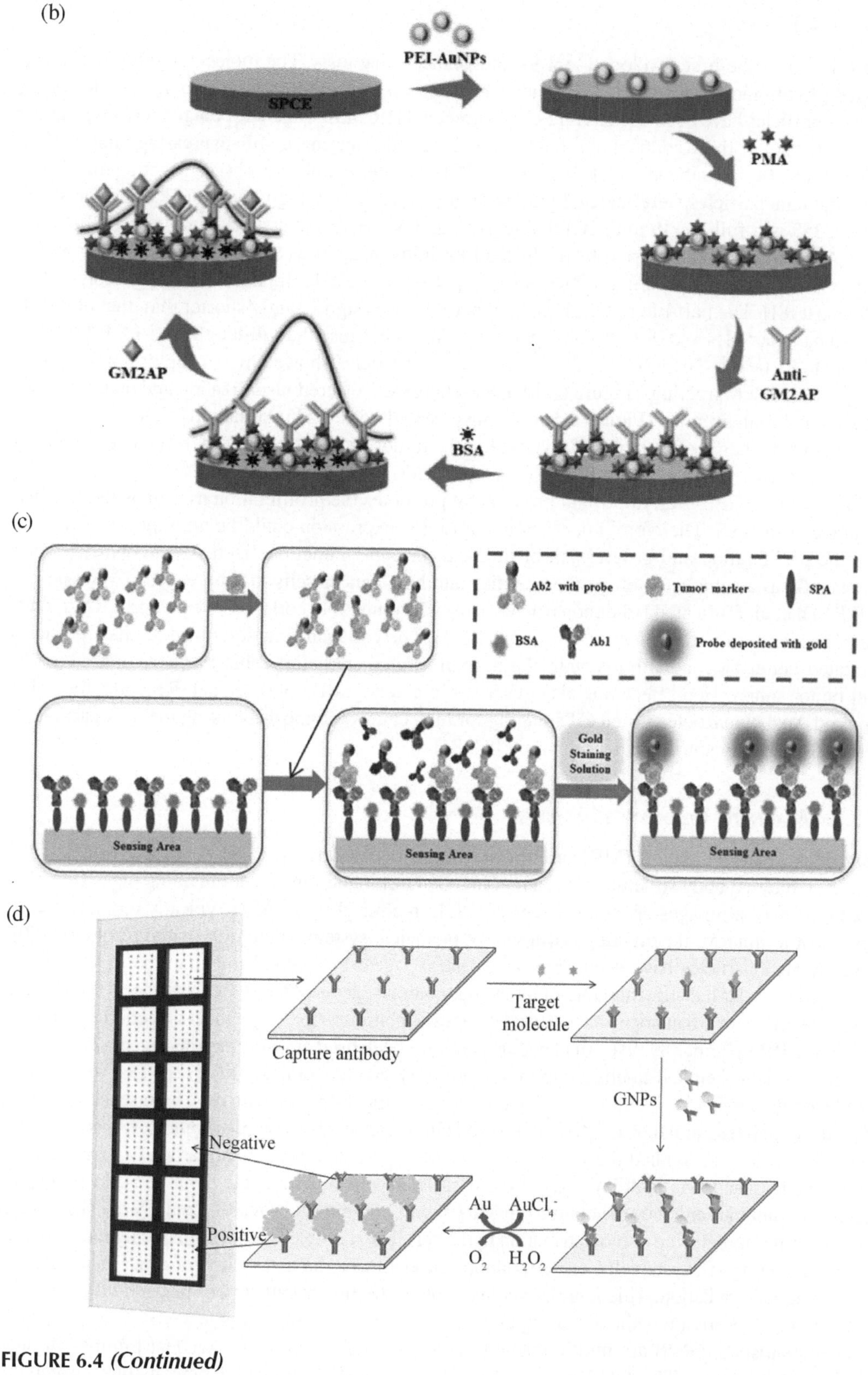

FIGURE 6.4 ***(Continued)***

6.5.2 CT Scans

The CT scan is the first developed tool for lung cancer diagnosis. The molecular study found epidermal growth factor receptor (EGFR) as an overexpressed protein in lung tumors [79]. Recently, Au nanoparticles have been used as an EGFR-directed CT contrast agent for diagnosing lung cancer [80]. In this study, three contrast agents were compared for imaging EGFR-expressing tumors. Both targeted (i.e., functionalized with C225 and VHH-122 protein) and control (i.e., functionalized with PEG) Au nanoparticles were injected into nude mice that display high levels of EGFR expression. Here, C255 is a full-length anti-EGFR antibody, and VHH-122 is a short-chain anti-EGFR antibody. All contrast agents were effectively used for both vascular as well as tumor imaging by CT (Figure 6.5**a**). The PEG coating allows Au nanoparticles to evade the immune recognition system of the host [81]. The half-life of C225 Au nanoparticles was significantly shorter than that of VHH-122 Au nanoparticles and PEG Au nanoparticles. This is because the full-length anti-EGFR is more specific to its target. Therefore, C225-coated Au nanoparticles showed higher tumor uptake due to the specific tumor targeting. (Figure 6.5**b**). Dual-energy CT offered visualization and quantification Au nanomaterials in mice (Figure 6.5**c**). Tumors treated with C225 Au nanoparticles appeared to have a much higher accumulation of gold within the tumor, showing a marked difference compared to tumors treated with VHH and PEG Au nanoparticles.

In addition, EGFR is an important protein that promotes the proliferation as well as the survival of lung cancer cells. Therefore, knocking down EGFR expression could be an important target to stop the proliferation and progression of lung cancer cells. Functionalized Au nanoparticles can act potently as a vehicle to deliver therapeutics into lung cancer cells to suppress the expression of EGFR. Yu et al. fabricated Au nanoparticles using biocompatible collagen to enhance siRNA loading efficiency for the treatment of lung cancer [82]. The collagen-functionalized Au nanoparticles exhibited better biocompatibility with the normal airway epithelial cells. Furthermore, a significant tumor suppression effect was also observed in a xenograft mouse model. Recently, they also reported Au nanoparticles-based CT imaging for lung cancer diagnosis, making the nanoparticles a potential contrast agent for cancer imaging [83, 84].

6.5.3 Magnetic Resonance Imaging

Magnetic resonance imaging (MRI) uses powerful magnetic fields and radio waves to create detailed images of internal body structures. It is a non-invasive technique capable of obtaining high-resolution three-dimensional images of living organisms [85]. In clinical trials, MRI typically requires a contrast agent to improve the imaging sensitivity of biological systems with high spatial resolution [86]. Recently, nanoparticles have been used as a contrast agent in medical imaging due to their more prolonged vascular half-life than that of molecular contrast agents [87, 88]. Recent reports support the beneficial effects of Au nanoparticles–based diagnostic platforms for lung cancer detection and *in vivo* real-time MRI imaging [89–91]. Au nanoparticles have been used as X-ray contrast agents *in vivo* due to their excellent chemical stability, superior EPR effect, and biocompatibility [92]. They were conjugated with fluorogenic molecules and were used for imaging purposes to achieve high sensitivity [93–95]. Functionalized Au nanoparticles can be used simultaneously as contrast agents for *in vivo* X-rays and MRI. Alric et al. studied the X-ray as well as MRI potential of Au nanoparticles functionalized with organic ligands via multilayer gadolinium chelate shells and disulfide bonds [96]. In principle, the gadolinium ions enhance the contrast in MRI stems, and the Au provides strong X-ray absorption for "real-time" monitoring of biodistribution in the cell. The multiple layers of organic ligands provide specific targeting of cancer cells, and Au nanoparticles induce the destruction of cancerous cells via electromagnetic radiation. This non-destructive tool allows tumor cells to be imaged and circulated freely in the bloodstream without causing unwanted accumulation in the organ.

Gold nanostars (GNS) are multibranched Au nanoparticles that hold excellent promise in biomedical applications. The GNS absorbs near-infrared (NIR) light and demonstrates excellent

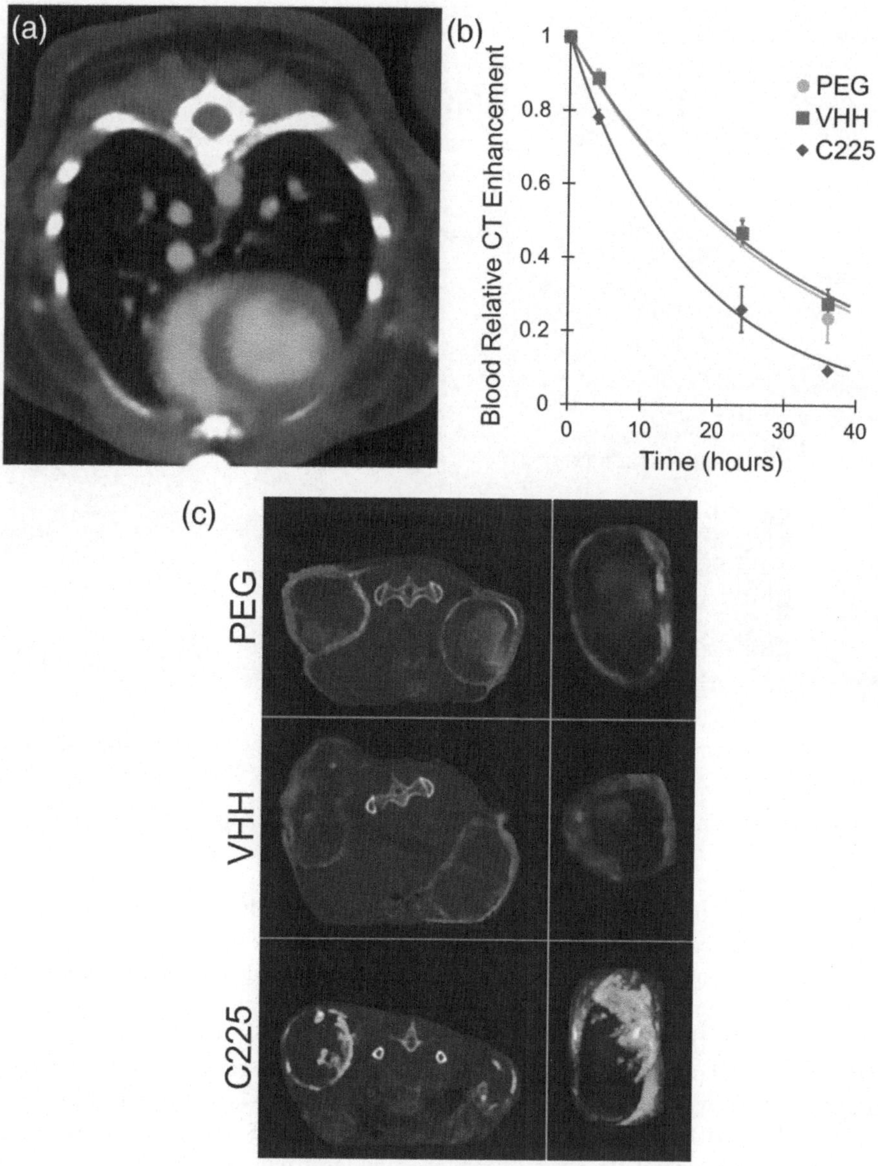

FIGURE 6.5 Effect of Au nanoparticle contrast agents in CT: (a) axial CT section of heart nearly after injection of VHH-GNPs; (b) estimation of contrast agents in the blood over time; (c) dual-energy CT sections of tumors 48 hours after injection.

Source: Adapted and Copyright © from Ashton et al. 2018 [80].

photothermal irradiation effect in the tumor microenvironment. Due to their large specific surface area and tunable plasma absorption, they can exhibit excellent photothermal effect and can be easily penetrated by external electromagnetic fields [97]. This photothermal irradiation brings about the death of lung cancer cells and has an insignificant effect on healthy epithelial lung cells. Therefore, GNS indicates a potential application of MRI-guided photothermal therapy for lung cancer. Liu et al. developed a novel theranostic nanoprobe by combining GNS with MnO_2 nanosheets [98]. GNS grew on the surface of 2D nanosheets of MnO_2 and demonstrated GSH-responsive MRI

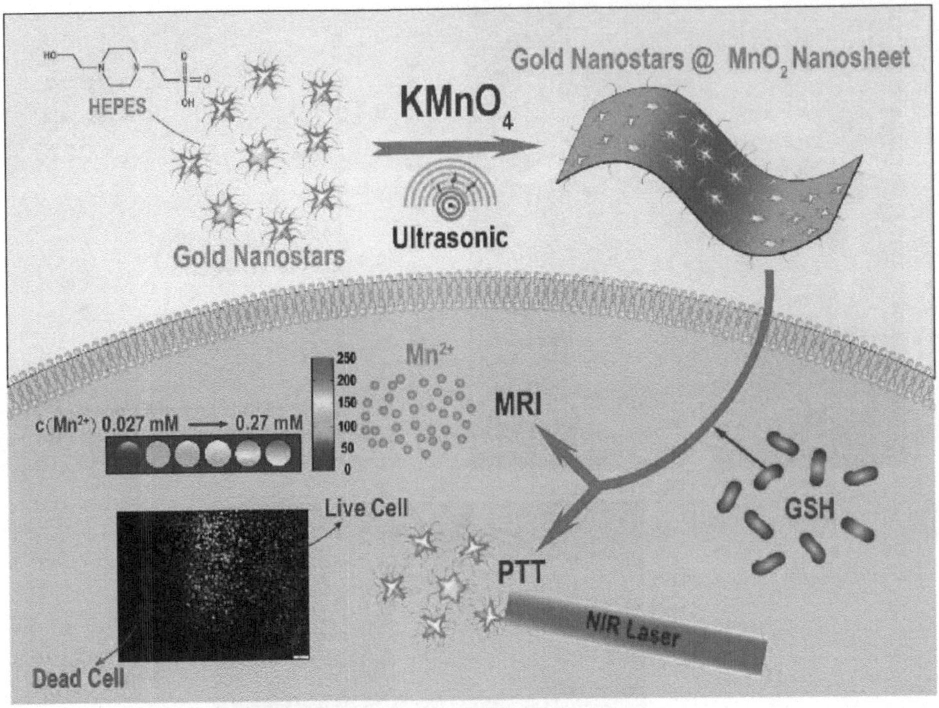

FIGURE 6.6 Schematic of possible theranostic nanoprobe mechanism, GSH-responsive MRI, and PTT in tumor cells.

Source: Adapted from Liu et al. 2018 [98]. Copyrights © 2018 Elsevier Ltd.

properties. Ultrathin MnO_2 nanosheets are readily degraded to Mn^{2+} in the tumor microenvironment, which typically contains high levels of H_2O_2, GSH (glutathione), and low pH [99–101]. The good biocompatibility of MnO_2 nanosheets makes them a potential candidate for non-invasive and stimuli-responsive imaging contrast agents that can improve imaging sensitivity and accuracy in early cancer diagnosis (Figure 6.6). A study demonstrated the generation of multifunctional nanomaterials for MRI and photothermal therapy [102]. Magnetic Au nanoshells were prepared by fabricating silica spheres with Au embedded in Fe_3O_4 nanoparticles. Magnetite nanoparticles and gold nanoshells used as MRI contrast enhancement agents absorb NIR radiation and produce heat to kill cancer cells. The synergistic effect of femtosecond laser pulses and near-infrared irradiation destroys cancer cells very effectively through localized heating.

6.5.4 Surface-Enhanced Raman Scattering

Surface-enhanced Raman scattering (SERS) is an interesting phenomenon in which the Raman signal of an analyte absorbed on a rough metal surface is greatly amplified even for single molecule detection. Recently, SERS technology has been extensively investigated for compound detection and biological sensing to provide a robust tool for biochemical analysis and disease diagnosis. SERS has recently emerged as a promising analytical tool with the compensation of excellent contrast and high spectral specificity. This is a non-invasive, highly sensitive spectroscopic technique that distinguishes tumor cells from healthy tissues by measuring changes in the vibrational state of molecules. However, due to the slow imaging speed, this leads to poor temporal resolution and hinders the rapid detection of a large number of target biomarkers [103].

Recently, Saranya et al. developed a smart programmable nanoarchitecture by integrating Raman-active fluorophores with spherical Au nanoparticles via peptide linkage [104]. In cancer cells, the pH sensitivity of the nanoprobes enables enzymatic cleavage at the tumor site, disrupting the Au nanoparticles complex and facilitating fluorophore activation (Figure 6.7**a**). Unique fingerprint at 837, 354, and 617 cm^{-1} from the three probes was used to simultaneously detect multiple biomarkers with high sensitivity and specificity (Figure 6.7**b**). Measured SERS spectra acquired from RhB-cathB@anti-EGFR- and Rh110-cathB@anti-CK-treated A549 cells showed distinct fingerprint signatures corresponding to each Raman tag. SERS and fluorescence imaging studies demonstrated specific recognition of respective protein-coding tags, highlighting the suitability of the assay (Figure 6.7**c**).

In another study, SERS-active magnetic nanoparticles (Fe_3O_4) were labeled with gold nanoflowers to detect carcinoembryonic antigens (CEA) [105]. CEA is not normally present in the blood of healthy adults; however, its level gets elevated in patients with lung, colon, stomach, pancreas, breast, and medullary thyroid cancers. Recently, CEA has been widely used for lung cancer screening, treatment efficacy evaluation, prognosis assessment, and recurrence and metastasis monitoring. The developed ultrasensitive tool has an LOD of 0.01 fg/mL.

Tian et al. exploited the plasmonic-enhanced Raman property of Au nanoparticles for imaging cancer cells as well as the delivery of anticancer drugs [106]. A stronger signal from methotrexate-bound (MTX) Au nanoparticles was observed in human Caucasian pulmonary epidermoid carcinoma cells (Calu-1) when compared to the signal in the extracellular medium. In addition, real-time intracellular SERS imaging allows us to investigate the intracellular drug release profiles as well as assessment of a spectrum of cellular structures characteristic of cell lineages (Figure 6.7**d**).

6.6 CONCLUSION

One in five cancer-related deaths worldwide is caused by lung cancer, making it a prominent cause of death. Most of the tools used to diagnose lung cancer are invasive and can detect lung cancer only in the advanced stage. The treatment of lung cancer is associated with high costs and significant side effects due to the administration of chemotherapy and radiotherapy. As a result, there is an immediate need for novel diagnostic techniques that can facilitate the early detection of lung cancer. Bioactive Au nanoparticles have several inherent advantages (e.g., inertness, large surface area,

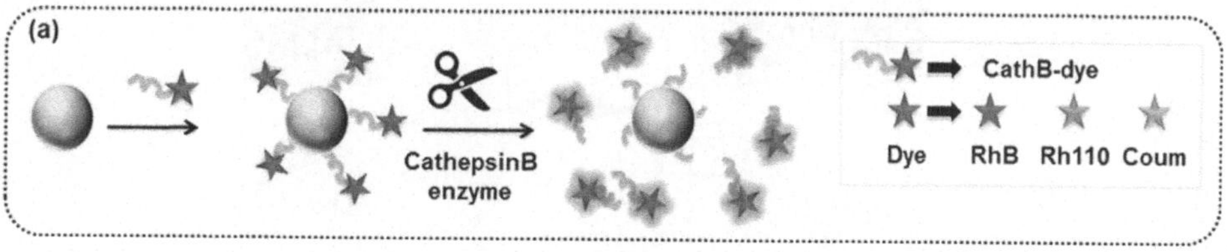

FIGURE 6.7 Lung cancer diagnosis by SERS-active Au nanoparticles: (a) switching SERS and fluorescence properties of fluorochrome-conjugated Au nanoparticles; (b) SERS spectral analysis of cathepsin B–tagged (cathB) 7-hydroxy-3-carboxycoumarin (Coum-cathB), rhodamine 110 (Rh110-cathB), and rhodamine B (RhB-cathB) in phosphate-buffered saline (pH 7.4) (SERS spectra were obtained by excitation using a 633 nm laser); (c) detection of lung cancer biomarkers with fluorescent SERS-guided multiplexed probes; (d) real-time absorbance of SERS map of Calu-1 cells incubated with drug-loaded gold nanostars for 48 h (the red color represents nanoparticle region, and green and light blue indicate the background; scale bars 20 μm; SERS spectra of cells incubated with drug-loaded gold nanostars were recorded after 1, 8, and 48 hr of treatment).

Source: (a–c) Adapted from Saranya et al. 2018 [104]. Copyright © 2018, American Chemical Society. (d) Adapted from Tian et al. 2016 [106]. Copyright © 2016 Elsevier Ltd.

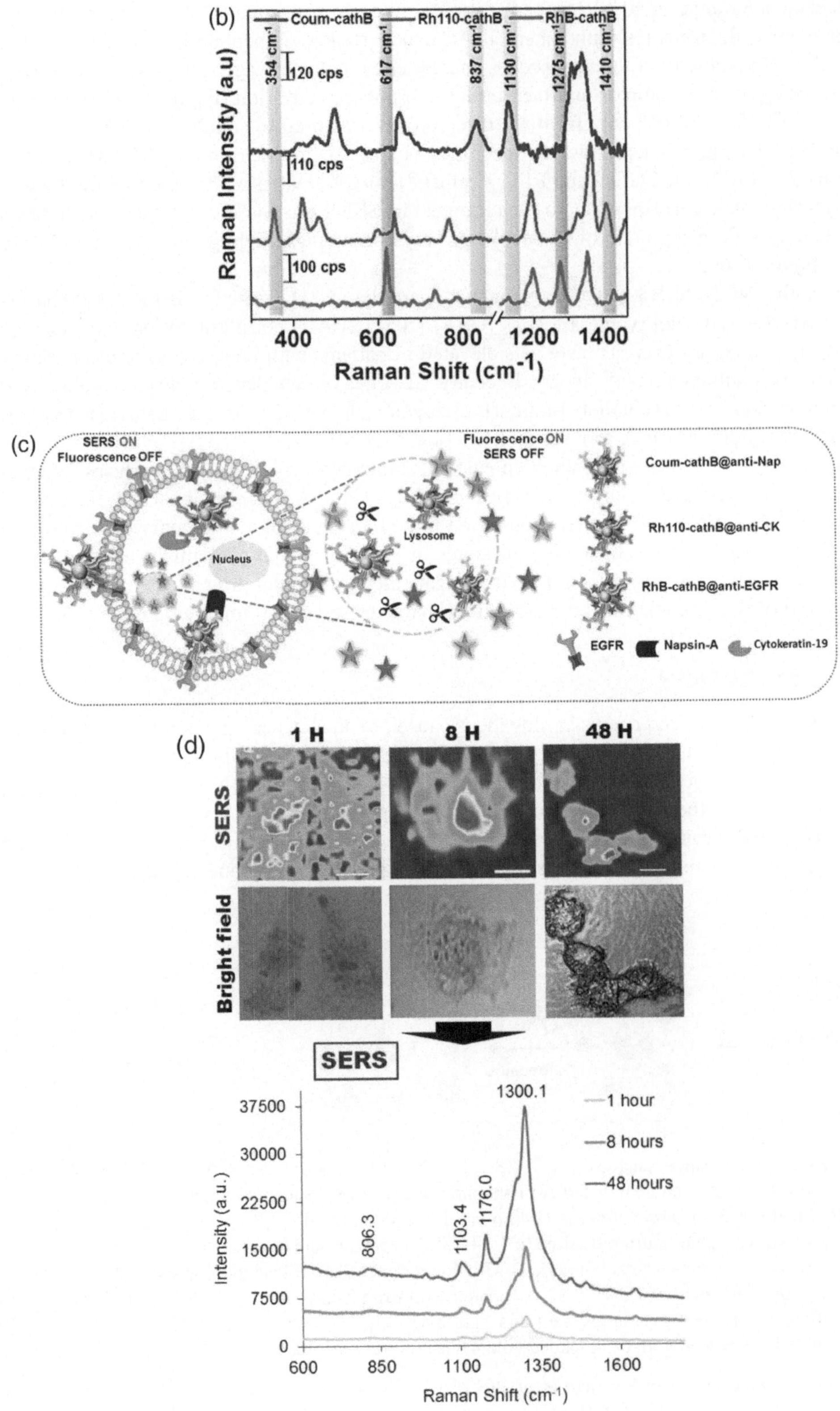

FIGURE 6.7 *(Continued)*

good biocompatibility, flexibility to functionalize with biomolecules, enhanced permeability and retention effect, etc.) to be utilized as a platform for the development of cancer diagnostic device to diagnose lung cancer at the early stage. In addition, bioactive gold nanomaterials–based biosensors are promising tools for noninvasive point-of-care diagnosis of lung cancer in the early stage.

REFERENCES

1. C.S.D. Cruz, L.T. Tanoue, R.A. Matthay, Lung cancer: Epidemiology, etiology, and prevention, *Clin Chest Med* 32(4) (2011) 605–644.
2 C.A. Ridge, A.M. McErlean, M.S. Ginsberg, *Epidemiology of Lung Cancer, Seminars in Interventional Radiology*, Thieme Medical Publishers, New York, 2013, pp. 93–98.
3 H. Lemjabbar-Alaoui, O.U. Hassan, Y.-W. Yang, P. Buchanan, Lung cancer: Biology and treatment options, *Biochim Biophys Acta* 1856(2) (2015) 189–210.
4. L. Kachuri, M. Johansson, S.R. Rashkin, R.E. Graff, Y. Bossé, V. Manem, N.E. Caporaso, M.T. Landi, D.C. Christiani, P. Vineis, Immune-mediated genetic pathways resulting in pulmonary function impairment increase lung cancer susceptibility, *Nat Commun* 11(1) (2020) 1–14.
5. W.D. Travis, E. Brambilla, G.J. Riely, New pathologic classification of lung cancer: Relevance for clinical practice and clinical trials, *J Clin Oncol* 31(8) (2013) 992–1001.
6. A.B. Mariotto, A.-M. Noone, N. Howlader, H. Cho, G.E. Keel, J. Garshell, S. Woloshin, L.M. Schwartz, Cancer survival: An overview of measures, uses, and interpretation, *J. Natl. Cancer Inst. Monographs* 2014(49) (2014) 145–186.
7. H.J. Park, S.H. Lee, Y.S. Chang, Recent advances in diagnostic technologies in lung cancer, *Korean J Intern Med* 35(2) (2020) 257.
8. C. Cainap, L.A. Pop, O. Balacescu, S.S. Cainap, Early diagnosis and screening in lung cancer, *Am J Cancer Res* 10(7) (2020) 1993.
9. A. Guinart, H.L. Perry, J.D. Wilton-Ely, T.D. Tetley, Gold nanomaterials in the management of lung cancer, *Emerg Top Life Sci* 4(6) (2020) 627–643.
10. G.F. Combes, A.-M. Vučković, M. Perić Bakulić, R. Antoine, V. Bonačić-Koutecky, K.J.C. Trajković, Nanotechnology in tumor biomarker detection: The potential of liganded nanoclusters as nonlinear optical contrast agents for molecular diagnostics of cancer, *Cancers (Basel)* 13(16) (2021) 4206.
11. A.M. Cryer, A.J. Thorley, Nanotechnology in the diagnosis and treatment of lung cancer, *Pharmacol Ther* 198 (2019) 189–205.
12. N.R.S. Sibuyi, K.L. Moabelo, A.O. Fadaka, S. Meyer, M.O. Onani, A.M. Madiehe, M. Meyer, Multifunctional gold nanoparticles for improved diagnostic and therapeutic applications: A review, *Nanoscale Res Lett* 16(1) (2021) 1–27.
13. C. Jin, K. Wang, A. Oppong-Gyebi, J. Hu, Application of nanotechnology in cancer diagnosis and therapy – a mini-review, *Int J Med Sci* 17(18) (2020) 2964.
14. M.M. Hossain, S.A. Polash, M. Takikawa, R.D. Shubhra, T. Saha, Z. Islam, S. Hossain, M.A. Hasan, S. Takeoka, S.R. Sarker, Investigation of the antibacterial activity and *in vivo* cytotoxicity of biogenic silver nanoparticles as potent therapeutics, *Front Bioeng Biotechnol* 7 (2019) 239.
15. M.S. Niloy, M.M. Hossain, M. Takikawa, M.S. Shakil, S.A. Polash, K.M. Mahmud, M.F. Uddin, M. Alam, R.D. Shubhra, M.M.A.K. Shawan, Synthesis of biogenic silver nanoparticles using Caesalpinia digyna and investigation of their antimicrobial activity and *in vivo* biocompatibility, *ACS Appl. Bio Mater* 3(11) (2020) 7722–7733.
16. S.A. Polash, M. Hossain, T. Saha, S.R. Sarker, *Biogenic Silver Nanoparticles: A Potent Therapeutic Agent, Emerging Trends in Nanomedicine*, Springer, Singapore, 2021, pp. 81–127.
17. K.M. Mahmud, M.M. Hossain, S.A. Polash, M. Takikawa, M.S. Shakil, M.F. Uddin, M. Alam, M.M. Ali Khan Shawan, T. Saha, S. Takeoka, Investigation of antimicrobial activity and biocompatibility of biogenic silver nanoparticles synthesized using Syzigyum cymosum extract, *ACS Omega* 7(31) (2022) 27216–27229.
18. S. Polash, A. Hamza, M. Hossain, M. Tushar, M. Takikawa, R. Shubhra, N. Saiara, T. Saha, S. Takeoka, S. Sarker, Diospyros malabarica fruit extract derived silver nanoparticles: A biocompatible antibacterial agent, *Front Nanotechnol* 4 (2022) 888444.
19. C. Kohout, C. Santi, L. Polito, Anisotropic gold nanoparticles in biomedical applications, *Int J Mol Sci* 19(11) (2018) 3385.

20. S.R. Sarker, S.A. Polash, M.N. Karim, T. Saha, C. Dekiwadia, V. Bansal, Y. Sabri, A.E. Kandjani, S.K. Bhargava, Functionalized concave cube gold nanoparticles as potent antimicrobial agents against pathogenic bacteria, *ACS Appl Bio Mater* 5(2) (2022) 492–503.
21. Y. Sabri, A.E. Kandjani, C.J. Harrison, S.R. Sarker, A. Chalkidis, V.E. Coyle, G. Matthews, S. Ippolito, K.M. Kabir, M. Srinivasan, Gold nanorod self-assembly on a quartz crystal microbalance: An enhanced mercury vapor sensor, *Environ Sci Nano* 8(11) (2021) 3273–3281.
22. S. Ranjan Sarker, S.A. Polash, J. Boath, A.E. Kandjani, A. Poddar, C. Dekiwadia, R. Shukla, Y. Sabri, S.K. Bhargava, interfaces, functionalization of elongated tetrahexahedral Au nanoparticles and their antimicrobial activity assay, *ACS Appl Mater Interfaces* 11(14) (2019) 13450–13459.
23. E. Oh, J.B. Delehanty, K.E. Sapsford, K. Susumu, R. Goswami, J.B. Blanco-Canosa, P.E. Dawson, J. Granek, M. Shoff, Q. Zhang, Cellular uptake and fate of PEGylated gold nanoparticles is dependent on both cell-penetration peptides and particle size, *ACS Nano* 5(8) (2011) 6434–6448.
24. Y.-C. Yeh, B. Creran, V.M. Rotello, Gold nanoparticles: Preparation, properties, and applications in bionanotechnology, *Nanoscale* 4(6) (2012) 1871–1880.
25. J.-L. Li, L. Wang, X.-Y. Liu, Z.-P. Zhang, H.-C. Guo, W.-M. Liu, S.-H. Tang, *In vitro* cancer cell imaging and therapy using transferrin-conjugated gold nanoparticles, *Cancer Lett* 274(2) (2009) 319–326.
26. H. Katas, N.Z. Moden, C.S. Lim, T. Celesistinus, J.Y. Chan, P. Ganasan, S. Suleman Ismail Abdalla, Biosynthesis and potential applications of silver and gold nanoparticles and their chitosan-based nanocomposites in nanomedicine, *J Nanotechnol* 2018 (2018).
27. P. Singh, S. Pandit, V. Mokkapati, A. Garg, V. Ravikumar, I. Mijakovic, Gold nanoparticles in diagnostics and therapeutics for human cancer, *Int J Mol Sci* 19(7) (2018) 1979.
28. L. Gharibvand, D. Shavlik, M. Ghamsary, W.L. Beeson, S. Soret, R. Knutsen, S.F. Knutsen, The association between ambient fine particulate air pollution and lung cancer incidence: Results from the AHSMOG-2 study, *Environ Health Perspect* 125(3) (2017) 378–384.
29. J.E. Larsen, J.D. Minna, Molecular biology of lung cancer: Clinical implications, *Clin Chest Med* 32(4) (2011) 703–740.
30. P. Tomasini, P. Walia, C. Labbe, K. Jao, N.B. Leighl, Targeting the KRAS pathway in non-small cell lung cancer, *Oncologist* 21(12) (2016) 1450–1460.
31. E. Brambilla, A. Gazdar, Pathogenesis of lung cancer signalling pathways: Roadmap for therapies, *Eur Respir J* 33(6) (2009) 1485–1497.
32. I.B Weinstein, Addiction to oncogenes – the Achilles heal of cancer, *Science* 297(5578) (2002) 63–64.
33. L. Sequist, R. Heist, A. Shaw, P. Fidias, R. Rosovsky, J. Temel, I. Lennes, S. Digumarthy, B. Waltman, E. Bast, Implementing multiplexed genotyping of non-small-cell lung cancers into routine clinical practice, *Ann Oncol* 22(12) (2011) 2616–2624.
34. P.Y. Yip, B. Yu, W.A. Cooper, C.I. Selinger, C.C. Ng, C.W. Kennedy, M.R. Kohonen-Corish, B.C. McCaughan, R.J. Trent, M.J. Boyer, Patterns of DNA mutations and ALK rearrangement in resected node negative lung adenocarcinoma, *J Thorac Oncol* 8(4) (2013) 408–414.
35. P. Liu, C. Morrison, L. Wang, D. Xiong, P. Vedell, P. Cui, X. Hua, F. Ding, Y. Lu, M. James, Identification of somatic mutations in non-small cell lung carcinomas using whole-exome sequencing, *Carcinogenesis* 33(7) (2012) 1270–1276.
36. L. Ding, G. Getz, D.A. Wheeler, E.R. Mardis, M.D. McLellan, K. Cibulskis, C. Sougnez, H. Greulich, D.M. Muzny, M.B. Morgan, Somatic mutations affect key pathways in lung adenocarcinoma, *Nature* 455(7216) (2008) 1069–1075.
37. R. Govindan, L. Ding, M. Griffith, J. Subramanian, N.D. Dees, K.L. Kanchi, C.A. Maher, R. Fulton, L. Fulton, J. Wallis, Genomic landscape of non-small cell lung cancer in smokers and never-smokers, *Cell* 150(6) (2012) 1121–1134.
38. Y.S. Ju, W.-C. Lee, J.-Y. Shin, S. Lee, T. Bleazard, J.-K. Won, Y.T. Kim, J.-I. Kim, J.-H. Kang, J.-S. Seo, A transforming KIF5B and RET gene fusion in lung adenocarcinoma revealed from whole-genome and transcriptome sequencing, *Genome Res* 22(3) (2012) 436–445.
39. T. Kohno, H. Ichikawa, Y. Totoki, K. Yasuda, M. Hiramoto, T. Nammo, H. Sakamoto, K. Tsuta, K. Furuta, Y. Shimada, KIF5B-RET fusions in lung adenocarcinoma, *Nat Med* 18(3) (2012) 375–377.
40. D. Lipson, M. Capelletti, R. Yelensky, G. Otto, A. Parker, M. Jarosz, J.A. Curran, S. Balasubramanian, T. Bloom, K.W. Brennan, Identification of new ALK and RET gene fusions from colorectal and lung cancer biopsies, *Nat Med* 18(3) (2012) 382–384.
41. Cancer Genome Atlas Research Network, Comprehensive genomic characterization of squamous cell lung cancers, *Nature* 489(7417) (2012) 519.

42. S. Stapley, D. Sharp, W. Hamilton, Negative chest X-rays in primary care patients with lung cancer, *Br J Gen Pract* 56(529) (2006) 570–573.
43. S.H. Bradley, S. Abraham, M.E. Callister, A. Grice, W.T. Hamilton, R.R. Lopez, B. Shinkins, R.D. Neal, Sensitivity of chest X-ray for detecting lung cancer in people presenting with symptoms: A systematic review, *Br J Gen Pract* 69(689) (2019) e827–e835.
44. M.C. Tammemagi, S. Lam, Screening for lung cancer using low dose computed tomography, *BMJ* 348 (2014).
45. S.A. Deppen, J.D. Blume, C.D. Kensinger, A.M. Morgan, M.C. Aldrich, P.P. Massion, R.C. Walker, M.L. McPheeters, J.B. Putnam, Jr., E.L. Grogan, Accuracy of FDG-PET to diagnose lung cancer in areas with infectious lung disease: A meta-analysis, *JAMA* 312(12) (2014) 1227–1236.
46. J.F. Bruzzi, R.F. Munden, PET/CT imaging of lung cancer, *J Thorac Imaging* 21(2) (2006) 123–136.
47. W. Jiao, T. Qiu, Necessity of preoperative bone scintigraphy for cT1N0 lung cancer: Evidence from retrospective to prospective study, *Thorac Cancer* 12(4) (2021) 413–414.
48. F.B. Thunnissen, Sputum examination for early detection of lung cancer, *J Clin Pathol* 56(11) (2003) 805–810.
49. X. Yao, M. Gomes, M. Tsao, C. Allen, W. Geddie, H. Sekhon, Fine-needle sspiration biopsy versus core-needle biopsy in diagnosing lung cancer: A systematic review, *Curr Oncol* 19(1) (2012) 16–27.
50. H. Carmona, S. Allison, M. Triplette, R. Adamson, Transthoracic needle biopsy, *Am J Respir Crit Care Med* 204(1) (2021) P1–P2.
51. N.I. Lindeman, P.T. Cagle, D.L. Aisner, M.E. Arcila, M.B. Beasley, E.H. Bernicker, C. Colasacco, S. Dacic, F.R. Hirsch, K. Kerr, D.J. Kwiatkowski, M. Ladanyi, J.A. Nowak, L. Sholl, R. Temple-Smolkin, B. Solomon, L.H. Souter, E. Thunnissen, M.S. Tsao, C.B. Ventura, M.W. Wynes, Y. Yatabe, Updated molecular testing guideline for the selection of lung cancer patients for treatment with targeted tyrosine kinase inhibitors: Guideline from the college of American pathologists, the international association for the study of lung cancer, and the association for molecular pathology, *Arch Pathol Lab Med* 142(3) (2018) 321–346.
52. N. Normanno, M.G. Denis, K.S. Thress, M. Ratcliffe, M. Reck, Guide to detecting epidermal growth factor receptor (EGFR) mutations in ctDNA of patients with advanced non-small-cell lung cancer, *Oncotarget* 8(7) (2017) 12501–12516.
53. C.R. Singh, K. Kathiresan, Molecular understanding of lung cancers – a review, *Asian Pac J Trop Biomed* 4(Suppl 1) (2014) S35–S41.
54. J.P. Gregg, T. Li, K.Y. Yoneda, Molecular testing strategies in non-small cell lung cancer: Optimizing the diagnostic journey, *Transl Lung Cancer Res* 8(3) (2019) 286–301.
55. G. Pinheiro, T. Pereira, C. Dias, C. Freitas, V. Hespanhol, J.L. Costa, A. Cunha, H.P. Oliveira, Identifying relationships between imaging phenotypes and lung cancer-related mutation status: EGFR and KRAS, *Sci Rep* 10(1) (2020) 3625.
56. K. Inamura, K. Takeuchi, Y. Togashi, S. Hatano, H. Ninomiya, N. Motoi, M.Y. Mun, Y. Sakao, S. Okumura, K. Nakagawa, M. Soda, Y.L. Choi, H. Mano, Y. Ishikawa, EML4-ALK lung cancers are characterized by rare other mutations, a TTF-1 cell lineage, an acinar histology, and young onset, *Mod Pathol* 22(4) (2009) 508–515.
57. J. Wu, Y. Qu, Q. Yu, H. Chen, Gold nanoparticle layer: A versatile nanostructured platform for biomedical applications, *Mater Chem Front* 2(12) (2018) 2175–2190.
58. J. Zhang, L. Mou, X. Jiang, Surface chemistry of gold nanoparticles for health-related applications, *J Chem Sci* 11(4) (2020) 923–936.
59. S. Zeng, K.-T. Yong, I. Roy, X.-Q. Dinh, X. Yu, F. Luan, A review on functionalized gold nanoparticles for biosensing applications, *Plasmonics* 6(3) (2011) 491–506.
60. B.N. Khlebtsov, R.S. Tumskiy, A.M. Burov, T.E. Pylaev, N.G. Khlebtsov, Quantifying the numbers of gold nanoparticles in the test zone of lateral flow immunoassay strips, *ACS Appl Nano Mater* 2(8) (2019) 5020–5028.
61. Z. Yang, D. Wang, C. Zhang, H. Liu, M. Hao, S. Kan, D. Liu, W. Liu, The applications of gold nanoparticles in the diagnosis and treatment of gastrointestinal cancer, *Front Oncol* 11 (2021) 819329.
62. Y. Yang, X. Zheng, L. Chen, X. Gong, H. Yang, X. Duan, Y. Zhu, Multifunctional gold nanoparticles in cancer diagnosis and treatment, *Int J Nanomedicine* 17 (2022) 2041–2067.
63. Y. Gu, G. Gong, Y. Jiang, J. Qin, Y. Mei, J. Han, Electrochemical immunosensor modified with nitrogen-doped reduced graphene oxide@ carboxylated multi-walled carbon nanotubes/chitosan@ gold nanoparticles for CA125 detection, *Chemosensors* 10(7) (2022) 272.

64. Y. Fan, S. Shi, J. Ma, Y. Guo, Smartphone-based electrochemical system with multi-walled carbon nanotubes/thionine/gold nanoparticles modified screen-printed immunosensor for cancer antigen 125 detection, *Microchem J* 174 (2022) 107044.
65. Y. Zeng, J. Bao, Y. Zhao, D. Huo, M. Chen, M. Yang, H. Fa, C. Hou, A sensitive label-free electrochemical immunosensor for detection of cytokeratin 19 fragment antigen 21–1 based on 3D graphene with gold nanoparticle modified electrode, *Talanta* 178 (2018) 122–128.
66. H. Zhang, H. Ke, Y. Wang, P. Li, C. Huang, N. Jia, 3D carbon nanosphere and gold nanoparticle-based voltammetric cytosensor for cell line A549 and for early diagnosis of non-small cell lung cancer cells, *Mikrochim Acta* 186(1) (2018) 39.
67. M. Amouzadeh Tabrizi, M. Shamsipur, L. Farzin, A high sensitive electrochemical aptasensor for the determination of VEGF(165) in serum of lung cancer patient, *Biosens Bioelectron* 74 (2015) 764–769.
68. O. Barash, N. Peled, U. Tisch, P.A. Bunn, Jr., F.R. Hirsch, H. Haick, Classification of lung cancer histology by gold nanoparticle sensors, *Nanomedicine* 8(5) (2012) 580–589.
69. D.M. Kim, H.B. Noh, D.S. Park, S.H. Ryu, J.S. Koo, Y.B. Shim, Immunosensors for detection of Annexin II and MUC5AC for early diagnosis of lung cancer, *Biosens Bioelectron* 25(2) (2009) 456–462.
70. A.K. Sharma, H. Kaurav, A. Sharma, R. Nagraik, D.N. Kapoor, Applications of nanotechnology in pulmonary disease diagnosis, In *Advanced Drug Delivery Strategies for Targeting Chronic Inflammatory Lung Diseases*, Springer, Singapore, 2022, pp. 171–193.
71. M. Hasan, M. Ahommed, M. Daizy, M. Bacchu, M. Ali, M. Al-Mamun, M.A.S. Aly, M. Khan, S.I. Hossain, Recent development in electrochemical biosensors for cancer biomarkers detection, *Biosens. Bioelectron* 8 (2021) 100075.
72. Y. Fan, S. Shi, J. Ma, Y. Guo, A paper-based electrochemical immunosensor with reduced graphene oxide/thionine/gold nanoparticles nanocomposites modification for the detection of cancer antigen 125, *Biosens Bioelectron* 135 (2019) 1–7.
73. C. Yang, Q. Guo, Y. Lu, B. Zhang, G. Nie, Ultrasensitive "signal-on" electrochemiluminescence immunosensor for prostate-specific antigen detection based on novel nanoprobe and poly (indole-6-carboxylic acid)/flower-like Au nanocomposite, *Sens Actuators B Chem* 303 (2020) 127246.
74. J.A. Ho, H.C. Chang, N.Y. Shih, L.C. Wu, Y.F. Chang, C.C. Chen, C. Chou, Diagnostic detection of human lung cancer-associated antigen using a gold nanoparticle-based electrochemical immunosensor, *Anal Chem* 82(14) (2010) 5944–5950.
75. Y.S. Gao, X.F. Zhu, J.K. Xu, L.M. Lu, W.M. Wang, T.T. Yang, H.K. Xing, Y.F. Yu, Label-free electrochemical immunosensor based on Nile blue A-reduced graphene oxide nanocomposites for carcinoembryonic antigen detection, *Anal Biochem* 500 (2016) 80–87.
76. K. Kuntamung, P. Sangthong, J. Jakmunee, K. Ounnunkad, A label-free immunosensor for the detection of a new lung cancer biomarker, GM2 activator protein, using a phosphomolybdic acid/polyethyleneimine coated gold nanoparticle composite, *Analyst* 146(7) (2021) 2203–2211.
77. Y. Zou, X. Zhang, C. An, C. Ran, K. Ying, P. Wang, A point-of-care testing system with Love-wave sensor and immunogold staining enhancement for early detection of lung cancer, *Biomed Microdevices* 16(6) (2014) 927–935.
78. W. Gao, W. Wang, S. Yao, S. Wu, H. Zhang, J. Zhang, F. Jing, H. Mao, Q. Jin, H. Cong, C. Jia, G. Zhang, J. Zhao, Highly sensitive detection of multiple tumor markers for lung cancer using gold nanoparticle probes and microarrays, *Anal Chim Acta* 958 (2017) 77–84.
79. D. Veale, T. Ashcroft, C. Marsh, G. Gibson, A. Harris, Epidermal growth factor receptors in non-small cell lung cancer, *Br J Cancer* 55(5) (1987) 513–516.
80. J.R. Ashton, E.B. Gottlin, E.F. Patz, Jr., J.L. West, C.T. Badea, A comparative analysis of EGFR-targeting antibodies for gold nanoparticle CT imaging of lung cancer, *PLoS One* 13(11) (2018) e0206950.
81. J.C. Zheng, N. Lei, Q.C. He, W. Hu, J.G. Jin, Y. Meng, N.H. Deng, Y.F. Meng, C.J. Zhang, F.B. Shen, PEGylation is effective in reducing immunogenicity, immunotoxicity, and hepatotoxicity of α-momorcharin *in vivo*, *Immunopharmacol Immunotoxicol* 34(5) (2012) 866–873.
82. A.Y.-H. Yu, R.-H. Fu, S.-H. Hsu, C.-F. Chiu, W.-H. Fang, C.-A. Yeh, C.-M. Tang, H.-H. Hsieh, H.-S. Hung, Epidermal growth factor receptors siRNA-conjugated collagen modified gold nanoparticles for targeted imaging and therapy of lung cancer, *Mater Today Adv* 12 (2021) 100191.
83. J. Liu, Z. Xiong, J. Zhang, C. Peng, B. Klajnert-Maculewicz, M. Shen, X. Shi, Zwitterionic Gadolinium(III)-complexed dendrimer-entrapped gold nanoparticles for enhanced computed tomography/magnetic resonance imaging of lung cancer metastasis, *ACS Appl Mater Interfaces* 11(17) (2019) 15212–15221.
84. W. Cai, T. Gao, H. Hong, J. Sun, Applications of gold nanoparticles in cancer nanotechnology, *Nanotechnol Sci Appl* 1 (2008) 17–32.

85. G.J. Soufi, A. Hekmatnia, S. Iravani, R.S. Varma, Nanoscale contrast agents for magnetic resonance imaging: A review, *ACS Appl Nano Mater* 5(8) (2022) 10151–10166.
86. D. Lima, A. Ribicki, L. Gonçalves, A.C.M. Hacke, L.C. Lopes, R.P. Pereira, K. Wohnrath, S.T. Fujiwara, C.A. Pessôa, Nanoconjugates based on a novel organic-inorganic hybrid silsesquioxane and gold nanoparticles as hemocompatible nanomaterials for promising biosensing applications, *Colloids Surf B Biointerfaces* 213 (2022) 112355.
87. L. Bondaz, P. Fontaine, F. Muller, N. Pantoustier, P. Perrin, I. Morfin, M. Goldmann, F. Cousin, Controlled synthesis of gold nanoparticles in copolymers nanomolds by X-ray radiolysis, *Langmuir* 36(22) (2020) 6132–6144.
88. O. Perlman, A. Borodetsky, Y. Kauffmann, Y. Shamay, H. Azhari, I.S. Weitz, Gold/Copper@ polydopamine nanocomposite for contrast-enhanced dual modal computed tomography–magnetic resonance imaging, *ACS Appl Nano Mater* 2(10) (2019) 6124–6134.
89. E.M. Hébert, P.J. Debouttière, M. Lepage, L. Sanche, D.J. Hunting, Preferential tumour accumulation of gold nanoparticles, visualised by magnetic resonance imaging: Radiosensitisation studies *in vivo* and *in vitro*, *Int J Radiat Biol* 86(8) (2010) 692–700.
90. S. Wan, F. Cui, B. Li, K. Zhao, H. He, Y. Zhang, J. Liu, L. Zhang, K. Liu, Dysprosium-modified gold nanoparticles as T 2ex contrast agents for magnetic resonance imaging, *ACS Appl Nano Mater* 3(9) (2020) 9433–9439.
91. D. Luo, X. Wang, C. Burda, J.P. Basilion, Recent development of gold nanoparticles as contrast agents for cancer diagnosis, *Cancers (Basel)* 13(8) (2021).
92. Y.J. Tseng, S.W. Chou, J.J. Shyue, S.Y. Lin, J.K. Hsiao, P.T. Chou, A versatile theranostic delivery platform integrating magnetic resonance imaging/computed tomography, pH/cis-Diol controlled release, and targeted therapy, *ACS Nano* 10(6) (2016) 5809–5822.
93. E.A. Daza, A.S. Schwartz-Duval, K. Volkman, D. Pan, Facile chemical strategy to hydrophobically modify solid nanoparticles using inverted micelle-based multicapsule for efficient intracellular delivery, *ACS Biomater Sci Eng* 4(4) (2018) 1357–1367.
94. S.K. Mishra, S. Kannan, Doxorubicin-conjugated bimetallic silver-gadolinium nanoalloy for multimodal MRI-CT-optical imaging and pH-responsive drug release, *ACS Biomater Sci Eng* 3(12) (2017) 3607–3619.
95. G. De Crozals, D. Kryza, G.J. Sánchez, S. Roux, D. Mathé, J. Taleb, C. Dumontet, M. Janier, C. Chaix, Granulocyte colony-stimulating factor nanocarriers for stimulation of the immune system (part I): Synthesis and biodistribution studies, *Bioconjug Chem* 29(3) (2018) 795–803.
96. C. Alric, J. Taleb, G. Le Duc, C. Mandon, C. Billotey, A. Le Meur-Herland, T. Brochard, F. Vocanson, M. Janier, P. Perriat, S. Roux, O. Tillement, Gadolinium chelate coated gold nanoparticles as contrast agents for both X-ray computed tomography and magnetic resonance imaging, *J Am Chem Soc* 130(18) (2008) 5908–5915.
97. S. Sasidharan, D. Bahadur, R. Srivastava, Protein-poly (amino acid) nanocore-shell mediated synthesis of branched gold nanostructures for computed tomographic imaging and photothermal therapy of cancer, *ACS Appl Mater Interfaces* 8(25) (2016) 15889–15903.
98. J. Liu, H. Cui, S. Yan, X. Jing, D. Wang, L. Meng, Gold nanostars decorated MnO_2 nanosheets for magnetic resonance imaging and photothermal erasion of lung cancer cell, *Mater Today Commun* 16 (2018) 97–104.
99. Z. Zhao, H. Fan, G. Zhou, H. Bai, H. Liang, R. Wang, X. Zhang, W. Tan, Activatable fluorescence/MRI bimodal platform for tumor cell imaging via MnO_2 nanosheet-aptamer nanoprobe, *J Am Chem Soc* 136(32) (2014) 11220–11223.
100. W. Fan, W. Bu, B. Shen, Q. He, Z. Cui, Y. Liu, X. Zheng, K. Zhao, J. Shi, Intelligent MnO_2 nanosheets anchored with upconversion nanoprobes for concurrent pH-/H_2O_2-responsive UCL imaging and oxygen-elevated synergetic therapy, *Adv Mater* 27(28) (2015) 4155–4161.
101. E. Ju, K. Dong, Z. Liu, F. Pu, J. Ren, X. Qu, Tumor microenvironment activated photothermal strategy for precisely controlled ablation of solid tumors upon NIR irradiation, *Adv Funct Mater* 25(10) (2015) 1574–1580.
102. J. Kim, S. Park, J.E. Lee, S.M. Jin, J.H. Lee, I.S. Lee, I. Yang, J.S. Kim, S.K. Kim, M.H. Cho, T. Hyeon, Designed fabrication of multifunctional magnetic gold nanoshells and their application to magnetic resonance imaging and photothermal therapy, *Angew Chem Int Ed Engl* 45(46) (2006) 7754–7758.
103. J. He, C. Li, L. Ding, Y. Huang, X. Yin, J. Zhang, J. Zhang, C. Yao, M. Liang, R.P. Pirraco, J. Chen, Q. Lu, R. Baldridge, Y. Zhang, M. Wu, R.L. Reis, Y. Wang, Tumor targeting strategies of smart fluorescent nanoparticles and their applications in cancer diagnosis and treatment, *Adv Mater* 31(40) (2019) e1902409.

104. G. Saranya, M.M. Joseph, V. Karunakaran, J.B. Nair, V.N. Saritha, V.S. Veena, K. Sujathan, A. Ajayaghosh, K.K. Maiti, Enzyme-driven switchable fluorescence-SERS diagnostic nanococktail for the multiplex detection of lung cancer biomarkers, *ACS Appl Mater Interfaces* 10(45) (2018) 38807–38818.
105. C. Song, L. Min, N. Zhou, Y. Yang, B. Yang, L. Zhang, S. Su, L. Wang, Ultrasensitive detection of carcino-embryonic antigen by using novel flower-like gold nanoparticle SERS tags and SERS-active magnetic nanoparticles, *RSC Adv* 4(78) (2014) 41666–41669.
106. F. Tian, J. Conde, C. Bao, Y. Chen, J. Curtin, D. Cui, Gold nanostars for efficient *in vitro* and *in vivo* real-time SERS detection and drug delivery via plasmonic-tunable Raman/FTIR imaging, *Biomaterials* 106 (2016) 87–97.

7 Biogenic Nanoparticles as a Strong Larvicidal Agent in Integrated Pest Management

Sunita Patil, Chandrasekaran Rajkuberan, and John Joseph

7.1 INTRODUCTION

Worldwide, vectoring of various diseases, including malaria, dengue, filariasis, yellow fever, Zika virus, Japanese encephalitis, etc., has been observed through the most common vector, mosquitoes (Figure 7.1). These deadly diseases are responsible for a significant number of global morbidity and mortality. Despite control, efforts that have been done for centuries, the disproportionate effects of these diseases on children and adolescents cannot be controlled [1]. Malaria claims almost 1 million children deaths, and half of the world's population is at risk of it, caused by the parasite *Plasmodium*. These parasites are vectored to the host by the bite of the infected mosquito species *Anopheles* [2].The report shows about 350 to 500 million clinical cases each year, with almost 1 million deaths. Of those, African children account for 80%, with the age group below 5 years [3]. In the year 1955, WHO set its aim to rid the world of the deadly malaria disease. The Global Malaria Eradication Programme campaign was set to control the malaria parasite vector, *Anopheles* [3]. But the program did not succeed practically due to insecticide-resistant development in mosquitoes and the emergence of drug-resistant *Plasmodium* [3]. Y.Tu discovered artemisinin, derived from the plant *Artemisia annua*, acts effectively against the parasite, and the vaccine development against *P. falciparum* has helped reduce malaria cases. But parasite strains with artemisinin resistance have been detected, and the RTS,S vaccine does not protect against malaria caused by *Plasmodium vivax* observed in many nations outside Africa [4]. At present, various vector control and malaria prevention tools are available and are pretty costly, thus can't be afforded by everyone. The efficient, economic vector control strategy could do better malaria control worldwide.

Dengue is a systemic viral disease that shows acute symptoms. Dengue has established both endemic and epidemic transmission cycles. Dengue virus infection in humans can cause a wide range of clinical symptoms, from placid fever to potentially fatal dengue shock syndrome. Till now, no effective antiviral agents have been discovered against the dengue virus, and thus, dengue treatment can be done only through supportive agents [5, 6]. Dengue is vectored mainly by *Aedes aegypti* and, to some extent, *Aedes albopictus*. In 128 countries, the dengue virus keeps 3,900 million people at risk of infection [7]. In 2010, there were 96 million apparent dengue infection cases reported. Among these, 70% of cases were reported in Asia, with high disease transmission in thickly populated places. From a global total, India contributed 34% of infections [7]. There is no effective treatment for dengue, so controlling the vector population can help control the spread of dengue.

Zika virus is a mosquito-borne flavivirus; it was first detected in monkeys in 1947, and later, in 1952, it was detected in humans. Zika virus is vectored by *Aedes* mosquitoes that bite during the daytime. Infants of Zika virus–infected pregnant ladies may be born with microcephaly and congenital Zika syndrome. Miscarriage and preterm birth may also be associated with Zika virus infection. Zika virus disease outbreaks have been recorded in the Americas, Asia, Africa, and the Pacific countries [8]. As per the report, till July 2019, 87 countries and territories have had evidence of indigenous transmission of the Zika virus via mosquito vectors [9, 10]. Globally, 61 countries

DOI: 10.1201/9781003430087-7

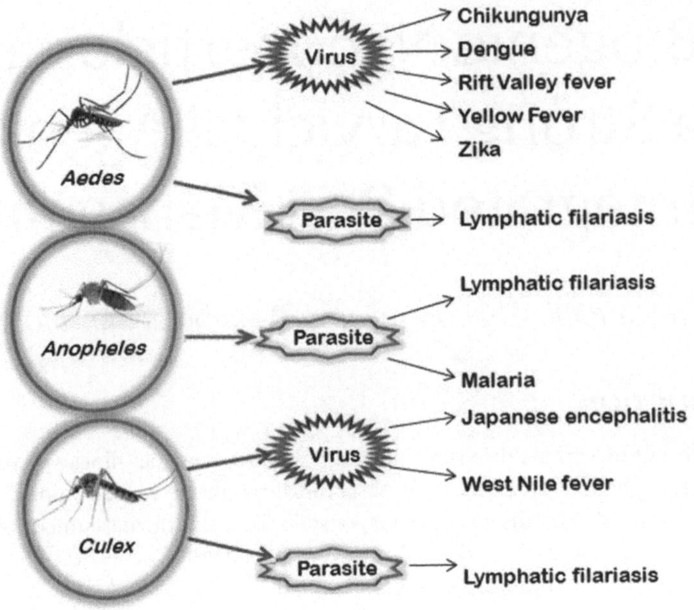

FIGURE 7.1 Mosquito-borne diseases.

and territories have evidence of *Ae. Aegypti* vectors. These regions are not yet documented for Zika virus but are at potential risks for Zika virus [9].

Yellow fever is an arbovirus of the flavivirus genus that causes acute hemorrhagic disease, transmitted by infected *Aedes* and *Haemogogus* species of mosquitoes. This arbovirus is endemic in tropical areas of African and Central and South American countries. The symptoms include fever, headache, muscle pain, jaundice, fatigue, nausea, and vomiting. Some patients die within 7 to 10 days due to life-threatening symptoms. The introduction of the virus into heavily populated areas with high mosquito density leads to epidemics of yellow fever, and infected mosquitoes of the *Ae. aegypti* species act as a vector. A study based on data sources from Africa in 2013 reported 84,000–170,000 severe cases and 29,000–60,000 deaths. Yellow fever can be prevented by a safe and affordable vaccine [11] and by the effective control of *Ae. aegypti*.

Lymphatic filariasis is a tropical disease known as elephantiasis. This infection is transmitted to humans through different types of mosquito vectors, like *Anopheles*, *Culex*, and *Aedes*. There are three types of filarial worms, *Wuchereria bancrofti*, *Brugia malayi*, and *Brugia timori*. *Wuchereria bancrofti* infection is most prevalent in 90% of the cases. Globally, 25 million men with hydrocele and over 15 million people with lymphoedema were estimated. Almost 36 million people remain with these chronic disease symptoms. In 2019, 859 million people from 50 countries were living in areas that required preventive chemotherapy to stop the transmission of filariasis infection [12].

West Nile virus is transmitted primarily through infected mosquito vector bites. It can cause a deadly neurological disease in humans. West Nile virus shows a mosquito-bird-mosquito transmission cycle. The principal vector of the West Nile virus is the *Culex* mosquito, particularly *Cx. pipiens*. West Nile virus is predominantly established in Africa, West Asia, Europe, the Middle East, and the North American regions. Along with humans, horses and other mammals can also get infected by this virus. West Nile virus shows symptoms like headache, fever, fatigue, and body ache, nausea, and vomiting; in some cases, skin rash on the trunk and swollen lymph glands may be observed. The spread of the West Nile virus disease can be controlled by acquiring efficient mosquito vector control [13].

Chikungunya is an arthritogenic viral disease caused by the virus chikungunya. The transmission of chikungunya is observed through *Ae. aegypti* or *Ae. albopictus* mosquito species. Symptoms of a chikungunya virus infection are fever and joint swelling, severe joint pain with muscle pain, headache, nausea, fatigue, and rashes. The disease is mostly observed in Africa, the Indian subcontinent, and other Asian countries [14]. Chikungunya virus vaccines are in the clinical pipeline [15], so preventive modalities should be applied to control the disease.

Japanese encephalitis virus is a flavivirus transmitted to humans mainly through infected *Culex tritaeniorhynchus* mosquito species. Every year, about 68,000 clinical viral encephalitis cases are found in Asia. Twenty-four countries from Southeast Asia and Western Pacific regions have transmission of endemic Japanese encephalitis virus, exposing around 3 billion people to the risk of viral infection. The disease can be cured, but efficient vaccines are there to prevent Japanese encephalitis [16].

7.2 PRESENT STRATEGIES TO CONTROL DISEASES

By considering the aforementioned diseases and the severe consequences they have caused now, there is a vital need for approaches to reducing the burden of mosquitoes to control the transmission of mosquito-borne pathogens. The adopted strategy should be safe, efficient, and sustainable. For the reduction and control of mosquito vectors, various tools have been adopted, like the use of insecticides, genetic alteration, use of mosquito bed nets, and environmental modification, but these strategies have shown setbacks after long-term application due to factors like the emergence of insecticide resistance (Figure 7.2).

Chemical insecticidal agents are most widely used for controlling the mosquito vector population. With the use of dichlorodiphenyltrichloroethane (DDT) insecticide during the 1940s, the beginning of the chemical era of DDT-based insecticides, against larvae and adult mosquitoes, has taken place [17]. Chemical insecticides such as organophosphates, organochlorines, carbamates, pyrethroids, and pyrethrins biorational have been used for controlling the vector population. Organochlorines are a widely used chlorinated pesticide group. These pesticides are persistent organic pollutants to the environment. The organophosphates show their effects through the inactivation of the acetylcholinesterase enzyme, which is important for proper nerve function. Carbamates insecticides are derivatives of carbamic acid, with carbamate esters as a functional group. They show reversible inactivation of the acetylcholinesterase enzyme. From the flowers of *C. cinerariaefolium*, pyrethroids and *Chrysanthemum coccineum* pyrethrins and pyrethroids are isolated. Pyrethroids act on the sodium channels, resulting in paralysis of the vector, while the insecticidal properties are shown by keto alcoholic esters of chrysanthemum and pyrethroid acids of pyrethrins [18, 19]. Initially, these insecticides have shown sound efficacy, but after long-term use, they are just useless, as vectors have adopted resistance against them [20]. Each of these insecticides triggers the mode of one or more mechanisms of resistance development. The evolution in mosquitoes is observed in the rate of

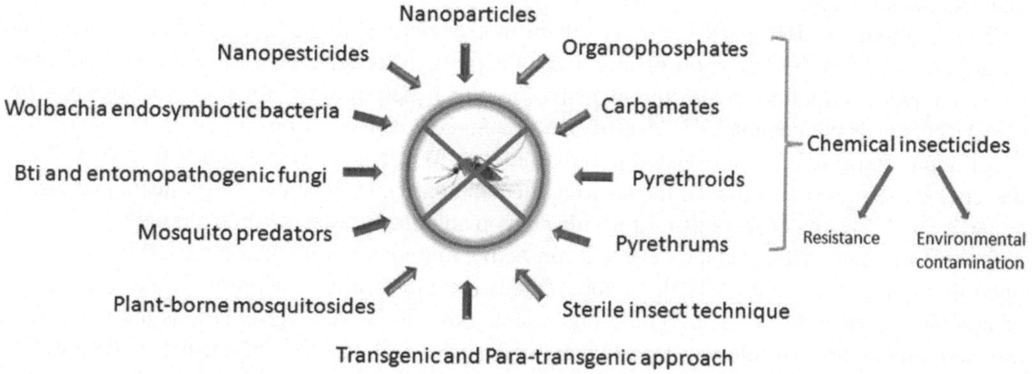

FIGURE 7.2 Various strategies used for mosquito control.

production of insecticidal detoxifying enzymes, such as glutathione-S-transferases, monooxygenases, and carboxyl-cholinesterase, which contribute to resistance development. Monooxygenases often produce metabolic resistance against pyrethroids and permethrin, while glutathione-S-transferases produce resistance toward organochlorides. Pyrethroids, organophosphates, and carbamates resistance are incurred by the increased activity of carboxyl-cholinesterase. Alteration in metabolism, esterase-based resistance, glutathione S-transferase resistance, monooxygenase-based resistance, target-site resistance, and alterations in acetylcholinesterase, GABA receptor, and sodium channels are some common insecticide mechanisms developed in vectors [21,22]. Moreover, environmental concerns caused due to the use of chemical insecticides are a matter of worry; almost 2.5 million tons of insecticides are used annually, contributing to global damage of amount $100 billion annually [20].

In response to this enzyme inactivation and resistance problems, current research on potential alternatives and balanced biological control strategies, like transgenic and paratransgenic approaches, has flourished. Biological control approaches are sustainable, targeting a variety of mosquito species and reducing the dependence on insecticide-based control of mosquitoes. Plant-derived mosquitocidal, repellents, mosquito predators, bti, and entomopathogenic fungi; application of the endosymbiotic bacteria *Wolbachia* that infect around 40% of different insect species; the sterile insect technique; and genetically modified mosquitoes are the commonly used biological control strategies [23].Biological control is usually more difficult to implement and maintain, as these controlling agents are very susceptible to changes in environmental parameters than chemical strategies. Consequently, it causes fluctuations in activity and control over the insect population even though these biological controlling approaches have a slow action that lacks rapid control [24]. Due to resistance developed from physiological changes and behavioral patterns like changes in their blood-feeding times in response to bed nets, controlling the mosquito population is still an immense challenge to us. Chemical control strategies also cause an unintentional effect on non-target environmentally friendly insects, like pollinators, that leads to harmful ecological consequences.

7.3 ROLE OF NANOTECHNOLOGY IN INTEGRATED PEST MANAGEMENT

As conventional mosquito control strategies have shown either moderate effectiveness or their efficiency reduced with time due to resistance, still we don't have a fully efficient mosquito control strategy. Lots of advances in science and technology have been made in the area of pest management in the last decades. Nanotechnology introduced novel nanobased pesticide formulations called nanopesticides. Nanopesticides can be prepared using emulsion, polymer, lipid, ceramics, proteins, and metals. Many researchers have reported the control of mosquito larvae using biologically fabricated nanoparticles that possess bioactive constituents. The development of nanoparticles (1–100 nm) and their application in pest management were increased using metals like silver, gold, zinc, copper, etc. [25].

The small size of nanoparticles makes them more mobile and enables better penetration into tissues and leads to the induction of insecticidal activity. Nanoparticles can be entered either by ingestion and absorption through the digestive tract or penetration by direct contact through the insect's cuticle. Nanotechnology is an effective strategy to conquer various problems like solubility, penetration, absorption, etc. as related to bulk material. While no effective alternative to chemical insecticide molecules is available, formulating the nanomaterial can enable a significant increase in aqueous solubility and ensures its uniform dispersion and distribution on application [26].

Nanoinsecticides are developed either as an active ingredient or as the carrier molecule developed through nanotechnology [27]. Nanopesticides include a variety of products that can consist of organic ingredients and/or inorganic ingredients [28]. These nanoformulations have selective, targeted, and controlled release of pesticides for long-term effects, which are more eco-friendly. The material can be injected into insects efficiently in the form of nanosized particles, capsules, gels, emulsions, etc.

During the last decade, various papers have been published on the green or biological synthesis of nanoparticles that have shown activity against all stages of mosquito life. They attribute a broad spectrum of activity against insects/pests. Moreover, these nanoparticles reduce a considerable extent of air, water, soil, and chemical pollution. Metallic nanopesticides such as Ag, Au, Cu, TiO_2, Fe, and ZnO and their formulations have been reported for their mosquitocidal activity [29]. Several plant-mediated nanoparticles have been studied as effective ovicides, larvicides, pupicides, adulticides, and ovideterrents against various mosquito species [30]. The nanotechnology produces insect-repellent, larvicidal, and pupicidal formulations that are proven more efficacious and less toxic [31].

7.4 LARVICIDAL ACTIVITY OF PLANT-MEDIATED METALLIC NANOPARTICLES

Plant-mediated nanoparticles demonstrate excellent mosquito larvicidal properties. Many studies have shown that plant-mediated nanoparticles are very toxic to mosquitoes. The plant extracts as reducers and stabilizers result in producing metal nanoparticles with varying sizes, types, and toxic properties against mosquitoes [32]. Recently, controlling mosquito vectors by using plant-mediated metal and metal oxide nanoparticles is reported from metallic nanoparticles like silver NPs, gold NPs, copper NPs, zinc oxide NPs, palladium NPs, cadmium NPs, titanium NPs, and others [33–37].

7.4.1 Silver Nanoparticles as a Larvicidal Agent

Silver NPs act as potential mosquitocidal agents. The assessment of larvicidal activity of AgNPs synthesized using leaf aqueous extracts of *A. arborescens*against larvae of *Ae. aegypti* mosquitos was evaluated at different concentrations (0.2–0.5 ppm). The synthesized AgNPs were more toxic to the *Ae. Aegypti* with LC_{50} 0.28 ppm [38].Govindan et al. [39] studied the larvicidal potential of *Plumbago auriculata*–mediated AgNPs against fourth instar larvae of *C. quinquefasciatus* and *Ae. aegypti* at concentrations of 20–100 μg/mL.The AgNPs trigger stringent cytotoxicity against *C. quinquefasciatus* with LC_{50} 41 μg/mL and *Ae. Aegypti* with LC_{50} 45 μg/mL. Further, the author demonstrated that plumbagin docked with salivary proteins of *C. quinquefasciatus* and *Ae. aegypti*. As a result, the plumbagin was effectively docked with proteins with a significant docking score. The mosquito larvicidal potential of different solvent extracts of *Nelumbo Nucifera* and synthesized AgNPs was evaluated against *C. quinquefasciatus* and *A. subpictus* larvae [40]. The results showed the promising effect of AgNPs at various dose concentrations against the larvae of *C. quinquefasciatus* (LC_{50} 1.10 mg/mL) and *A. subpictus* (LC_{50} 0.78 mg/mL). Likewise, Shelar et al. [41] investigated the larvicidal properties of *Momordica charantia*–mediated AgNPs. The phytochemicals in the aqueous fruit peel extract were found to be essential for the formation of AgNPs under 100 nm. At 20 ppm concentration, these nanoparticles inhibited 85% of *Ae. aegypti* and *A. albopictus* larvae.

Fruit extracts of *Solanum mammosum* were utilized as a reducing and stabilizing agent for the synthesis of AgNPs [42]. The synthesized AgNPs exhibited remarkable physicochemical properties, and it reflected in the larvicidal activity. The assay was investigated against the third instar larvae of *A. aegypti* with LC_{50} 0.06 ppm in 24 h. Likewise, flower-mediated synthesis of silver nanoparticles was accomplished by using the extract of *Hibiscus vitifolius* [43]. The authors demonstrated that larvicidal activity was assessed against fourth instar larvae of *A. aegypti*. The assay revealed that floral AgNPs at a concentration of 5 mg/L cause 100% mortality within 3 h. Murugan et al. [44] derived marine weed silver nanoparticles from the sponge *Codium tomentosum*. The author investigated various bioassays against the *A. stephensi* larvae and pupa. The LC_{50} 18.1–29.6 ppm was determined for the acute toxicity assay for larvae I–IV of *A. stephensi*. In addition, the nanoparticles also exert a toxic effect against the pupae.

7.4.2 Gold Nanoparticles as a Larvicidal Agent

Plant-mediated gold nanoparticle synthesis has recently been offered as a novel and eco-friendly mosquito control method. To combat the mosquito problem, varieties of strategies have been devised.

One method for preventing mosquito-borne disease is to kill mosquito larvae. Sundararajan and Ranjitha Kumari [45] evaluated the mosquito larvicidal potential against third and fourth instar mosquito larval stages of *Aedes aegypti* gold nanoparticles synthesized from *Artemisia vulgaris*. The AuNPs had potential larvicidal activity in both 12 h and 24 h exposure periods of the third instar and fourth instar larvae stages of mosquito when compared to essential oil. In another investigation, the AuNPs and AgNPs were synthesized using *A. fruticosa* leaf extracts to inhibit the larvae of *C. pipiens*. In this study, results confirm that AuNPs had a high efficacy compared to the plant extracts and AgNPs [46]. *Cymbopogon citratus* leaf extract fabricate spherical cubic crystalline gold nanoparticles and assessed its larvicidal properties against the *Anopheles stephensi* and *A. aegypti* [47]. The assay had an outcome that gold nanoparticles are highly toxic to the larvae and pupae of *A. stephensi* and *A. aegypti*. At the same time, the gold nanoparticles were assessed for the predation efficiency in crustacean *Mesocyclops aspericornis*. The gold nanoparticles significantly increase the predation efficiency in gold nanoparticles–contaminated environment. Paramasivam Deepak et al. [48] formulated seaweed *Turbinaria ornate*–mediated gold nanoparticles and studied the larvicidal properties against the malarial vector *Anopheles stephensi*. The assay depicts that seaweed gold nanoparticles trigger cytotoxic to the larvae of *A. stephensi* with LC_{50} 12.79 ppm and cause 100% mortality in 24 h.

7.4.3 Copper Oxide Nanoparticles as Larvicidal Agent

The leaf extract of *T. procumbens* was used to synthesize copper oxide NPs for the control of the mosquito *A. aegypti*. The CuONPs had more activity compared to leaf extract [49]. In another study, the copper nanoparticles (CuNPs) synthesized from *Artocarpus heterophyllus* were evaluated against *A. aegypti*. The nanoparticles were highly effective against all the juvenile larvae stages compared to leaf extracts [50].

7.4.4 Titanium Dioxide Nanoparticles as Larvicidal Agent

Gandhi et al. [51] synthesized TiO_2 nanoparticles (NPs) using the aqueous extract of *Momordica charantia* against *A. stephensi*. TiO_2 NPs were highly toxic to the larvae and pupae of *A. stephensi* even at low dosages. Similarly, the TiO_2 NPs synthesized from *Pouteria campechiana* were evaluated for their larvicidal and pupicidal activities against *A. aegypti* [52]. Upon 48 h of exposure to TiO_2NPs, tremendous mortality was noticed at LC_{50} at 570.47 ppm.

7.4.5 Zinc Oxide Nanoparticles as Larvicidal Agent

Zinc oxide nanoparticles have a variety of uses, including in the biological and biomedical sectors [53]. Insecticidal properties of zinc oxide nanoparticles had been found by many researchers. *Syzgium cumini* seed extract was used to synthesize zinc oxide nanoparticles to evaluate the larvicidal and ovicidal activity against *A. aegypti* with LC_{50} at 51.94 ppm, and a maximum mortality rate of 80% at maximum concentrations was achieved [54]. Loganathan et al. [55] synthesized the zinc oxide nanoparticle from *Knoxia sumatrensis* for mosquito larvicidal activity against *C. quinquefasciatus* (LC_{50} 0.08 ppm)

7.4.6 Iron Oxide Nanoparticles as Larvicidal Agent

Celosia argentea–mediated iron oxide nanoparticles trigger the toxic effects in the larvae of *A. subpictus* [56]. With respect to the increase in time and dosage, the nanoparticles cause 100% mortality in 48 h. Likewise, floral-mediated iron oxide nanoparticles also showcased 100% mortality in the dengue vector *A. aegypti* [57]. The iron oxide nanoparticles also proved that the biogenic nanoparticles are a safe larvicidal agent.

7.4.7 Manganese Oxide Nanoparticles as Larvicidal Agent

Aqueous leaf extract of *Hibiscus rosa-sinensis* reduced and stabilized the manganese oxide nanoparticles with spherical and cubic crystalline nature [58]. The synthesized nanoparticles exhibited dose-dependent activity with 50% mortality occurring at 400 ppm in the third instar larvae of *A. aegypti. Melia azedarach* seed extract fabricated polydispersed MgO nanoparticles and also exhibited a dose-dependent activity with 86% mortality occurring in 24 h at the concentration of 90 µg/mL in the third instar larvae of *A. aegypti* [59].

7.4.8 Other Metal/Metal Oxide Nanoparticles as Larvicidal Agent

Palladium nanoparticles were synthesized from the leaf aqueous extract of *Cymodocea serrulata* [60].The synthesized nanoparticles displayed strong mosquitocidal activity against *A. stephensi* with LC_{50} 4 µg/mL. The leaf extract of *Dillenia indica* was reduced and stabilized to produce oval-shaped selenium nanoparticles [61]. The nanoparticles induce growth inhibition in the *A. aegypti* and *C. quinquefasciatus* larvae. Sowndarya et al. [62] synthesized selenium nanoparticles from leaf extracts of *Clausena dentate*. The nanoparticles strongly inhibit the growth of the fourth instar larvae of *A. aegypti, A. stephensi*, and *C. quinquefasciatus*. Thus, with the preceding scientific evidence, it is now well established that plant-mediated nanoparticles can effectively inhibit the growth of mosquitoes with respect to genus and species. The efficacy of the nanoparticles depends on the phytoconstituents that coat the surface of the nanoparticles (Table 7.1)

7.5 GREEN SYNTHESIS OF MICROORGANISMS-MEDIATED NANOPARTICLES

The synthesis of microbial-mediated nanoparticles is now a revolutionary way to control mosquitos. Conspicuously, the use of microorganisms to synthesize nanoparticles is based on their interactions with metal bases, and they are also commercially used in biotechnological progressions [73]. The metabolites were produced by microbes either intracellularly or extracellularly [74]. Synthesis of nanoparticles by extracellular metabolites nanoparticles is less expensive and time-consuming than intracellular. As a result, most investigations on microbial nanoparticle synthesis are focused on this approach [75].

Microbial-mediated nanoparticles had high efficiency of mosquito larvicidal activity compared to plant-mediated nanoparticles. Because the size of nanoparticles was very tiny, time-consuming, cheap, and eco-friendly, the microbial-mediated nanoparticles are less in size, leading to easily penetrating the plasma membrane of the mosquito and damaging the central dogma.

Bacteria are the most important microbial bases for fabricating nanoparticles. The AgNPs synthesized from the *Bacillus marisflavi* bacteria exposed potent mosquitocidal activity against *Anopheles stephensi, Aedes aegypti*, and *Culex quinquefasciatus*. The low concentration of nanoparticles shows high efficiency [76]. Similarly, different metallic nanoparticles such as ZnONPs, CuNPs, SeNPs, and TiO_2NPs were synthesized by using microorganisms for controlling mosquitoes.

Mycosynthesis of nanoparticles is very efficient in mosquito control. Fungi can produce more amount of primary metabolites compared to other microbes, especially protein; therefore, fungal extracts easily and rapidly synthesize the metallic nanoparticle by reducing the metallic bases [77]. In recent years, researchers have focused on the insecticidal potential of fungal-mediated nanoparticles. The copper nanoparticles synthesized from entomopathogenic fungi *Metarhizium robertsii*had potent larvicidal activity on fourth instar larvae of *Cx. quinquefasciatus, A. aegypti, A. stephensi*, and *T. molitor* and were less toxic to *A. nauplii, A. salina, E. eugeniae*, and *E. andrei* [78]. Seetharaman et al. [79] reported that silver nanoparticle was synthesized from the endophytic fungi *Phomopsis liquidambaris* and evaluated the efficacy against larvae of *Cx. quinquefasciatus* and *A. aegypti* mortality. The result confirms that silver nanoparticles trigger 100% mortality achieved at 10 ppm concentrations in mosquito species.

TABLE 7.1
Larvicidal Activity of Biogenic Nanoparticles

Plant Names	Plant/Microbe Source	Type of Metallic NPs	Size(nm)	Type of Mosquito	LC_{50}	References
Cinnamomum zeylanicum	Bark extract	AuNPs	46.48 nm	*A. stephensi*	2 ppm	[63]
Ficus natalensis	Leaf extract	FeNPs	35–40 nm	*C. quinquefasciatus*	35.9 ppm	[64]
Jasminum nervosum	Leaf extract	AgNPs	9.4 nm	*C. quinquefasciatus*	144.36 μg/mL	[65]
Tridaxprocumbens	Leaf extract	CuONPs	16	*Aedes aegypti*	4.20 ppm	[66]
Ocimumbasillicum	Leaf extract	CdS NPs	3.84	*Culex pipiens*	0.011 ppm	[67]
S.zaomyceticus Oc-5	Extracellular	–	–	–	–	[68]
Bacillus marisflavi	Extracellular	AgNPs	78.77	*A. aegypti*	13.96 ppm	[69]
Bacillus thuringiensis	Extracellular	AgNPs	43.5–142.9	*A. aegypti*	0.10 ppm	[70]
Penicillium corylophilum	Extracellular	Se-NPs	29.1–48.9	*A. stephensi*	25 ppm	[71]
Amphiroa rigida	Intracellular	AgNPs	25	*A. aegypti*	7 ppm	[72]

Algae are the primary producers of organic resources, which include minerals, proteins, and a variety of natural products [80]. These organic sources are able to reduce the metal base into nanoparticles. Many researchers synthesized algal-mediated nanoparticles for different biomedical applications. However, very few findings are available for mosquito larvicidal efficacy of algal-mediated nanoparticles. Ellumalai et al. [81] reported that silver nanoparticles were synthesized from an aqueous extract of algae *Oscillatoria santa* and evaluated its mosquito larvicidal potential against the fourth instar of *Aedes aegypti* and *Aedes stephensi*. The synthesized nanoparticles can kill at low concentrations.

7.6 MODE OF LARVICIDAL ACTIVITY BY NANOPARTICLES

The particular mechanism of larvicidal activity is not known clearly until now. However, some researchers propose the mechanism of nanoparticles for killing a mosquito. They hypothesize that nanoparticles pierce the shell membrane and bind to the DNA of the larvae. Eventually, it disturbs DNA activity, promotes fast protein breakdown, and significantly reduces insect membrane permeability. With the massive generation of ROS in the mosquito, all these things disrupt cell function and cause cell death [82].

In a detailed study, some reports have documented the impact of nanoparticles on the larvae of mosquitoes. In a study reported by Banumathi et al. [83], the biogenic nanoparticles against the larvae are studied. The nanoparticles cause loss of hair and damage to midgut epithelium, which eventually leads to death. Ishwarya et al. [84]reported that biogenic silver nanoparticles cause cuticular and abdomen damage in the larvae of *A. aegypti*.

In response to the treatment of nanoparticles, there were significant changes in the cytokine levels, protein calexcitin, ROS, lipids, and carbohydrates levels [85]. Fouad et al. [86]reported that AgNPs cause a drastic reduction in the protein content and enzymes acetylcholinesterase and carboxylesterase in *Aedes albopictus* and *Culex pipiens*.

Esterase is an important enzyme responsible for the development of resistance against synthetic insecticides by breaking carboxyl ester and phosphor diester bonds [87]. Salicylic acid–mediated AgNPs cause decreased levels of esterase enzyme in fourth instar larvae of *Aedes albopictus* [87]. Also, the silver nanoparticles cause the demolting process and temperature regulation in the insects. Other observations such as lesions, vacuolization of the cell, broken border brush, rupture of cells, and nuclei degradation were observed in response to the treatment of nanoparticles [88]. Nalini et al. [89] reported that nanoparticles disrupt the gonads, resulting in defective oogenesis and infertility also.

Molecular changes associated with nanoparticles are very important to understand in the context of drug development. In certain reports, there was a significant increase in the gamma-H2AX gene, which is closely related to apoptosis. In addition, increased levels of CAT and SOD expressions and a decrease of GPX indicate the production of ROS [90]. The efficacy of nanoparticles depends on the cells, phytoconstituents, and other factors. Overall, it is concluded that the nanoparticles breach the cuticle layer and disrupt the internal midgut membrane then react with biomolecules (DNA/RNA/proteins) to induce ROS and cause cell death.

7.7 CONCLUSION

The development of nanobased pesticides is indeed needed for the present scenario; a myriad of research is persuading in the development of green pesticides. But the lack of technologies in developing bioresources into pesticides is still challenging and difficult also. Altogether, the toxicity of nanobased pesticides should be extensively studied in the aquatic ecosystem. The accumulation of nanoparticles and their consequence in non-target organisms is a much-needed study for the development of nanopesticides. Thus, this chapter substantially covered all the aspects of biogenic nanoparticles and their prospective features as the larvicidal agent.

REFERENCES

1. Tolle, M.A. Mosquito-borne diseases, *Current Problems in Pediatric and Adolescent Health Care*. 2009, *39*, 97–140.
2. Church, K.E.M.; Smith, R.J. Comparing malaria surveillance with periodic spraying in the presence of insecticide-resistant mosquitoes: Should we spray regularly or based on human infections? *Mathematical Biosciences*. 2016, *276*, 145–161.
3. Aponte, J.J.; Aide, P.; Renom, M.; et al. Safety of the RTSS/AS02D candidate malaria vaccine in infants living in a highly endemic area of Mozambique: A double blind randomised controlled phase I/IIb trial, *Lancet*. 2007, *370*, 1543–1551.
4. Benelli, G.; Mehlhorn, H. Declining malaria; rising of dengue and Zika virus: Insights for mosquito vector control, *Parasitology Research*. 2016, *115*, 1747–1754.
5. Nathan, M. B.; Dayal-Drager, R.; Guzman, M. Epidemiology, burden of disease and transmission. In *Dengue Guidelines for Diagnosis, Treatment, Prevention and Control: New Edition*. World Health Organization, Geneva, 2009, pp. 1–21.
6. Bhatt, S.; Gething, P.W.; Brady, O.J.; et al. The global distribution and burden of dengue, *Nature*. 2013, *496*, 504–507.
7. Becker, N.; Pluskota, B.; Kaiser, A.; Schaffner, F. Exotic mosquitoes conquer the world. In *Arthropods as Vectors of Emerging Diseases*. Springer, Verlag Berlin Heidelberg, 2012, pp. 31–60.
8. Zika virus, 2018. www.who.int/news-room/fact-sheets/detail/zika-virus
9. Countries and territories with current or previous Zika virus transmission, World health organization, 2019. www.who.int/emergencies/diseases/zika/countries-with-zika-and-vectors-table.pdf
10. Zika epidemiology update, World health organization, 2019. www.who.int/publications/m/item/zika-epidemiology-update–february-2022
11. Yellow fever, World health organization, 2019.www.who.int/news-room/fact-sheets/detail/yellow-fever
12. Lymphatic filariasis, World health organization, 2022. www.who.int/news-room/fact-sheets/detail/lymphatic-filariasis
13. West Nile virus, World health organization, 2017. www.who.int/news-room/fact-sheets/detail/west-nile-virus
14. Chikungunya, World health organization, 2020. www.who.int/news-room/fact-sheets/detail/chikungunya
15. Goya, M.; Chauhan, A.; Goyal, V.; Jaiswal, N.; Singh, S.; Singh, M. Recent development in the strategies projected for chikungunya vaccine in humans. *Drug Design, Development and Therapy*. 2018, *12*, 4195–4206.
16. Japanese encephalitis, World health organization, 2019. www.who.int/news-room/fact-sheets/detail/japanese-encephalitis
17. Chanda, E.; Govere, J. M.; Macdonald, M. B.; Lako, R. L.; Haque, U.; Baba, S. P.; Mnzava, A. Integrated vector management: A critical strategy for combating vector-borne diseases in South Sudan. *Malaria Journal*. 2013, *12*(1), 1–9.
18. Jayaraj, R.; Megha, P.; Sreedev, P.; Review article: Organochlorine pesticides, their toxic effects on living organisms and their fate in the environment. *Interdisciplinary Toxicology*. 2016, *9*, 90–100.
19. Reigart, R. J.; Roberts, J. R. *Recognition and Management of Pesticide Poisonings*. 5th ed. U.S. Environmental Protection Agency, Washington, DC, 1999, pp. 34–93.
20. Cuervo-Parra, J.A.; Cortés, T.R.; Ramirez-Lepe, M. Mosquito-borne diseases, pesticides used for mosquito control, and development of resistance to insecticides. *Insecticide Resistance*. 2016, 111–134.
21. Hemingway, J.; Ranson, H. Insecticide resistance in insect vectors of human disease. *Annual Review of Entomology*. 2000, *45*, 371–391.
22. Logan, J.G.; Birkett, M.A. Juvenile hormonal activity of mono- and polyenehomobenzenes in *Drosophila virilis*, *Spodoptera frugiperda* and *Araschnia levana*. *Pest Management Science*. 2007, *63*, 809–814.
23. Benelli, G.; Jeffries, C.L.; Walker, T. Biological control of mosquito vectors: Past, present, and future. *Insects*. 2016, *7(4)*, 1–18.
24. Kendie, F.A. Potential biological control agents against mosquito vector in the case of larvae stage: A review. *World News of Natural Sciences*. 2020, *28*, 34–50.
25. Santhosh, S.B.; Yuvarajan, R.; Natarajan, D. *Annona muricata* leaf extract-mediated silver nanoparticles synthesis and its larvicidal potential against dengue, malaria and filariasis vector. *Parasitology Research*. 2015, *114(8)*, 3087–3096.
26. Margulis-Goshen, K.; Magdassi, S. Nanotechnology: An advanced approach to the development of potent insecticides. In *Advanced Technologies for Managing Insect Pests*. Springer, Dordrecht, 2012, pp. 295–314.

27. Zannat, R.; Rahman, M.M.; Afroz, M. Application of nanotechnology in insect pest management: A review. *SAARC Journal of Agriculture*. 2021, *19(2)*, 1–11.
28. Ragaei, M.; Sabry, A.H. Nanotechnology for insect pest control, *International Journal of Scientific & Technology*. 2014, *3(2)*, 2278–3687.
29. Bhattacharya, A.; Epidi, T.T.;Kannan, M.Nano-technology applications in pest management. In*Innovative Pest Management Approaches for the 21st Century Harnessing Automated Unmanned*. Springer, Singapore, 2020, pp. 391–401.
30. Benelli, G. Plant-mediated biosynthesis of nanoparticles as an emerging tool against mosquitoes of medical and veterinary importance: A review. *Parasitology Research*. 2016, *115(1)*, 23–34.
31. Vangelie E.; Campos, R.; Oliveira, J.L.; et al. Recent developments in nanotechnology for detection and control of *Aedes aegypti*-borne diseases, *Asian Pacific Journal of Tropical Disease*. 2020, *8*, 1–17.
32. Benelli, G.; Caselli, A.; Canale, A. Nanoparticles for mosquito control: Challenges and constraints. *Journal of King Saud University-Science*. 2017, *29(4)*, 424–435.
33. Idowu, E.T.; Adeogun, A.O.; Adams, L.A.; et al. Larvicidal potential of two silver nano-particles (*Moringa oleifera* and *Ficus exasperata*) against laboratory and field strains of Anopheles gambiae (Diptera: Culicidae) in Lagos, Nigeria. *The Journal of Basic and Applied Zoology*. 2021, *82(1)*, 1–9.
34. Alhag, S.K.; Al-Mekhlafi, F.A.; Abutaha, N.; Abd, A.; Galil, F.M.; Wadaan, M.A. Larvicidal potential of gold and silver nanoparticles synthesized using *Acalypha fruticosa* leaf extracts against *Culex pipiens* (Culicidae: Diptera). *Journal of Asia-Pacific Entomology*. 2021, *24(1)*, 184–189.
35. Rajagopal, G.; Nivetha, A.; Sundar, M.; et al. Mixed phytochemicals mediated synthesis of copper nanoparticles for anticancer and larvicidal applications. *Heliyon*. 2021, *7(6)*, e07360.
36. Nityasree; B.R.; Chalannavar, R.K.; Kouser, S.; et al. Bioinspired synthesis of zinc oxide nanoparticles by using leaf extract of *Solanum lycopersicum* L. for larvicidal activity of *Aedes aegypti* L. *Advances in Natural Sciences: Nanoscience and Nanotechnology*. 2021, *12(1)*, 015009.
37. Bhakyaraj, K.; Kumaraguru, S.; Gopinath, K.; et al. Eco-friendly synthesis of palladium nanoparticles using *Melia azedarach* leaf extract and their evaluation for antimicrobial and larvicidal activities. *Journal of Cluster Science*.2017, *28(1)*, 463–476.
38. Morejón, B.; Pilaquinga, F.; Domenech, F.; Ganchala, D.; Debut, A.; Neira, M. Larvicidal activity of silver nanoparticles synthesized using extracts of *Ambrosia arborescens* (Asteraceae) to control *Aedes aegypti* L. (Diptera: Culicidae). *Journal of Nanotechnology*. 2018, *2018*.
39. Govindan, L.; Anbazhagan, S.; Altemimi, A.B; et al. Efficacy of antimicrobial and larvicidal activities of green synthesized silver nanoparticles using leaf extract of *Plumbago auriculata* lam. *Plants*.2020, *9(11)*, 1577.
40. Santhoshkumar, T.; Rahuman, A.A.; Rajakumar, G.; et al. Synthesis of silver nanoparticles using *Nelumbo nucifera* leaf extract and its larvicidal activity against malaria and filariasis vectors. *Parasitology Research*. 2011, *108(3)*, 693–702.
41. Shelar, A.; Sangshetti, J.; Chakraborti, S.; Singh, A.V.; Patil R.; Gosavi, S. Helminthicidal and larvicidal potentials of biogenic silver nanoparticles synthesized from medicinal plant *Momordica charantia*. *Journal of Medicinal Chemistry*. 2019, *15(7)*, 781–789.
42. Pilaquinga, F.; Morejón, B.; Ganchala, D.; et al. Green synthesis of silver nanoparticles using Solanum mammosum L. (Solanaceae) fruit extract and their larvicidal activity against Aedes aegypti L. (Diptera: Culicidae). *PLoS One*. 2019,*14(10)*, e0224109.
43. Kumar, P.; Begum, S.; Sunil, G.; Devi, P.; Kesana, N. Screening of larvicidal activity of nanoparticles synthesized from flower extracts of Hibiscus vitifolius. *Journal of Applied Biotechnology and Bioengineering*. 2018, 157. https://doi. org/10.15406/jabb
44. Murugan, K.; Panneerselvam, C.; Subramaniam, J.; et al. Eco-friendly drugs from the marine environment: Spongeweed-synthesized silver nanoparticles are highly effective on Plasmodium falciparum and its vector Anopheles stephensi, with little non-target effects on predatory copepods. *Environmental Science and Pollution Research*. 2016, *23(16)*, 16671–16685.
45. Sundararajan, B.; Kumari, B.R. Novel synthesis of gold nanoparticles using Artemisia vulgaris L. leaf extract and their efficacy of larvicidal activity against dengue fever vector *Aedes aegypti* L. *Journal of Trace Elements in Medicine and Biology*. 2017, *43*, 187–196.
46. Alhag, S.K.; Al-Mekhlafi, F.A.; Abutaha, N.; Abd Al Galil, F.M.; Wadaan, M.A. Larvicidal potential of gold and silver nanoparticles synthesized using Acalypha fruticosa leaf extracts against Culex pipiens (Culicidae: Diptera). *Journal of Asia-Pacific Entomology*. 2021, *24(1)*, 184–189.
47. Murugan, K.; Benelli, G.; Panneerselvam, C.;et al. Cymbopogon citratus-synthesized gold nanoparticles boost the predation efficiency of copepod Mesocyclops aspericornis against malaria and dengue mosquitoes. *Experimental Parasitology*. 2015,*153*, 129–138.

48. Deepak, P.; Sowmiya, R.; Balasubramani, G.; Aiswarya, D.; Arul, D.; Josebin, M.P.D.; Perumal, P. Mosquito-larvicidal efficacy of gold nanoparticles synthesized from the seaweed, Turbinaria ornata (Turner) J. Agardh 1848. *Particulate Science and Technology.* 2018, *36(8)*, 974–980.
49. Selvan, S.M.; Anand, K.V.; Govindaraju, K.; et al. Green synthesis of copper oxide nanoparticles and mosquito larvicidal activity against dengue, zika and chikungunya causing vector Aedes aegypti. *IET Nanobiotechnology.*2018, *12(8)*, 1042–1046.
50. Sharon, E.A.; Velayutham, K.; Ramanibai, R. Biosynthesis of copper nanoparticles using Artocarpus heterophyllus against dengue vector Aedes aegypti. *International Journal of Life-Sciences Scientific Research.* 2018, *1912(1716)*, 1872–1879.
51. Gandhi, P.R.; Jayaseelan, C.; Kamaraj, C.; Rajasree, SR; Mary, R.R. *In vitro* antimalarial activity of synthesized TiO_2 nanoparticles using Momordica charantia leaf extract against Plasmodium falciparum. *Journal of Applied Biomedicine.*2018, *16(4)*, 378–386.
52. Narayanan, M.; Devi, P.G.; Natarajan, D.;et al. Green synthesis and characterization of titanium dioxide nanoparticles using leaf extract of Pouteria campechiana and larvicidal and pupicidal activity on Aedes aegypti. *Environmental Research.* 2021, *200*, 111333.
53. Bhadra, J.; Alkareem, A.; Al-Thani, N.A. Review of advances in the preparation and application of polyaniline based thermoset blends and composites. *Journal of Polymer Research.*2020, *27(5)*, 1–20.
54. Roopan, S.M.; Mathew, R.S.; Mahesh, S.S.; et al. Environmental friendly synthesis of zinc oxide nanoparticles and estimation of its larvicidal activity against Aedes aegypti. *International Journal of Environmental Science and Technology.*2019, *16*(*12*), 8053–8060.
55. Loganathan, S.; Shivakumar, M.S.; Karthi, S.; Nathan, S.S.; Selvam, K. Metal oxide nanoparticle synthesis (ZnO-NPs) of Knoxia sumatrensis (Retz.) DC: Aqueous leaf extract and It's evaluation of their antioxidant, anti-proliferative and larvicidal activities. *Toxicology Reports.* 2021, *8*, 64–72.
56. Velsankar, K.; Parvathy, G.; Mohandoss, S.; Krishna Kumar, M.; Sudhahar, S. Celosia argentea leaf extract-mediated green synthesized iron oxide nanoparticles for bio-applications. *Journal of Nanostructure in Chemistry.* 2021, 1–16.
57. Sudhakar, C.; Poonkothai, M.; Selvankumar, T.; Selvam, K. Facile synthesis of iron oxide nanoparticles using Cassia auriculata flower extract and accessing their photocatalytic degradation and larvicidal effect. *Journal of Materials Science: Materials in Electronics.* 2022, *33(14)*, 11434–11445.
58. Khan, M.A.; Ali, F.; Faisal, S.;et al. Exploring the therapeutic potential of Hibiscus rosa sinensis synthesized cobalt oxide (Co_3O_4-NPs) and magnesium oxide nanoparticles (MgO-NPs). *Saudi Journal of Biological Sciences.* 2021, *28(9)*, 5157–5167.
59. Dinga, E.; Ekennia, A.; Ogbonna, C.U.; Udu, D.A.; Mthiyane, D.M.N.; Marume, U.; Onwudiwe, D.C. Phyto-mediated synthesis of MgO nanoparticles using Melia azedarach seed extract: Larvicidal and antioxidant activities. *Scientific African.* 2022, *17*, e01366.
60. Amutha, V.; Deepak, P.; Kamaraj, C.; et al. Mosquito-larvicidal potential of metal and oxide nanoparticles synthesized from aqueous extract of the seagrass, Cymodocea serrulata. *Journal of Cluster Science.* 2019, *30(3)*, 797–812.
61. Krishnan, M.; Ranganathan, K.; Maadhu, P.; Thangavelu, P.; Kundan, S.; Arjunan, N. Leaf extract of Dillenia indica as a source of selenium nanoparticles with larvicidal and antimicrobial potential toward vector mosquitoes and pathogenic microbes. *Coatings.* 2020, *10(7)*, 626.
62. Sowndarya, P.; Ramkumar, G.; Shivakumar, M.S. Green synthesis of selenium nanoparticles conjugated Clausena dentata plant leaf extract and their insecticidal potential against mosquito vectors. *Artificial Cells, Nanomedicine, and Biotechnology.* 2017, *45(8)*, 1490–1495.
63. Elumalai, D.; Hemavathi, M.; Rekha, G.S.; et al. Photochemical synthesizes of silver nanoparticles using Oscillatoria sancta micro algae against mosquito vectors Aedes aegypti and Anopheles stephensi. *Sensing and Bio-Sensing Research.*2021, *24*,100457.
64. Murugan, K.; Samidoss, C.M.; Panneerselvam, C.; et al. Seaweed-synthesized silver nanoparticles: An eco-friendly tool in the fight against Plasmodium falciparum and its vector Anopheles stephensi? *Parasitology Research.* 2015, *114*, 4087–4097.
65. Lallawmawma, H.; Sathishkumar, G.; Sarathbabu, S.; et al. Synthesis of silver and gold nanoparticles using *Jasminum nervosum* leaf extract and its larvicidal activity against filarial and arboviral vector *Culex quinquefasciatus*say (Diptera: Culicidae). *Environmental Science and Pollution Research.* 2015, *22(22)*, 17753–17768.
66. Selvan, S.M.; Anand, K.V.; Govindaraju, K.; et al. Green synthesis of copper oxide nanoparticles and mosquito larvicidal activity against dengue, zika and chikungunya causing vector Aedes aegypti. *IET Nanobiotechnology.*2018, *12(8)*, 1042–1046.

67. Mohammed, Z.Y.A. Recent, eco-friendly approach for controlling *Culex pipiens* (L.) using novel synthesized cadmium sulphide nanoparticles of *Ocimum basillicum*extract. *Egyptian Journal of Aquatic Biology and Fisheries.* 2021,*25(4)*, 359–377.
68. Fouda, A.; Hassan, S.E.D.; Abdo, A.M.; El-Gamal, M.S. Antimicrobial, antioxidant and larvicidal activities of spherical silver nanoparticles synthesized by endophytic *Streptomyces* spp. *Biological Trace Element Research.* 2020, *195(2)*, 707–724.
69. Thelma, J.; Balasubramanian, C. Ovicidal, larvicidal and pupicidal efficacy of silver nanoparticles synthesized by *Bacillus marisflavi* against the chosen mosquito species. *PLoS One.* 2021, *16(12)*, e0260253.
70. Najitha Banu, A.; Balasubramanian, C.; Moorthi, P.V. Biosynthesis of silver nanoparticles using Bacillus thuringiensis against dengue vector, Aedes aegypti (Diptera: Culicidae). *Parasitology Research.* 2014, *113(1)*, 311–316.
71. Salem, S.S.; Fouda, M.M.; Fouda, A.; Awad, M.A.; Al-Olayan, E.M.; Allam, A.A.; Shaheen, T.I. Antibacterial, cytotoxicity and larvicidal activity of green synthesized selenium nanoparticles using Penicillium corylophilum. *Journal of Cluster Science.* 2021, *32(2)*, 351–361.
72. Gopu, M.; Kumar, P.; Selvankumar, T.;et al. Green biomimetic silver nanoparticles utilizing the red algae Amphiroa rigida and its potent antibacterial, cytotoxicity and larvicidal efficiency. *Bioprocess and Biosystems Engineering.* 2021, *44(2)*, 217–223.
73. Gericke, M.; Pinches, A. Biological synthesis of metal nanoparticles. *Hydrometallurgy.* 2006, *83*,132–140.
74. Pirzadah, T. B., Pirzadah, B., Jan, A., Dar, F. A., Hakeem, K. R., Rashid, S., ... & Fazili, M. A. (2020). Development of nano-formulations via green synthesis approach. *Nanobiotechnology in Agriculture: An Approach Towards Sustainability*, 171–183.
75. Srivastava, S.; Usmani, Z.; Atanasov, A.G.; et al. Biological nanofactories: Using living forms for metal nanoparticle synthesis. *Mini-Reviews in Medicinal Chemistry.* 2021, *21(2)*, 245–265.
76. Anthony, K.J.P.; Murugan, M.; Gurunathan, S. Biosynthesis of silver nanoparticles from the culture supernatant of Bacillus marisflavi and their potential antibacterial activity. *Journal of Industrial and Engineering Chemistry.* 2014, *20(4)*, 1505–1510.
77. Loshchinina, E.A.; Vetchinkina, E.P.; Kupryashina, M.A. Diversity of biogenic nanoparticles obtained by the fungi-MEdiated synthesis: A review. *Biomimetics.* 2022, *8(1)*, 1.
78. Vivekanandhan, P.; Swathy, K.; Thomas, A.; Kweka, E.J.; Rahman, A.; Pittarate, S.; Krutmuang, P. Insecticidal efficacy of microbial-mediated synthesized copper nano-pesticide against insect pests and non-target organisms. *International Journal of Environmental Research and Public Health.* 2021, *18(19)*, 10536.
79. Seetharaman, P.K.; Chandrasekaran, R.; Gnanasekar, S.; Chandrakasan, G.; Gupta, M.; Manikandan, D.B.; Sivaperumal, S. Antimicrobial and larvicidal activity of eco-friendly silver nanoparticles synthesized from endophytic fungi Phomopsis liquidambaris. *Biocatalysis and Agricultural Biotechnology.* 2018, *16*, 22–30.
80. Wells, M.L.; Potin, P.; Craigie, J.S.;et al. Algae as nutritional and functional food sources: Revisiting our understanding. *Journal of Applied Phycology.* 2017, *29(2)*, 949–982.
81. Elumalai, D.; Hemavathi, M.; Rekha, G.S.;et al. Photochemical synthesizes of silver nanoparticles using Oscillatoria sancta micro algae against mosquito vectors Aedes aegypti and Anopheles stephensi. *Sensing and Bio-Sensing Research.* 2021, *34*, 100457.
82. Ishwarya, R.; Vaseeharan, B.; Anuradha, R.;et al. Eco-friendly fabrication of Ag nanostructures using the seed extract of Pedalium murex, an ancient Indian medicinal plant: Histopathological effects on the Zika virus vector Aedes aegypti and inhibition of biofilm-forming pathogenic bacteria. *Journal of Photochemistry and Photobiology B: Biology.* 2017, *174*, 133–143.
83. Meng, X.; Abdlli, N.; Wang, N.;et al. Effects of Ag nanoparticles on growth and fat body proteins in silkworms (Bombyx mori). *Biological Trace Element Research.* 2017, *180(2)*, 327–337.
84. Fouad, H.; Hongjie, L.; Hosni, D.; Wei, J.; Abbas, G.; Ga'al, H.; Jianchu, M. Controlling Aedes albopictus and Culex pipiens pallens using silver nanoparticles synthesized from aqueous extract of Cassia fistula fruit pulp and its mode of action. *Artificial Cells, Nanomedicine, and Biotechnology.* 2018, *46(3)*, 558–567.
85. Kumar, D.; Kumar, P.; Singh, H.; Agrawal, V. Biocontrol of mosquito vectors through herbal-derived silver nanoparticles: Prospects and challenges. *Environmental Science and Pollution Research.* 2020, *27(21)*, 25987–26024.
86. William, G.B.; Janet, C. Heme peroxidase activity measured in single mosquitoes identifies individuals expressing an elevated oxidase for insecticide resistance. *Journal of the American Mosquito Control.* 1997,*13*, 233–237.

87. Ga'al, H.; Fouad, H.; Mao, G.; Tian, J.; Jianchu, M. Larvicidal and pupicidal evaluation of silver nanoparticles synthesized using Aquilaria sinensis and Pogostemon cablin essential oils against dengue and zika viruses vector Aedes albopictus mosquito and its histopathological analysis. *Artificial Cells, Nanomedicine, and Biotechnology.* 2018, *46(6)*, 1171–1179.
98. Nalini, M.; Lena, M.; Sumathi, P.; Sundaravadivelan, C. Effect of phyto-synthesized silver nanoparticles on developmental stages of malaria vector, Anopheles stephensi and dengue vector, Aedes aegypti. *Egyptian Journal of Basic and Applied Sciences.* 2017, *4(3)*, 212–218.
89. Mao, B.H.; Chen, Z.Y.; Wang, Y.J.; Yan, S.J. Silver nanoparticles have lethal and sublethal adverse effects on development and longevity by inducing ROS-mediated stress responses. *Scientific Reports.* 2018, *8(1)*, 1–16.
90. Araújo, P.S.; Caixeta, M.B.; Canedo, A.; da Silva Nunes, E.; Monteiro, C.; Rocha, T.L. Toxicity of plant-based silver nanoparticles to vectors and intermediate hosts: Historical review and trends. *Science of the Total Environment.* 2022, 155299.

8 Diagnostic and Therapeutic Applications of Biogenic Metal Nanoparticles

Govindan Nadar Rajivgandhi, Muthuchamy Maruthupandy, Franck Quero, Rajamanickam Mohanraj, and Balasubramanian Mythili Gnanamangai

8.1 INTRODUCTION

Nanotechnology has become a revolutionary area of research interest in recent years, with applications in almost all industries, including agriculture, food, biomedicine, and pharmaceuticals. It is a rapidly growing field that is connected to a variety of multidisciplinary sciences, including chemistry, physics, biology, and material sciences (Saif et al., 2016). It is a hot topic in the fields of basic and applied sciences research. Nanotechnology focuses on creating and using materials that range in size from 1 to 100 nm. It involves the control, manipulation, and creation of matter at the nanometer level. The nanoparticles (NPs) offer a wide surface area. And because of larger surface-area-to-volume ratio, they exhibit excellent physiochemical, biopharmaceutical, and biological properties (Subhaschandrabose et al., 2022), which have been exploited in various fields, such as electronics, mechanical, optics, biotechnology, environmental remediation, material science, and medicine (Hina et al., 2023). It is a bulk material derived by nanoscale size, functionalized surface moieties, larger surface of volume zero properties, with both hydrophobic and hydrophilic nature (Jamkhande et al., 2019). The NPs are prepared by different ways, such as physical, chemical, and biological (Castro et al., 2013). More researchers are choosing the biological-mediated nanoparticle synthesis for their applications due to their low cost, less time, and minimized manpower. Among the nanomaterials, metal oxide nanoparticles, including silver, copper, gold, zinc, iron, nickel, and some other nanoparticles, are very important.

8.2 SYNTHESIS OF METAL NANOPARTICLES

Metal nanoparticles are being synthesized by conventional methods and green synthesis methods. Toxicity and the process cost of conventional methods are very high when compared to the green synthesis method. In the green synthesis method, extracts or whole cells of living organisms, such as plants, algae, fungus, bacteria, etc., are used to prepare the nanoparticles from various substrates by means of their oxido-reduction reactions. Generally, nanoparticles are synthesized by two major fundamental approaches, that is, bottom-up approach and top-down approach. These two approaches can be subcategorized further based on the reaction, process conditions, and steps. The bottom-up approach involves building up of NPs by aggregating smaller units, whereas in the top-down approach, bulk materials are reduced into nanoscale level (El-Kassas et al., 2016). Green synthesis of nanoparticles is a challenging and essential research area to reduce the utilization of toxic chemicals and substrates in the synthesis process (Momeni and Nabipour, 2015). The present review is a detailed discussion of biomedical applications of metal, metal oxide, or composite

DOI: 10.1201/9781003430087-8

TABLE 8.1
Advantages and Disadvantages of Conventional Nanoparticles Synthesis

S. No.	Method	Advantages	Disadvantages
1	Ball milling	Large-scale production and enhanced solubility. It imparts new characteristics to nanoparticles.	Requires high energy for the process. Process time is too long, with possibility of contamination due to the milling process.
2	Mechanochemical	Simple and efficient method to prepare the nanoparticles.	Contact time is very high to prepare nanoscale particles, and also, contamination may occur due to process condition.
3	Laser ablation	Efficient method to prepare large quantity of nanoparticles. Nanoparticles can be prepared without solvents and surfactants.	Process time continues for the long time to prepare bulk quantity of nanoparticles.
4	Ion sputtering	This is an economical method, as the process cost is very low. Contaminations are very low.	Sputtering gases may cause some effects on the thin films.
5	Physical vapor deposition	Simple method to prepare the thin film of metals.	Process cost is very high. Product yield is also very low when compared to other methods.
6	Chemical vapor deposition	Surface morphology of synthesized nanoparticles can be controlled. Crystal structure of nanoparticles can be controlled.	Chemical changes may cause toxic particles and exhaust hazardous gases.
7	Solgel method	Particle size and morphology of nanoparticles can be controlled.	Cost of raw materials is high.
8	Chemical reduction	Simple and the most common method to prepare nanoparticles.	Some limitations for this method include the usage of very toxic components with less intensity of reducing capability.
9	Spray pyrolysis	Simple and low-cost process method.	Requires high precision.
10	Laser pyrolysis	Particles prepared are uniform and can be controlled for its morphology. Simple method to prepare large quantity of nanoparticles.	Skilled operations and meticulous planning is required.

nanomaterials. The merits and demerits of the methods involved in synthesis of metal NPs are presented in Table 8.1.

8.3 APPLICATIONS OF METAL NANOPARTICLES

8.3.1 Metal Nanoparticles in Drug Delivery System

The growing needs of mankind related to the longevity of life, the fight against devastating diseases, and the existing side effects of drugs targeting the host's immunity can only be countered by developing highly efficient and specific drug delivery systems (DDS). There are many scopes of nanoparticles to act as potential DDS. However, some of the existing nanoparticles that can act as potential drug delivery systems lack precision in their toxicity. This drawback may be tackled by the NPs synthesized using green methodologies. NPs of biological origin are reported to possess unique and synergetic properties, with reduced toxicity and higher therapeutic potential, at low cost. Nanomedicines are well established with targeted therapy, and nanobased drug delivery systems. Nanotechnology-based drug delivery systems were successfully launched in the nanomedicine era with liposomes and micelles. These DDS systems were commercially used to improve the therapeutic potential of metal NPs/inorganic NPs. The improved bioavailability of drug targeted with nanodrug delivery systems were accorded by preventing sensitive and soluble drugs that face many hindrances at the drug adsorption or drug metabolism site, like gastrointestinal. The controlled release and targeting of the sparingly soluble drugs using nano-DDS is one milestone for DDS, wherein the longevity of drugs in the blood circulation reduces the need for frequent dosing. Although natural chemistry diminished with numerous disadvantages in bioavailability, adsorption, and elimination, it was boosted in the nano era by conjugating nanoparticles with natural compounds. Based on their capacity to enhance the pharmaceuticals and pharmacodynamics of natural or chemical drugs, different types of nanoparticles are rendered suitable for use as DDS.

8.3.1.1 Various Drug Delivery Approaches

In recent years, the failure of the chemotherapeutic design of metal oxide NPs' (MONPs) usages is an important point in the research field due to continuous side effects, lower adsorption efficiency, decreased solubility, more drug requirements, uncontrolled or continuous usage of irrelevant drugs, and increased cytotoxicity (Marzieh Samiei et al., 2021). To overcome these problems, precise and improved drug delivery systems must be developed. The merits of drug delivery system include increased biodegradability, improved bioavailability and biocompatibility, controlled drug-delivering capacity with lesser side effects (Sanaz and Homayon, 2017). Owing to these advantages, various drug delivery systems, particularly liposomes, dendrimers, and some polymeric NPs, have been designed. However, none of them could work independently (Chandan et al., 2018). The drug delivery approaches are used to deliver the drug, enzymes, proteins, or antibiotics or other drug molecules in an effective way. This kind of drug delivery approach not only helps deliver the drugs but helps target site drug delivery, increased bioavailability, and highest degradability, and to diminish the toxicity levels. Further, the engineering and construction of smart nanocarriers are the alternative choice to carry large amounts of the drug or therapeutic molecules to the desired place. It also helps deliver drugs into target sites, controlled drug release, and acts as a diagnostic agent, which has been considered a great concentration in recent years. Therefore, drug delivery approaches are the best choice instead of existing drugs, and they can be used as an ideal drug delivery system (Jie et al., 2023). Importantly, polymeric materials with MONPs are clean, neat, and green approaches in the research field (Xiaolu et al., 2022). The involvement of various NPs in drug delivery approaches is presented in Figure 8.1.

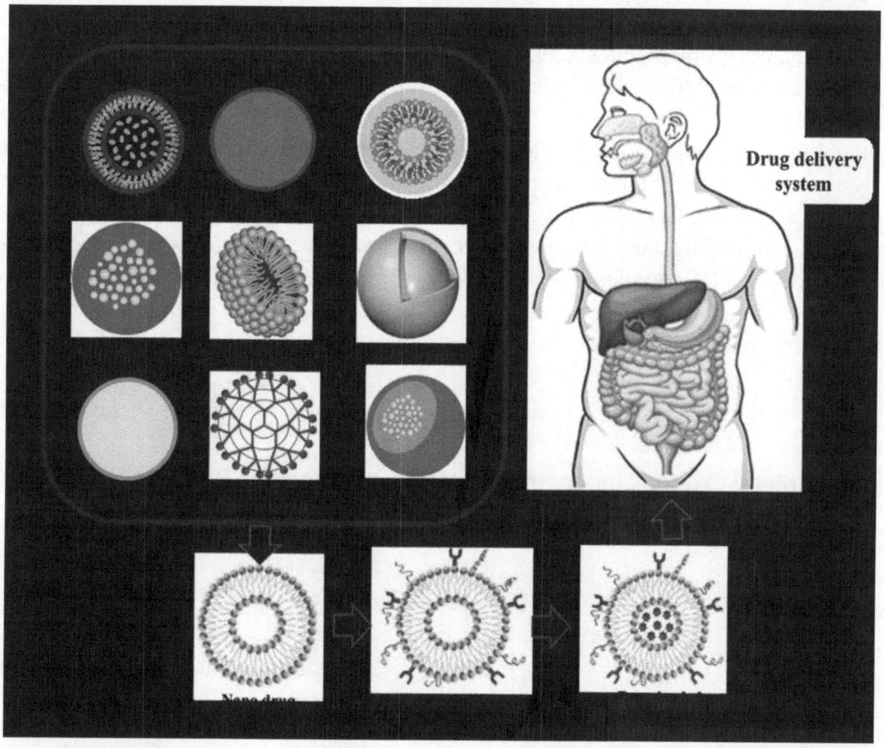

FIGURE 8.1 Applications of various nanoparticles in drug delivery systems.

Chemically synthesized drugs are excellent chemical derivatives, except for their toxicity and side effects. The NPs are being claimed for their ability to serve as potential DDS systems by seeming as biocompatible, coatings, cross-linked, composites, functionalized polymers to accord with various functional requisites, like size, surface properties, uptake, and release capabilities (Schulte, 2007). Gold NPs are widely known nanoparticles for thermal ablation of cancer cells wherein they were earlier claimed for treating tumor by the mode of photon therapy, where AuNPs convert light and radio frequencies into heat. Yet many novel polymer nanomaterial cross-linked with glutaraldehyde are claimed to treat cardiovascular diseases (Woei and Cheng, 2010).

Sensitiveness of the NP can be used for triggering the NP to deliver the "cargo" active ingredient. There are numerous drugs that are claimed for the sensitive property to trigger for release or to direct to site-specific organ or region that requires medications. Different NPs get triggered with different agents, like high, acoustic, ultrasound, thermal, photothermal, magnetic, pH-sensitive, temperature, radiation, receptor, and aptamers, whereas most of the NPs are claimed for aptamer receptors, followed by receptors triggered by irradiation. Sensitization or stimuli with other condition are still under the nascent stage and not claimed as much. Most of the sensitization are claimed for tumor treatment (Myolar and Toledo, 2008).

Nearly all major and minor routes of administration are addressed in terms of drug delivery, and the parenteral route (IM, IV, SC) is the most suitable method for administering these nanodrug delivery systems because it can maximize the delivery and availability of NPs. The parenteral route of administering nanovesicles via intravenous mode for carcinoma and sarcoma types of cancer is covered by US20130195983A1 (Desai et al., 2013). For the liposome type, gel and hydrogel for NP delivery are preferred. These routes of delivery take up various formulations of NPs.

To overcome the infectious issue of the biomedical field, different strategies help eradicate bacterial infection, viral infection, fungal infections, as well as cancer, larvicidal infections, and immunodeficiency (Negin et al., 2022). Importantly, the functionalization of silica plus biopolymer with MONPs combination is very efficient against all infectious diseases. It carries drugs, including penicillin, cefixime, ceftazidime/clavulanic acid, fluconazole, amphotericin, clotrimazole camptothecin, paclitaxel, doxorubicin, decetaxel, etc. (Sahar et al., 2023). These kinds of drug formulation methods were very useful to modern industry, but drug release efficiency is a big challenge because of controlled release of drug molecules of high molecular weights. The MONPs not only performed in biomedical field but also worked in various other applications, including gas storage, gas sensing, catalysis, and other environmental applications. The various applications of biopolymeric NPs combined with various metal oxide NPs are available in Table 8.2.

In addition, bioimaging and biosensing are also important applications in polymeric with metal oxide nanoparticles complex in drug delivery. Exclusively, the biodegradability, biocompatibility, and bioavailability of polymeric with metal oxide nanocompositions make them an excellent drug delivery material and produce a revolutionary drug delivery process. In addition, these

TABLE 8.2
Biopolymeric/Metal Oxide Nanocomposites against Pathogenic Bacteria

S. No.	Polymeric/Metal Nanoparticles	Biological Activity	References
1	Graphene/chitosan	Antibiofilm	(Maruthupandy et al., 2020)
2	Chitosan/beta lactam antibiotics	Antimicrobial	(Manpreet et al., 2022)
3	Doxorubicin and SAR405-loaded chitosan nanoparticles	Anticancer	(Mohammed et al., 2023)
4	Chitosan/silver	Antibiofilm	(Ramachandran et al., 2023)
5	Graphene/chitosan/magnetite nanocomposites	Anti-ESBL	(Maruthupandy et al., 2021)
6	Chitosan/graphene	Anticancer	(Bruna et al., 2022)
7	CS-AMP-P-ZnO	Antimicrobial	(Sathiyaseelan et al., 2023)

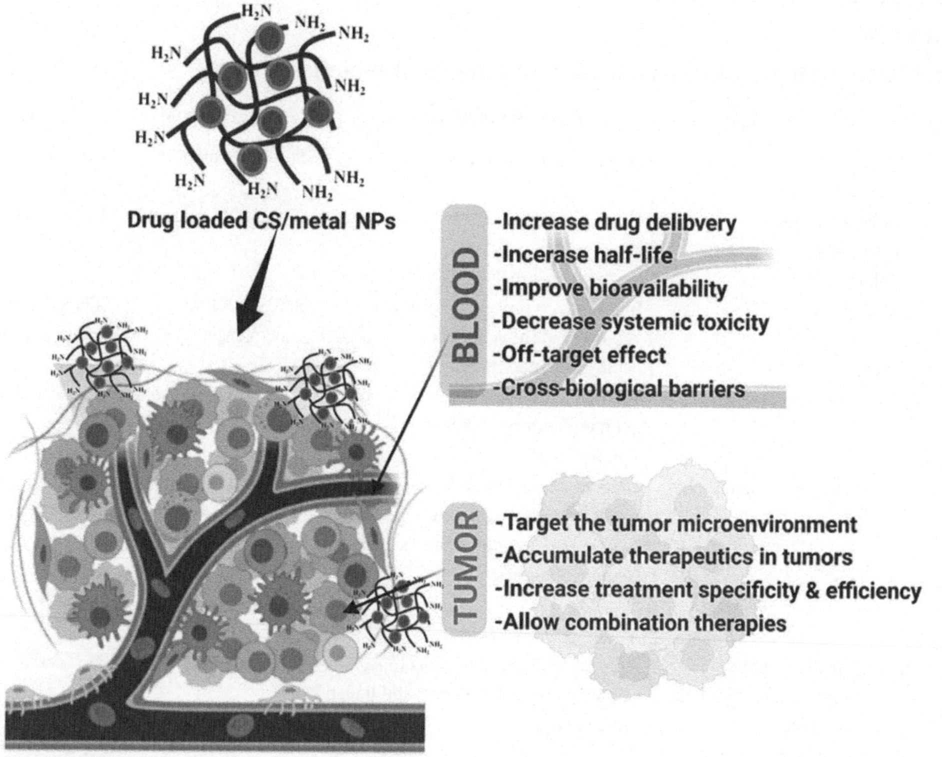

FIGURE 8.2 Biopolymeric nanoparticles with metal oxide nanoparticle combinations for various applications.

kinds of approaches helped avoid chemical usage, more time consumption, and toxic deficiency (Ramachandran et al., 2023). The schematic representation of Figure 8.2 exposes the polymeric nanoparticles/metal oxide nanocomposites for various applications.

8.3.2 Metal Nanoparticles in Diagnosis

Diagnosis is one potential application area of green synthesized metal nanoparticles. Although the organic and inorganic nanoparticles have expanded their role as therapeutic agents, there are some unique properties attributed to metal nanoparticles to serve as both diagnostic and therapeutic elements. In certain cases, the research hot spot was laid on Ag and Au, which made them employed as diagnostic devices for their unique localized surface plasmin resonance (LSPR) effect and physicochemical properties (Zhao et al., 2012). These metal nanoparticles served as biomarkers due to the fact that the reduced salts of gold and silver can be functionalized through the metal affinity to the thiol groups, which form part of proteins (Storhoff et al., 2004).

The major advantage of green synthesized metal nanoparticles is the use of cellular processes, like adhesion, growth, differentiation, interatomic forces, H-bonds, etc., by their nanostructures to mimic biological entities. Similarly, metal nanoparticles are used to detect qualitative changes through developing various points of case devices due to the LSRP in the visible region. These points of core devices screen analyses, such as nucleic acids, enzymes, proteins, aptamers, oligonucleotides, etc. This ability was widely commercialized for screening infection using colorimetric assay, as Au and Ag nanoparticles exhibit visual color change from red to blue and yellow to brown, respectively, in the test. The Au NPs are widely applied in bacterial detection due to their inherent properties, like excellent biocompatibility, functionalization, good solubility, and tunable synthesis (Table 8.3). Due to the surface charge conferring, gram-positive and gram-negative bacteria can

TABLE 8.3
Metal Nanoparticles for the Detection of Biological Moieties

S. No.	Nanoparticle	Principle of Detection	Target Signal	References
1	AuNPs–biotin-streptavidin conjugate	DNA binding transcription factor	Specific sequence	(Chan and Chen, 2012)
2	Au-Fe_3O_4 magnetic composites with silver enhancement	Glassy carbon as electrode as surface	27 mer DNA target	(Li et al., 2020)
3	Au nanoprobe	Immobilized probe changes in conformation on perfect complementary target	SNPS is KRAS oncogene	(Girigoswami et al., 2009)
4	Nanoporous surface	Wave voltammetry differential pulse voltammetry by redox indicator $[Fe(CN)_6]^{4-/3-}$	SNPS from TNF α gene	(Radwan et al., 2019)
5	AuNPs as tag for nanoprobe	Glassy carbon electrode, deposition of AuNPs, followed by immobilization of oligonucleotide probe on electrode	Specific sequence of PCR product of LTA-α gene	(Pan and Lu, 2008)
6	AgNPs	Cyclic voltammetry	Interaction of luteolin copper (II) and ds DNA	(Bai et al., 2010)
7	PH-controllable poly electrolyte shells; sShell as encapsulating AgNPs-biotin-avidin complex	Acid dissolution of Ag ions to AgNPs by stripping voltammetry and linear sweep voltammetry	DNA labelling	(Noor et al., 2013)
8	Au NR monolayer spanning over Au electrode	Current versus applied measurements	Bacterial detection	(Alfredo and Arben, 2010)

be distinguished easily, which facilitates the interaction of gold nanoparticles with cells and their aggregation for calorimetric assay detection.

Fluorescence assays are frequently used to detect fluorophores both quantitatively and qualitatively. It is actually believed that increasing fluorescence will increase sensitivity for high-throughput screening assays. Bacteria like *E. coli*, *MRSA*, *S. aureus*, and *Salmonella enteritidis* have all been subject to fluorescence assay screening. Separating different bacterial strains using fluorescence gradient response has also been reported (Refera et al., 2011). Both silver and gold metal nanoparticles were reported to have enhanced fluorescence system for detection of single-base change, RNA sensing, and double-stranded DNA sensing, through which bacterial population in low fold can be sensed and single-base change in RNA of viruses is detected.

The output correlation with target identification and quantification is made possible with accelerated signal processing system for transduction, enhanced intensity, and microwave assistance to promote accuracy in the detection of biomolecules (oligonucleotides of RNA). The mechanism behind the use of these metal NPs in improving the fluorescence, quenching phenomenon, and conjugation of metal nanoparticles with polymers for detecting the analyte in various application was reported earlier to be due to electrostatic interaction controllable extinction spectra and molar extinction coefficients (Niu et al., 2009).

As of now, AuNPs-catalyzed silver enhancement was reported by Girigoswami (Rijiravanich et al., 2008), which is used to detect specific sequence identification and quantification of BRCA 1 mutations. The AgNPs dissolutions in colloidal consumer product effects of particle size and capping agent are available (Berry et al., 2005). Sickle cell anemia associated mutation was reported to have noteworthy detection using AgNPs.

Metal nanoparticles like AuNPs and AgNPs with high SPR and surface enhancement of Raman scattering (SERS) were reported to overcome the spectral defects and lack of sensitivity in biomolecular detection. The studies done so far on these SERS-based detection projects have focused on the applications of nanoparticles like gold and silver that have 30- or 106-times stronger signals. The manifold increase in signal was proportional to the bacterial biomolecule detection limit of 0.1 ppb in AuNPs for SERS effect to detect multiple strains. These applications of metal NPs are not only limited to infections by bacterial/viral cells (Bruna et al., 2022; Sathiyaseelan et al., 2023; Ramachandran et al., 2023; Zhao et al., 2012). Electrochemical methods are also applied to the diagnosis of various electrochemical signals raised by the target. The electrodes used are surface-coated with various types of NP probes to detect the electrochemical signal with different underlying principles (Storhoff et al., 2004).

8.3.3 Metal Nanoparticles as Biosensors

In electrochemical sensing system, the different metal and metal oxide NPs play a variable role, notably as biosensors, enzyme sensors, protein sensors, immunosensors, or DNA sensors (Afonso et al., 2013). Previously, Hameed et al. (2022) reported that metal oxide nanoparticles have excellent conductivity and catalytic properties, which help them act as an electronic wire to improve the transfer of electrons between redox centers in proteins and electrode surfaces and as catalysts to increase the rates of electrochemical reactions. In some cases, the metal-oxide NPs have occasionally served as immobilized molecules, while their semiconductor nanoparticle counterparts have served as labels or as a means of transfer for electrochemical analysis. Recent reports stated that metal and metal oxide NPs are the sole sources used in electrochemical sensors and biosensors (Pedrosa and Baptista, 2015).

Metal NPs are involved in the biosensing applications too (Figure 8.3). They are involved in early disease diagnosis through biomarker detection with extremely high sensitivity (Labib et al., 2012). In biosensing method, metal oxide NPs is used, owing to their unique surface plasmon resonance. Surface plasmon resonance is an important behavioral approach designed by local charge oscillation in resonance with incident light when the size of the particles exhibited smaller than light wavelength. Metal oxide NPs can occasionally be affected by etching, growth, or other changes in their local environment as they aggregate or disaggregate, which can affect their optical properties. Sometimes, capping agents of metal oxide NPs exposed the excellent versatility in many molecular scenarios, which has the catalytic properties and involves the signal amplification process by precursor molecules (Sidhu et al., 2022).

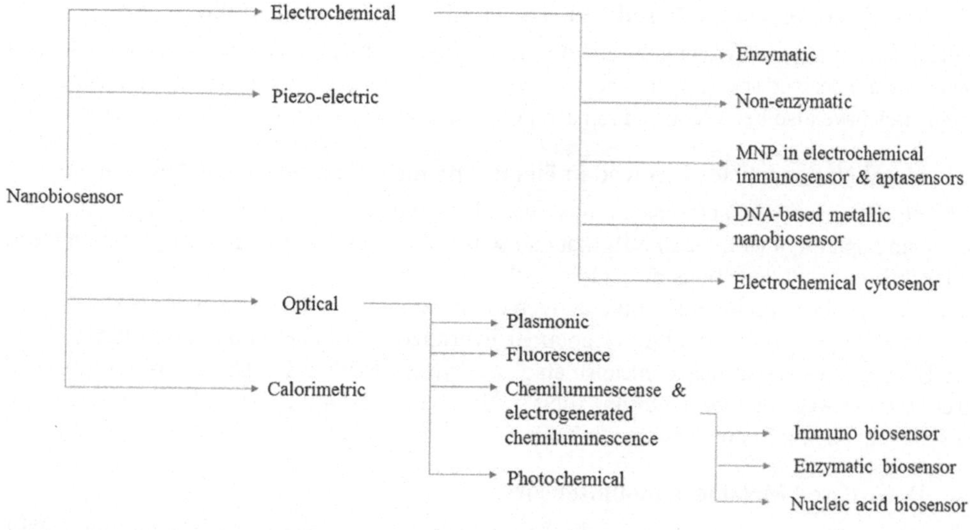

FIGURE 8.3. Taxonomic group of nanobiosensor as various fields.

Biosensors are a potential area where green metal nanoparticles which are still not taken up to secondary research. The biosensors have the biological component, physical component that convert the biological event into a detectable signal, and the signal display that reads out the concentration of the analyte that has taken up a role in the binding event with biological elements. Due to conferring of specificity, selectivity, and sensitivity, the metal NPs are used as one of the transducers to promote signal detection at very low levels.

8.3.3.1 Electrochemical Nanobiosensors

Biosensor is an excellent tool subjected to apply in the physical, chemical, and biological field identified based on the signal proportional to detect the concentration of electroanalyte under study (Choi et al., 2021). The biological element interacts with the analyte and produces physical or chemical changes which can be sensed by the transducer or detector, producing a signal proportional to the concentration of this analyte (Mohammad and Yen, 2020). The use of metal oxide nanoparticles (MONPs) in a variety of fields, including the optical, electrical, and piezoelectric sensing of the biorecognition process, has increased recently. These metal NPs exhibit a wide range of color based on the surface plasmon resonance phenomenon, and these surface plasmon resonances are strongly dependent on their size and shape (Białas et al., 2022). In this application, the results appear in the form of adsorption peaks applied for colorimetric detection that determine the color changes due to the aggregation of metal NPs. In addition, it was used in the enzyme degradation process due to biocatalysis, which influences color changes and the results taken by the adsorption spectrum. The enzyme inhibitions by metal oxide NPs are the major challenges in the electrochemical nanobiosensor. The smart metal oxide nanomaterials have excellent electroconductivity and photoactivity, high sensitivity, and signal amplifications. Among the various signal amplification strategies, the biocatalytic precipitation is an important application and has heightened great interest recent years (Yeşeren et al., 2022).

Recently, Chen et al. (2014) reported electrochemical-mediated bioassays as very important and an essential demand in the field of biological sciences. In addition, ultrasensitive techniques of immunobioassay based on the antigen–antibody recognition method is very important. Importantly, electrochemical bioassay methods are also very important due to the high sensitivity, intrinsic simplicity, field portability, and inexpensive instrumentation. Further, the immunoassay of DNA quantification is another milestone and measured by specific activity of the label-based method. In this method, the metal NPs act as a key factor and develop the signal-amplified electrochemical bioaffinity assay of protein and nucleic acids (Tingting et al., 2023; Caixin et al., 2020).

8.3.3.2 Metal Nanoparticles–Based Non-Enzymatic Electrochemical Biosensor

Due to the disadvantages inferred by the enzymatic reactions, non-enzymatic reactions for detection of molecules were derived specially to detect glucose. Metal NPs and their hybrids with other alloys and composites have also been reported to have precise sensitivity to glucose (Zhang et al., 2014).

8.3.3.3 Metal Nanoparticles Involved in Electrochemical, Immuno-, and Aptasensors

The immune biosensors and aptasensors make use of tagging, capping, or modifying of antigens, antibodies, and aptamers with metal NPs that can sensitize the analyte to a greater extent and lead to high-level throughput detections by exploiting the transduction and amplification of signals. The most studied is against carcinoembryonic antigen alpha fetoprotein. The detection of Alzheimer's disease is made easier by the binding of aptamer hybridized with Au nanoparticles (Choi et al., 2021) and using Au nanoparticles–encapsulated dendrimers MWCNTs (Dan et al., 2007), gold nanostructures, Au-Ag-Au core, chitosan AuNPs (Yang et al., 2014). The metal NPs explored for nanobiosensor applications are available in Table 8.4.

8.3.3.4 DNA-Based Metallic Nanobiosensors

These biosensors are a highly miniature sensor used to detect DNA mutations. The modified DNA biosensors with Au nanoparticles augmented to electrode surface were practiced and reported to reproduce accurate diagnosis of leukemia and dengue (Anees and Bansi, 2022).

TABLE 8.4
Applications of Metal Nanoparticles as Nanobiosensors

S. No.	Metal Nanoparticles	Type of Sensing	References
1	Gold nanoparticles	Catalytic activity of copper (II) ions using colorimetric sensing	(Yuan et al., 2023)
2	Au and chitosan as a template	Thiocholine inhibition sensor-based biocatalytic growth	(Zhao-Peng et al., 2007)
3	Iron nanoparticles encapsulated in boron–nitrogen Co-doped carbon nanotubes	Enzyme for electrochemical monitoring of dopamine and uric acid	(Jeong et al., 2013)
4	Au nanowire assays	Glucose detector amperometer	(Yang et al., 2014)
5	Platinum NPs over immobilized glucose oxidate	H_2O_2 by-product voltammetry measurement	(Ananda Murthy et al., 2021)
6	Pd-Co alloys in carbon nanofibers	H_2O_2 detection moving cyclic voltammograms	(Bhawna et al., 2020)

8.3.3.5 Electrochemical Cytosensors

The cell responses to stimuli can be recorded with electrochemical cytosensors. Cytosensing is beneficial for nanoinvasive diagnosis of diseases like cancer using Au nanoparticles, which amplify the signal (Singh et al., 2022).

8.3.3.6 Optical Biosensors

The localized surface plasmon resonance property of NPs of Au, Ag, and aluminum was applied to detect cancer (Tai et al., 2022), explosive vapors (Shubhanshi et al., 2020), and live sensing (Hayley et al., 2023).

8.3.3.7 Fluorescence Biosensors

Based on the fluorescence assays and due to the surface plasmon resonance features, quenching features, fluorescence emissions associated with Au nanoparticles, the AuNPs-functionalized DNA (Yuan et al., 2023), Go/AuNPs (Heredia et al., 2020), and AgNPs on CdSe QDs have been used for imaging applications in DNA hybridization (Meng et al., 2017), mercury pollutant detection concentration of CEA, and glucose sensing (Malekzad et al., 2017).

8.3.3.8 Chemiluminescence and Electrogenerated Chemiluminescence Biosensor

The concentration of the agents to be detected through chemiluminescence is directly proportional to the reaction between the reactants and the analyte in a chemical reaction. Colloidal AuNPs have a strong track record of being used to detect genes and proteins using chemiluminescence-based immunoassays, enzyme–CL sensors, and nucleic acid–CL sensors (Aiping et al., 2019).

8.3.3.9 Photochemical Nanobiosensors

For photochemical biosensing, the exciton–plasmon interaction and LSPR (localized surface plasmon resonance) of QDS and metal NPs were used to sense the detection elements. Some of the conjugated metal and quantum dots used are AuNPs-capped cysteamine carbon dots for Hela cell sensing (Liu et al., 2013).

8.3.3.10 Colorimetric Biosensors

Detection of specific analyte based on visual color change on association with metal NPs is the principle behind colorimetric biosensors. Despite the sensitivity, the simple design and requirement have led to the use of these types of biosensors for wide applications, like environmental pollutant screening for Hg, clinical diagnosis for glucose, and cancer biomarkers (Yao et al., 2015).

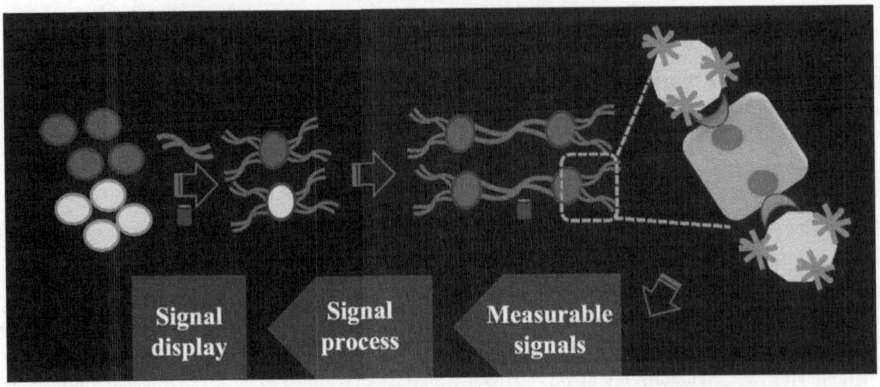

FIGURE 8.4 The role of noble metal nanoparticles for piezoelectric-based biosensor.

8.3.3.11 Piezoelectric Nanobiosensors

This type of nanobiosensor detects and amplifies signal when metal NPs biorecognize an element due to mass changes produced at the surface of piezoelectric crystals. Based on this, the AuNPs were reported for sensing *E. coli* in food samples by forming AuNPs–antibody conjugate (Chen et al., 2014). With similar mechanism, AuNPs were also applied to detect leukemia (Wenli et al., 2021). The piezoelectric-based biosensor and their mechanisms are available in Figure 8.4.

8.4 CONCLUSION

In recent years, the synthesis of biological-mediated metal and metal oxide nanoparticle has been heightened worldwide due to broad-spectrum antibiotics in the clinical field. In addition, it is a potential factor to be applied in almost all industries. Researchers are trying to find different kinds of innovation study to be used in the biopharmaceutical industries. In addition, those innovative approaches are more helpful in stressful environments in the medical field. Likewise, metal and metal oxide nanoparticles are very important materials which have been applied in all fields worldwide. In particular, the low toxicity of synthesized nanoparticles is heightened in the biomedical field as a format of new drugs. Supportively, biosynthesized metal nanoparticles have minimum toxicity level with the highest potential against various infectious diseases in biomedical applications. Among the metal nanoparticles studied, polymer-mediated metal nanoparticles or polymer/metal nanoparticle combination are very effective in drug delivery approaches. Based on the high bioavailability, biodegradability, biocompatibility, and minimized toxicity of the polymer/metal nanocomposites, they are the best choice to completely eradicate various diseases. Based on recently reported research work, polymer/metal nanoparticle combination as an alternative drug delivery molecule is used in the drug delivery approaches for various infections.

8.5 CONFLICT OF INTEREST

The authors have no conflict of interest.

ACKNOWLEDGMENTS

G. Rajivagndhi and Franck Quero acknowledge the financial support from ANID-FONDECYT (Chile) under the Postdoctoral Fellowship No. 3220019. Foremost thanks to the K. S. Rangasamy College of Technology Institutional Management, the principal, and the CEO. Word of acknowledgment to our national funders from the government of India (DST-FIST-SR/FST/College-235/2014 and DBT-STAR- PG Schemes (BT/HRD/11/09/2018; BT/HRD/01/07/2020).

REFERENCES

Afonso, S.; Pe, B.; Costa, M.M. et al. 2013. Electrochemical detection of Salmonella using gold nanoparticles. *Biosensors and Bioelectronics* 40: 121–126.

Aiping, C.; Jianwei, Z.; Wanqiao, B. et al. 2019. Signal-on electrogenerated chemiluminescence biosensor for ultrasensitive detection of microRNA-21 based on isothermal strand-displacement polymerase reaction and bridge DNA-gold nanoparticles. *Biosensors and Bioelectronics* 144: 111664.

Alfredo, E.-M.; Arben, M. 2010. Nanoparticle based enhancement of electrochemical DNA hybridization signal using nanoporous electrodes. *Chemical Communication* 47: 9007–9009.

Ananda Murthy, H.C.; Kiflom, G.K.; Ravikumar, C.R.; Nagaswarupa, H.P.; Tadesse, A.; Tegene, D. 2021. Graphene-supported nanomaterials as electrochemical sensors: A mini review. *Results in Chemistry* 3: 100131.

Anees, A.A.; Bansi, D.M. 2022. Current progress in organic–inorganic hetero-nano-interfaces based electrochemical biosensors for healthcare monitoring. *Coordination Chemistry Reviews* 452: 214282.

Bai, Y.; Li, J.; Xu, J. et al. 2010. Ultrasensitive electrochemical detection of DNA hybridization using Au/Fe_3O_4 magnetic composites combined with silver enhancement. *Analyst* 1672–1679.

Berry, V.; Gole, A.; Kundu, S. et al. 2005. Deposition of CTAB-terminated nanorods on bacteria to form highly conducting hybrid systems. *Journal of the American Chemical Society* 17600–17601.

Bhawna, B.; Samiksha, S.; Jyoti, A.; Minakshi, S. 2020. Electrochemical sensing of cytochrome *c* using graphene oxide nanoparticles as platform. *International Journal of Biological Macromolecules* 165: 1455–1462.

Białas, K.; Moschou, D.; Marken, F. et al. 2022. Electrochemical sensors based on metal nanoparticles with biocatalytic activity. *Microchimica Acta* 189: 172.

Bruna, L.M.; Rita, L.S.; Cátia, G.A.; André, F.M.; Ilídio, J.C.; Duarte, M. 2022. Chitosan-based injectable *in situ* forming hydrogels containing dopamine-reduced graphene oxide and resveratrol for breast cancer chemo-photothermal therapy. *Biochemical Engineering Journal* 185: 108529.

Caixin, G.; Jinliang, W.; Jing, C. 2020. Colorimetric sensing of copper (II) ions based on the inhibition of biocatalytic growth of gold nanoparticles. *Microchemical Journal* 157: 105015.

Castro, L.; Blázquez, M.L.; Muñoz, J.A.; González, F.; Ballester, A. 2013. Biological synthesis of metallic nanoparticles using algae. *IET Nanobiotechnology* 7: 109–116.

Chan, P.; Chen, Y. 2012. Human serum albumin stabilized gold nanoclusters as selective luminescent probes for staphylococcus aureus and methicillin resistant staphylococcus aureus. *Analytical Chemistry* 84: 8952–8956.

Chandan, A.; Anurag, M.; Debasis, N.; Anjan, C. 2018. Metal organic frameworks modified mesoporous silica nanoparticles (MSN): A nano-composite system to inhibit uncontrolled chemotherapeutic drug delivery from Bare-MSN. *Journal of Drug Delivery Science and Technology* 47: 1–11.

Chen, S.; Hai, X.; Chen, X.W. 2014. In Situ growth of silver nanoparticles on graphene quantum dots for ultrasensitive colorimetric detection of H_2O_2 and glucose. *Analytical Chemistry* 86(13): 6689–6694.

Chen, X., Wu, G., Cai, Z., Oyama, M., Chen, X. 2014. Advances in enzyme-free electrochemical sensors for hydrogen peroxide, glucose, and uric acid. *Microchimica Acta* 181: 689–705.

Choi, H.K.; Lee, M.-J.; Lee, S.N.; Kim, T.-H.; Oh, B.-K. 2021. Noble metal nanomaterial-based biosensors for electrochemical and optical detection of viruses causing respiratory illnesses. *Frontiers in Chemistry* 9: 672739.

Dan, D.; Jiawang, D.; Jie, C.; Aidong, Z. 2007. Electrochemical thiocholine inhibition sensor based on biocatalytic growth of Au nanoparticles using chitosan as template. *Sensors and Actuators B: Chemical* 127: 317–322.

Desai, N.P.; Angeles, L.; Tao, C.; De, T.; Ci, S.X.; Trieu, V. 2013. Nanoparticle formulations and uses thereof. US 2013/0195983A1.

El-Kassas, H.Y.; Aly-Eldeen, M.A.; Gharib, S.M. 2016. Green synthesis of iron oxide (Fe_3O_4) nanoparticles using two selected brown seaweeds: Characterization and application for lead bioremediation. *Acta Oceanologica Sinica* 35: 89–98.

Girigoswami, A.; Li, T.; Jung, C.; Young Mun, H.; Gyu Park, H. 2009. Gold nanoparticle-based label-free detection of BRCA1 mutations utilizing DNA ligation on DNA microarray. *Journal of Nanoscience and Nanotechnology* 9: 1019–1024.

Hameed, S.T.; Talal, F.Q.; Abdelghany, A.M.; Oraby, A.H. 2022. Effect of zinc oxide nanoparticles on physical properties of carboxymethyl cellulose/poly (ethylene oxide) matrix. *Physica B: Condensed Matter* 633: 413771.

Hayley, C.; James, G.; Nunzianda, F. 2023. Nanomaterials for optical biosensors in forensic analysis. *Talanta* 253: 123945.

Heredia, F.L.; Resto, P.J.; Parés-Matos, E.I. 2020. Fast adhesion of gold nanoparticles (AuNPs) to a surface using starch hydrogels for characterization of biomolecules in biosensor applications. *Biosensors* 10: 99.

Hina, R.; Waqar, A.; Nadir, Z.K.; Muhammad, A.;Tariq, K.; Ayaz, A.K. 2023. *Delphinium uncinatum* mediated biosynthesis of zinc oxide nanoparticles and *in-vitro* evaluation of their antioxidant, cytotoxic, antimicrobial, anti-diabetic, anti-inflammatory, and anti-aging activities. *Saudi Journal of Biological Sciences* 30: 103485.

Jamkhande, P.G.; Ghule, N.W.; Bamer, A.H.; Kalaskar, M.G. 2019. Metal nanoparticles synthesis: An overview on methods of preparation, advantages and disadvantages, and applications. *Journal of Drug Delivery Science and Technology* 53: 101174.

Jeong, B.; Akter, R.; Han, O.H. 2013. Increased electrocatalyzed performance through dendrimer-encapsulated gold nanoparticles and carbon nanotube-assisted multiple bienzymatic labels: Highly sensitive electrochemical immunosensor for protein detection. *Analytical Chemistry* 85(3): 1784–1791.

Jie, Y.; Shuaishuai, J.; Tian, C. et al. 2023. Construction of bionic nanoparticles camouflaged with macrophage membranes for drug delivery in breast cancer. *Journal of Drug Delivery Science and Technology* 84: 104433.

Labib, M.; Zamay, A.S.; Kolovskaya, O.S. et al. 2012. Aptamer-based viability impedimetric sensor for bacteria. *Analytical Chemistry* 84(21): 8966–8969.

Li, S.; He, J.; Xu, Q.H. 2020. Aggregation of metal-nanoparticle-induced fluorescence enhancement and its application in sensing. *ACS Omega* 5(1): 41–48.

Liu, X.; Zhang, J.; Liu, S. et al. 2013. Gold nanoparticle encapsulated-tubular TiO_2 nanocluster as a scaffold for development of thiolated enzyme biosensors. *Analytical Chemistry* 85(9): 4350–4356.

Malekzad, H.; Zangabad, P.S.; Mirshekari, H.; Karimi, M.; Hamblin, M.R. 2017. Noble metal nanoparticles in biosensors: Recent studies and applications. *Nanotechnology Reviews* 6: 301–329.

Manpreet, K.; Yael, C.; Elena, P.; Evgeni, E. 2022. Synergistic antimicrobial effect of the combination of beta-lactam antibiotics and chitosan derivative on multidrug-resistant bacteria. *International Journal of Biological Macromolecules* 223: 1107–1114.

Maruthupandy, M.; Rajivgandhi, G.; Muneeswaran, M.; Anand, M.; Quero, F. 2021. Highly efficient antibacterial activity of graphene/chitosan/magnetite nanocomposites against ESBL-producing *Pseudomonas aeruginosa* and *Klebsiella pneumoniae*. *Colloids and Surfaces B: Biointerfaces* 202: 111690.

Maruthupandy, M.; Rajivgandhi, G.; Shine, K.; Veeramani, T.; Naiyf, S.A.; Muneeswaran, T. 2020. Antibiofilm investigation of graphene/chitosan nanocomposites against biofilm producing *P. aeruginosa* and *K. pneumonia*. *Carbohydrate Polymers* 230: 115646.

Marzieh Samiei, F.; Ali, Z.; Hassan, A.; Reza, K.; Amirhosein, K. 2021. A drug delivery system based on nanocomposites constructed from metal-organic frameworks and Mn_3O_4 nanoparticles: Preparation and physicochemical characterization for BT-474 and MCF-7 cancer cells. *Colloids and Surfaces B: Biointerfaces* 202: 111712.

Meng, J.; Guifen, J.; Lu, T.; Shuyan, N. 2017. AgNPs-3D nanostructure enhanced electrochemiluminescence of CdSe quantum dot coupled with strand displacement amplification for sensitive biosensing of DNA. *Analytica Chimica Acta* 983: 166–172.

Mohammad, T.Y.; Yen, N.T. 2020. Metal nanoparticles-enhanced biosensors: Synthesis, design and applications in fluorescence enhancement and surface-enhanced Raman scattering. *Chemistry an Asian Journal* 15: 3180–3208.

Mohammed, N.A.; Suet, L.C.; Che Azurahanim, C.A.; Mas, J.M. 2023. Increased efficacy of biologics following inhibition of autophagy in A549 lung cancer cells in bimodal treatment of doxorubicin and SAR405-loaded chitosan nanoparticles. *Open Nano* 11: 100142.

Momeni, S.; Nabipour, I. 2015. A simple green synthesis of palladium nanoparticles with sargassum alga and their electrocatalytic activities towards hydrogen peroxide. *Applied Biochemistry and Biotechnology* 176: 1937–1949.

Myolar, R.; Toledo, O. 2008. Folate receptor gene modulation for cancer dagnosis and therapy. US 2008/0131366A1.

Negin, B.; Shadab, S.; Iman, A.D.; Leila, R.S. 2022. Design of slow-release methotrexate drug delivery system using PHBV magnetic nanoparticles and evaluation of its cytotoxicity. *Journal of Drug Delivery Science and Technology* 77: 103854.

Niu, S.; Han, B.; Cao, W. et al. 2009. Analytica chimica acta sensitive DNA biosensor improved by Luteolin copper (II) as indicator based on silver nanoparticles and carbon nanotubes modified electrode. *Analytica Chimica Acta* 651: 42–47.

Noor, M.R.; Goyal, S.; Christensen, S.M. et al. 2013. Electrical detection of single-base DNA mutation using functionalized nanoparticles electrical detection of single-base DNA mutation using functionalized nanoparticles. *Applied Physics Letters* 073703: 1–4.

Pan, Q.; Lu, Z. 2008. An electrochemical approach for detection of specific DNA-binding protein by gold nanoparticle-catalyzed silver enhancement. *Analytical Biochemistry* 375: 179–186.

Pedrosa, P.; Baptista, P.V. 2015. Gold and silver nanoparticles for diagnostics of infection. *Nanotechnology in Diagnosis, Treatment and Prophylaxis of Infectious Diseases* 1–18.

Radwan, I.M.; Gitipour, A.; Potter, P.M.; Dionysios D.D.; Al-Abed, S.R. 2019. Dissolution of silver nanoparticles in colloidal consumer products: Effects of particle size and capping agent. *Journal of Nanoparticle Research* 21: 1–155.

Ramachandran, G.; Alharbi, N.S.; Chackaravarthy, G. et al. 2023. Chitosan/silver nanocomposites enhanced the biofilm eradication in biofilm forming Gram positive S. aureus. *Journal of King Saud University – Science* 35: 102597.

Refera, T.; Henry, O.Y.F.; Ciara, K.Ó. 2011. Electrode surface nanostructuring via nanoparticle electronucleation for signal enhancement in electrochemical genosensors. *Biosensors and Bioelectronics* 26: 3962–3966.

Rijiravanich, P.; Somasundrum, M.; Surareungchai, W. 2008. Femtomolar electrochemical detection of DNA hybridization using hollow polyelectrolyte shells. *Analytical Chemistry* 80: 3904–3909.

Sahar, M.A.D.; Saeed, B.; Hevi, S.G.; Iskander, T.; Quynh, H.L. 2023. Drug release using nanoparticles in the cancer cells on 2-D materials in order to target drug delivery: A numerical simulation via molecular dynamics method. *Engineering Analysis with Boundary Elements* 148: 34–40.

Saif, S.; Tahir, A.; Chen, Y. 2016. Green synthesis of iron nanoparticles and their environmental applications and implications. *Nanomaterials* 6: 1–26.

Sanaz, H.; Homayon, A.P. 2017. Fabrication of magnetite nanoparticles modified with copper based metal organic framework for drug delivery system of letrozole. *Journal of Molecular Liquids* 243: 102–107.

Sathiyaseelan, A.; Saravanakumar, K.; Zhang, X.; Vishven Naveen, K.; Myeong-Hyeon, W. 2023. Ampicillin-resistant bacterial pathogens targeted chitosan nano-drug delivery system (CS-AMP-P-ZnO) for combinational antibacterial treatment. *International Journal of Biological Macromolecules* 237: 124129.

Schulte, R. 2007. Nanoparticle enhanced proton computed tomography and proton therapy. US 2007/0031337A1.

Shubhanshi, S.; Rashmi, K.; Shailendra, K.V.; Basudev, L. 2020. Optical biosensing with electromagnetic nanostructures. *Reviews in Physics* 5: 100044.

Sidhu, A.K.; Verma, N.; Kaushal, P. 2022. Role of biogenic capping agents in the synthesis of metallic nanoparticles and evaluation of their therapeutic potential. *Frontiers in Nanotechnology* 3: 801620.

Singh, R.; Sharma, A.; Saji, J. 2022. Smart nanomaterials for cancer diagnosis and treatment. *Nano Convergence* 9: 21.

Storhoff, J.J.; Lucas, A.D.; Garimella, V. et al. 2004. Homogeneous detection of unamplified genomic DNA sequences based on colorimetric scatter of gold nanoparticle probes. *Nature Biotechnology* 22: 883–887.

Subhaschandrabose, J.; Suresh, N.; Subburaman, C.; Narayanan, V.; Mohamed Ragab, A.G.; Mohamed S.E. et al. 2022. Biological synthesis of zinc oxide nanoparticles from the plant extract, *Wattakaka volubilis* showed anti-microbial and anti-hyperglycemic effects. *Journal of King Saud University – Science* 34: 101881.

Tai, J.; Fan, S.; Ding, S.; Ren, L. 2022. Gold nanoparticles based optical biosensors for cancer biomarker proteins: A review of the current practices. *Frontiers in Bioengineering and Biotechnology* 10: 877193.

Tingting, W.; Lei, S.; Yifei, H.; Yanqing, R.; Baizhao, Z.; Faqiong, Z. 2023. Highly sensitive photoelectrochemical and electrochemical dual-mode biosensing of polynucleotide kinase based on multifunctional $BiOBr_{0.8}I_{0.2}$/CuSCN composite and biocatalytic precipitation. *Sensors and Actuators B: Chemical* 388: 133818.

Wenli, Z.; Marjan, S.; Maryam, S.; Mohammad, J.; Hanieh, M.; Naser, A. 2021. Synthesis and application of Au NPs-chitosan nanocomposite in the treatment of acute myeloid leukemia *in vitro* and *in vivo*. *Arabian Journal of Chemistry* 14: 102929.

Woei, P.; Cheng, C. 2010. Composition. US 2010/0029544A1.

Xiaolu, C.; Rongying, M.; Zhaoming, F.; Qianhong, S.; Xingyu, L.; Yingchao, H. 2022. Metal-phenolic networks-encapsulated cascade amplification delivery nanoparticles overcoming cancer drug resistance via combined starvation/chemodynamic/chemo therapy. *Chemical Engineering Journal* 442: 136221.

Yang, C.; Wang, Q.; Xiang, Y. 2014. Target-induced strand release and thionine-decorated gold nanoparticle amplification labels for sensitive electrochemical aptamer-based sensing of small molecules. *Sensors and Actuators B: Chemical* 197: 149–154.

Yao, L.Y.; Yu, X.Q.; Zhao, Y.J. 2015. An aptamer-based chemiluminescence method for ultrasensitive detection of platelet-derived growth factor by cascade amplification combining rolling circle amplification with hydroxylamine-enlarged gold nanoparticles. *Analytical Methods* 7: 8786–8792.

Yeşeren, S.; Semra, A.; Adil, D. 2022. Preparation of magnetic nanoparticles-assisted plasmonic biosensors with metal affinity for interferon-α detection. *Materials Science and Engineering: B* 280: 115687.

Yuan, L.; Jie, L.; Min, L.; Lingling, L.; Xicun, W.; Zhengjun, Q. 2023. Iron nanoparticles encapsulated in boron-nitrogen Co-doped carbon nanotubes biomimetic enzyme for electrochemical monitoring of dopamine and uric acid in human serum. *Microchemical Journal* 185: 108184.

Zhang, Z.; Liu, S.; Shi, Y.; Peacock, D.; Yan, F. 2014. Label-free aptamer biosensor for thrombin detection on a nanocomposite of graphene and plasma polymerized allylamine. *Journal of Materials Chemistry B* 2: 1530–1538.

Zhao, P.; Li, N.; Astruc, D. 2012. State of the art in gold nanoparticle synthesis. *Coordination Chemistry Reviews* 257: 638–665.

Zhao-Peng, C.; Zhao-Feng, P.; Yan, L. et al. 2007. Successively amplified electrochemical immunoassay based on biocatalytic deposition of silver nanoparticles and silver enhancement. *Biosensors and Bioelectronics* 23: 485–491.

9 Biogenic Metals/Metal Oxides Nanocomposites as Eco-Friendly Antagonistic Agents

Marwa M. Eltarahony and Ayman Kamal

9.1 INTRODUCTION

During the last epoch, the prevalence of microbial infections has increased in a dramatic manner. To treat them, continuous misuses of antimicrobial drugs were applied, which ultimately led to the emergence of resistance phenomenon among different strains of microbes. As reported by WHO, these resistant microbes, such as bacteria, molds, yeasts, viruses, and parasites, have the capability to struggle against antimicrobial drug attacks through various metabolic resistance mechanisms. By such way, ineffective therapy results and causes persistence and the development of hospital-acquired infectious diseases (HAID). According to WHO reports, multidrug resistance (MDR) is commonly disseminated among various bacterial genera, such as *Klebsiella pneumoniae*, *Escherichia coli*, *Staphylococcus aureus*, *Neisseria gonorrhoeae*, *Salmonella* sp., *Streptococcus pneumoniae*, *Shigella* sp., and *Mycobacterium tuberculosis*. Nonetheless, it also exists among strains of *Aspergillus* spp., *Trichosporon beigelii*, *Candida* spp., *Scopulariopsis* spp., *Cryptococcus neoformans*, and *Pseudallescheria boydii*. Notably, parasitic MDR has been recorded in *Plasmodia*, *Trichomonas vaginalis*, schistosomes, *Leishmania*, *Entamoeba*, and *Toxoplasma gondii* [1]. Several adverse implications are associated with the MDR phenomena, such as elevation of mortality rates, increasing medical costs, and longer treatment periods. All such outcomes provoked scientists to develop, modify, and fabricate novel, sustainable, and effective medications.

Recently, nanoscience became an inescapable tool in modern technology that translates the traditional scientific concepts into a new, robust technological invention in day-to-day applications. Recently, inorganic metals and their oxides (MNPs/MONPs) are categorized among brilliant nanomaterials that prove to be of great interest on both scientific and technology levels. The conversion of metal from their bulk into nanoscale enhances their properties relative to the original state. Their peculiar features give them a higher significance by possessing distinct catalytic, optical, electrical, magnetic, and biological performances. However, such traits could be improved more and more by the incorporation of MNPs/MONPs in composites, which yield merits of being superior in performance, an unusual eminent combination of structural properties, as well as unique hybrid functional possibilities. Remarkably, the hybridization of MNPs/MONPs with other nanomaterial is considered as sort of modification that furnishes new hybrid nanocomposite with more stability during the processing stage and also increases their safety, which is urgently required for the ecosystem [2]. All such characteristic combinations enhance the output of any application by the dint of synergistic and mutual performance, as will be explained in this chapter [3]. Such hybrid material could be other MNPs/MONPs, organic linker, enzyme, polymer, or carbon matrix. Therefore, the main highlight of this chapter is to present a general glance on metals/oxides-based nanocomposites (MNCs) as effective biocidal agents.

9.2 METAL ORGANIC FRAMEWORKS (MOFS)

MOFs are a promising line in nanoporous science and could be described as a self-assembly of metallic ions and organic/inorganic-based ligands that act as linkers in metallic centers, whereas

DOI: 10.1201/9781003430087-9

metallic ions act as coordination centers. This configuration generates a relatively novel crystalline nanoporous matter. Remarkably, this hybrid organic–inorganic coordination networks possess outstanding traits, such as informal functionalization, considerable biodegradability, and unique biocompatibility as well, let alone their higher porosity and surface area compared to other conventional porous materials, which trigger employing them easily in drug delivery systems, as listed in extensive studies [2]. In recent years, MOFs were utilized as antimicrobial agents, as pointed out by [4], who prepared an Mg(H4gal) MOF material and employed it as antioxidant carrier and antipathogenic drug too. In the same sense, [5] studied the entrapment of AgNPs within the nanoscaled porous photoactive titanium (IV) aminoterephthalate MIL-125(Ti)NH_2 and denoted the possibility to be applied as active antibiofilm agent for coating high-touch surfaces (e.g., surgical instruments, tools, and doorknobs). Meanwhile, [6] immobilized CuNPs over cellulosic fibers and used such Cu-MOFs in eliminating the development of *E. coli* culture on agar plates and liquid medium. As observed, despite their antagonistic effectiveness, they are commonly and widely synthesized via physicochemical means. No study, till our knowledge, was reported regarding biosynthesis of MOFs. That encourages for more endeavors to optimize biological means for their fabrication. Owing to the significance of MOFs and their prominent antimicrobial capacity, it is noteworthy to give fast insinuation herein despite a nonbiological synthesis origin.

9.3 BIMETALLIC NPS (BMNPS)

Metal-based nanostructures can be classified into monometallic (one metal), bimetallic (two metals), trimetallic (three metals), and multimetallic (more than three metals), reckoning the number of metallic constituents. The enhanced functionality of bimetallic nanoformulations due to characteristic mixing patterns and geometrical architecture have gained attention in research and technological sectors. In comparison to monometallic counterparts, the particular combination of two metals triggers bimetallic NPs (BMNPs) that exhibit superior stability, catalytic activity, selectivity, electronic/optical performance, and biological efficacy than that showed by monometallic NPs [7]. In BMNPs, each metal with its outshining properties exerts particular function in this combination to implement the overall reaction mechanism efficiently. Therefore, the concurrent and synergistic impacts of two metals offer new properties with improved potentiality. Tuning of traits and performance can be accomplished by the proper selection of both metal combination and optimizing the reaction condition as well. In general, the most commonly fabricated BMNPs encompass noble transition metals, especially Ag, as shown in Table 9.1. Other metals, such as Pd, Au, and Pt, have acquired a great interest in industrial-scale manufacturing [8]. However, to obtain the required functional performance, reaction solution pH, temperature, reaction time, metal salt concentration, and reducing/capping agent concentrations must be optimized, where all these parameters considerably influence the size, architecture, morphology, and quality of as-prepared BMNPs [9].

Interestingly, different morphologies and also varied architectures could be generated from bimetallic nanocombinations [20]. According to the atom's configuration (i.e., random and ordered arrangement of atoms), BMNPs can be classified into several categories, namely, mixed, alloy, crown jewel, heterostructure, subcluster, core–shell, and multishell core–shell structures. However, based on the desired type of BMNPs, the production process can be categorized into sequential and concurrent reduction. Broadly, BMNPs as nanocomposites are prepared by physical and chemical means; nonetheless, biosynthesis of BMNPs by natural extracts has been preferred recently due to the advantages that were mentioned tremendously. Notably, the phyto-synthesis of BMNPs or metal nanocomposites is the most commonly applied biological method.

Actually, biosynthesis using microbial entities is rare due to several difficulties that were recorded by [7]. Nevertheless, [21] succeeded in the synthesis of biogenic PdPt BMNPs by *Shewanella oneidensis* under anaerobic conditions and applied in elimination of 4-nitrophenol (4-NP). On the other hand, [22] fabricated palladium/ruthenium (bio-Pd@Ru) core–shell NPs by *B. benzeovorans* using sequential reduction strategy. The catalytic activity of Pd@Ru core–shell BMNPs was examined

TABLE 9.1
Examples on Bimetallic NPs Produced by Biological Means, with Their Characteristic Features in Restricting Growth of Different Microbial Species

BMNPs Type	Biological Reducing Agent	Treated Organism	Characteristic Features	References
Ag/Cu	*Ricinus Communis* leaf extract	*S. aureus*, *E. coli*, and *A. niger*	Spheroid, 10–25 nm, SPR at 407 nm	[10]
	Majorana hortensis leaf extract	*Pseudomonas putida*, *Bacillus subtilis*, *Klebsiella pneumonia*, *S. aureus*, and *E. coli*	Spherical, 3 nm, SPR at 280 nm	[11]
Ag/Fe	*Gardenia jasminoides* leaf extract	*C. albicans*, *P. aeruginosa*, *S. aureus*, *E. coli*	Homogenously distributed core–shell, spherical, 13 nm, SPR at 425 nm	[12]
Ag-ZnO	*Mirabilis jalapa* leaf extract	*P. aeruginosa*, *S. aureus*, *E. coli*, *K. pneumoniae*, *B. subtilis*, and *leishmania*	Plates, sheets, and spherical, 19.3–67.4 nm	[13]
Ag/Co	Root extract of palmyra palm	*Culex quinquefasciatus*	SPR at 420 nm	[14]
Ag/TiO$_2$	*Origanum majorana* leaf extract	*E. coli*, *B. subtilis*, and *A. niger*	Spherical, 20–50 nm, SPR at 380 and 570 nm	[15]
Cu/Mn	*Vinca rosea* leaf extract	*K. pneumonia*, *E. coli*, *S. typhi*, and *S. aureus*	Monodispersed and spherical, 412 nm, SPR at 269 and 372 nm	[16]
Cu/Zn	*Borassus flabellifer*	*Alcaligenes faecalis*, *S. aureus*, *Citrobacter freundii*, *K. pneumoniae*, and *C. perfringens*	Spherical, 100 nm, SPR at 338 nm	[17]
ZnSe	*Gracilaria corticata*	*P. aeruginosa*, *K. pneumoniae*, *S. aureus*, *E. coli*, *S. typhi*, and *B. subtilis*	Spherical with the smooth surface, 55.5 nm, SPR at 350–400 nm	[18]
SeAg	Quercetin and gallic acid	*E. coli*, *B. subtilis*	Spherical, 30–35 nm, SPR at 400–500 nm	[19]

in converting 5-hydroxymethyl furfural (5-HMF) to 2,5-dimethyl furan (2,5-DMF), which is the fuel precursor. Interestingly, [7] produced biogenic functionalized CuFe hybrid nanocomposites by *Streptomyces cyaneofuscatus*, utilizing them as biosafe antimicrobial, antibiofilm, anticancer, and dye biosorpent agent. Several and various combinations of metals in BMNPs that were synthesized biologically, characterized and employed as biocidal agents, were tabulated in Table 9.1.

9.4 POLYMER-BASED METAL NANOCOMPOSITES (PMNCS)

Polymer/metal nanocomposite (PMNCs) can be described as an assortment of organic–inorganic hybrid nanocombinations which show innovative features generating from the synergism between the traits of the components. PMNCs are the composites that consist of polymer matrix ornamented with metal/metal oxide particles in the nanoscale, one dimension at least [23]. However, the aggregation and agglomeration are considered the major problems associated with the polymer nanocomposite. Therefore, the doping with metals in the nanorange is symbolized as a key to address functionalization and stabilization problems, which subsequently enhance polymer and metal characteristics [24]. Interestingly, the main application of polymer/metal nanocomposites emerges mainly in medical and pharmaceuticals domains, especially to keep medical tools, packaging products, surgical devices, and delivery systems safe away from hospital-acquired infections (HAIs). Generally, the utilization of metals with biocidal potency is wildly applied; nevertheless, the employment of polymers with antimicrobial efficiency has been highly urgent recently.

Remarkably, polymer/metal nanocomposite preparations can be implemented via two general approaches based on the location of the synthesized nanoparticles. The first one is *in situ*, where the polymer matrix serves as the reaction medium and the biocidal MNPs/MONPs are embedded into the polymer bulk; it is common in the preparation of polymer hydrogel nanocomposites. The second one is *ex situ*, where nanometals are synthesized firstly and thereafter incorporated into the polymer matrix as a surface coating. The polymer matrix acts in this case as the dispersion medium. This method is highly recommended in thermoplastic composites [25]. Broadly, co-evaporation and co-sputtering approaches are the most and wildly used protocols to synthesize PMNCs, as denoted by [24]. Notably, polypropylene, polyethylene, polyamide, polypyrole, polyhydroxy alkonat, polystyrene, polyvinyl alcohol, polymethyl methacrylate, polyhydroxyl butyrate, polylactic acid, and polyglycolic acid are general synthetic polymers harnessed in PMNCs for healthcare applications. On the other hand, natural polymers, or biopolymers, such as cellulose, chitosan, heparin, starch, Ɛ-Polylysine., alginate, gelatin, pectin, dextran, rubber, fibrin, and guar gum, are mostly utilized currently depending on their low cytotoxicity, biodegradability, and biocompatibility [26, 27]. As well-known, for environmental considerations, all attention is directed toward prevailing natural, renewable, eco-friendly biodegradable substances. Interestingly, the antimicrobial activity of such polymers, either synthetic or natural, can be summarized based on their charge, hydrophilic/hydrophobic property, chelating capacity, molecular weight, density, and structural constituents (e.g., amine groups). It is plausible to display different examples on PMNCs that are characterized by their antagonistic potency (Table 9.2). Noteworthy is the synthesis of such PMNCs by biological means, which seems to be rare.

9.5 ENZYME–METAL NANOCOMPOSITES (EMNCS)

In spite of the brilliant efficiency of antibiotics, they recently represent a dilemma. Therefore, effective alternatives have emerged for antibiotic-free therapies. MNPs/MONPs are not only the magic

TABLE 9.2
Examples of Different Polymers, Either Natural or Synthetic, in Their Combination with Different Metals in PMNCs in Controlling the Growth of Various Microbial Groups

Polymer Type in PMNCs	Metal NPs in PMNCs	Biocidal Activity	References
Sodium alginate and CMC	Ag	*S. aureus MTCC 740, E. coli MTCC 9492*	[28]
Chitosan-PVP	Ag-curcumin	*E. coli, P. aeruginosa, Staphylococcus* sp., *Micrococcus* sp., *C. albicans*	[29]
CMC	Ag, ZnO, CuO	*S. aureus* and *E. coli*	[30]
Starch	Ag	*E. coli, S. aureus, C. albicans*	[31]
Cellulose	Ag	*B. stearothermophilus*	[32]
Gelatin	Ag	Airborne microbes	[33]
Pectin	Ag	*E. coli, S. aureus, A. japonicus*	[34]
Natural rubber latex	Ag	*E. coli, S. aureus*	[35]
Polyacrylic acid	Zn	*S. aureus, P. aeruginosa*	[36]
Galactomannan and carrageenan	Ag	*S. aureus ATCC 25923, P. aeruginosa ATCC 27853, E. coli ATCC 25922, B. subtilis*	[37]
Carrageenan	TiO_2	*E. coli, S. aureus*	[38]
Carrageenan	ZnO, CuO	*E. coli, L. monocytogenes.*	[39]
Chitosan	Fe	*E. coli, S. aureus, P. aeruginosa*	[40]
Chitosan	Zn-Cu	*R. solani, A. alternata, B. cinerea*	[41]
Chitosan	TiO_2	*Colletotrichum gloeosporioides, Cladosporium oxysporum, P. steckii*	[42]

key for replacing antibiotics, but enzymes also currently prove their catalytic function in pathogen conquering. Enzymes possess a prominent biocompatibility and outstanding catalytic efficiency. As found by [43, 44], they show antimicrobial and antiviral properties, owing to their synergistic effect and diminishing toxicity influence. However, to efficiently express their role as bifunctional materials, enzyme functionality must not be hindered by metal coupling, where in some cases the presence of MNPs/MONPs in the active site of enzymes causes deactivation. Thus, the preparation method of these nanocomposites is critical, and a balance should be implemented between metals and enzymes. Nowadays, the co-immobilization and *in situ* reduction are the two main synthesis means of EMNCs to achieve better catalytic performance with maintaining enzyme stability and functionality. In the co-immobilization approach, both enzymes and metal nanoparticles are combined on the same carrier, which could be metal–organic frameworks (MOFs), polymeric matrices, mesoporous silica, etc. On the other hand, enzymes mediate reduction and stabilization of MNPs by acting as templates or reductants in *in situ* approach [45]. Interestingly, [46] utilized lysozyme-Ag-NCs as antimicrobial agent to suppress the growth of silver-resistant *P. mirabilis* strains. In addition, [47] prepared proteinase K-Au-NCs, investing the antimicrobial properties of proteinase K in inducing proteolytic digestion of extracellular macromolecules and causing physicochemical alterations in bacterial cell walls. The functionalization of AuNPs with proteinase K yields lethal dispersion in *Pseudomonas* biofilms, relative to each sole component. In the same sense, bovine lactoferrin (LTF) immobilized with AgNPs in EMNCs manifested eminent antagonistic functionality in opposing *P. aeruginosa*, *S. aureus*, and MRSA [48].

9.6 CARBON-BASED METALS/METAL OXIDES NANOCOMPOSITES (CMNCS)

Carbon-based nanomaterials (CBN) are gaining a lot of momentum in both sectors of fundamental theory and technological applications. That could be assigned to their role as a template or supporting material for metal deposition, besides showing superior high surface area, mechanical strength, advantageous heat transport, high electrical conductivity, and biocidal potency as well. Interestingly, nanodiamonds, carbon nanotubes (CNTs), fullerenes, graphene (G) with its derivatives of reduced graphene oxide (rGO), graphene oxide (GO), and carbon-based quantum dots, are categorized in the carbon family [49, 50]. Conclusively, a plethora of studies published on databases were conducted in graphene with its derivatives, owing to their superior biocompatibility conjugated with its strongest structure of two-dimensional one-atom-thick sheet of carbon. Moreover, the antagonistic potency of them was recorded tremendously. The studies ascribed such potency to several physical and chemical inhibition scenarios. Regarding physical inhibition mechanisms, it could be achieved by the direct interaction of the G sharp boundaries with microbial membranes, which is called "sharp edge–mediated insertion" or "insertion mode of action". Via this pathway, the sharp edges of graphene nanosheets serve as knives or blade to injure the bacterial cell, which is followed by intracellular material leakage or lipid extraction. Interestingly, the lipophilic surface of graphene edges break the protein–protein bonding of membranes, leading to functional failure [51].

Generally, graphene and its derivatives mediate as excellent support material for a vast array of metals, such as Cu, Pt, Ag, Ni, Au, and Pd, in several applications. The synergism and mutual association of characteristics derived from both CBN and MNPs/MONPs in CMNCs undoubtedly improve their catalytic activity and boost their functionality. Albeit physicochemical methods are widely applied, certain studies have achieved biological fabrication of CMNCs successfully. Furthermore, [52] impregnated an rGO matrix with NiO, AgO, and ZnO to fabricate rGO-AgO, rGO-NiO, and rGO-ZnO nanocomposites and test their biocidal potency against a wide spectrum of planktonic and biofilm pathogens. The ultrastructure studies of *Cryptococcus neoformans* before and after rGO-NiO-NCs denoted lysis of the outer cellular surface, with clear damage of the fungal cells, as illustrated by SEM. Moreover, [49] summarized various researches addressing the hybridization of polymers (natural or synthetic) with graphene immobilized with metals, yielding innovative polymer-graphene-metals nanocomposites with promising features and outstanding applications.

9.7 ANTIMICROBIAL MECHANISMS OF NPS

Despite the promising results of metal-based NPs, there is a debate regarding the exact mechanisms involved in this process. The extensive utilization of metal-based NPs has led to a high number of literatures interpreting the biocidal potentiality of NPs, where metal NPs can alter the metabolic activity of microbes and also change their resistance capacity to biocidal agents [53], let alone their capacity to suppress biofilm generation based on blocking the expression of biofilm metabolic genes [54]. Furthermore, alterations in microbial membrane permeability, protein deactivation, electrolyte balance disorders, enzyme prohibition, lipid peroxidation, and heterogeneous alterations in algae, viruses, and parasites were also recorded. Therefore, different instrumentations were used to study how inhibition occurs, like electron spin resonance, proteomics tools, EDAX, FTIR analysis, liquid chromatography–mass spectrometry, and electron microscopy transmission (TEM)/scanning (SEM). Herein the antagonistic strategies against various microbial taxa, followed by metal-based NPs, will be discussed. The scenario begins by the contact and adsorption of NPs on microbial cells by various means, such as electrostatic attraction [55], hydrophobic interactions [56], receptor–ligand [57], and van der Waals forces [58]. Thereafter, the metal ions are slowly released from NPs and bind with the functional groups of the cell membrane carrying a negative charge (e.g., carboxyl, hydroxyl, phosphate, amino, mercapto groups, etc.), which is Coulomb gravity–dependent [59]. After binding and diffusion, release and penetration occur, and the deterioration tactic initiates via either non-oxidative mechanisms or oxidative stress induction, as demonstrated graphically in Figure 9.1.

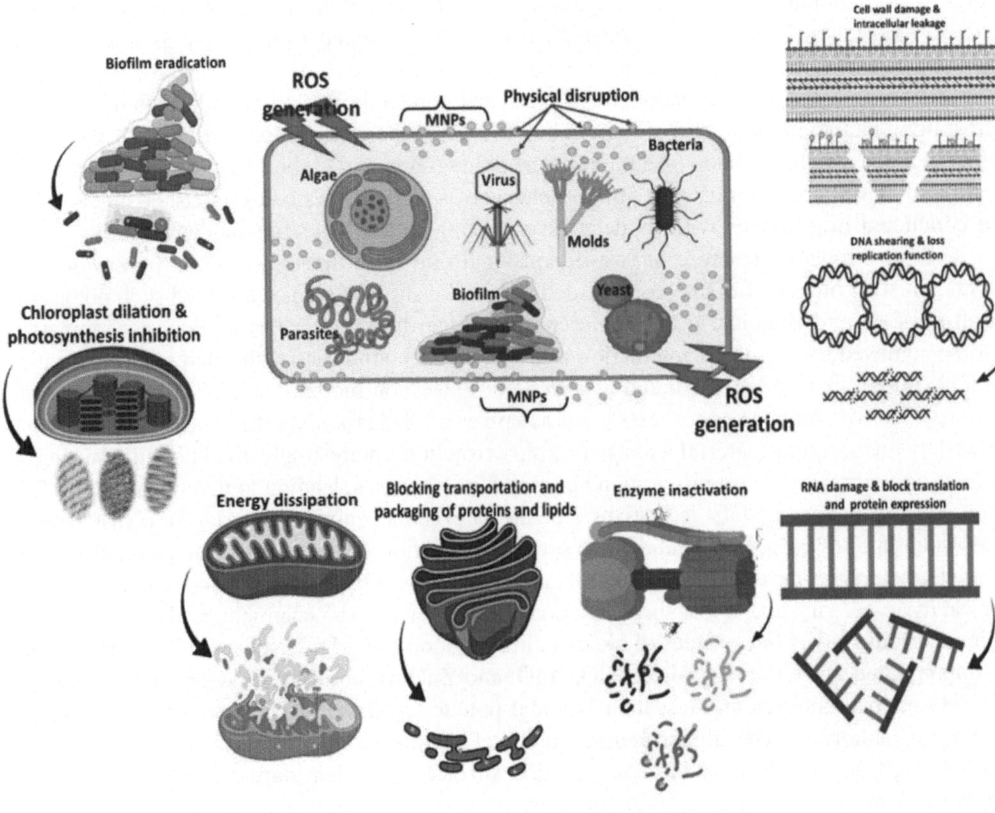

FIGURE 9.1 Schematic diagram representing the biocidal effects of MNPs/MONPs on different microbial forms.

9.7.1 Non-Oxidative Mechanisms

9.7.1.1 Destruction of Cell Barriers (Cell Wall and Cell Membrane)

The most vital defensive barriers for any cell are the cell wall and membrane, which protect the cells against the external environment and maintain the cell's natural shape [60]. Interestingly, the cell wall of gram-negative bacteria consists of a small layer of peptidoglycan which is covered by a large external layer of negatively charged lipopolysaccharide (LPS), lipoproteins, and phospholipids, preventing the penetration of only macromolecules, whereas the cell wall of gram-positive bacteria is mainly of peptidoglycan that is decorated with glycopolymers, characterized by the presence of pores [61]. In some NP treatments, the LPS of gram-negative bacteria attracts a lot of NPs, causing several damages to the cell wall, followed by cell death. In contrast, the abundance of pores in teichoic acid permits the entrance of foreign molecules, and in other cases, NPs bind to the molecular chain of phosphate, blocking their aggregation. Enormous studies showed that NPs exhibit greater biocidal potency against gram-positive bacteria than that in gram-negative bacteria, as reflected by [61]. In parallel, NPs also disturb both composition and structure of the cell membrane, thereby lowering membrane integrity and further deteriorating the respiratory performance of the cell membrane, which is considered to be its main function. Additionally, leakage of cellular contents, dissipation of proton motive force, generation of osmotic imbalance, and the diminishing of ATP intracellular levels were also detected. [62] studied the effect of NPs on *E. coli* growth by measuring the reduction in 2,3,5-triphenyltetrazolium chloride or the uptake of O_2, affirming the lethal effect of AgNPs by the elevation of membrane permeability and damaging of respiratory chain. In the same extent, X-ray photoelectron spectroscopy and attenuated total reflection (ATR)-FTIR analysis were used to determine the saturated oxidation of fatty acid and unsaturated fatty acid in the membrane after NP treatments [63, 64].

9.7.1.2 Interaction of NPs with Intracellular Biomolecules and Gene Expression Regulation

Once NPs pass through cell barriers into the cell internally, NPs interact with functional groups of amino acids, especially SH group of amino acids, and formation of S–S bonds, which induce molecular changes that lead to protein inactivation and ribosomes denaturation [65]. Besides, NPs could replace or interact with essential ions of biomolecules (e.g., oxygen, nitrogen, and sulfur) at their active sites, either specifically and/or selectively, in a process called ionic/molecular mimicry. In this process, NP–protein complexes are generated in an altered conformational structure, leading to protein malfunction. CuO-NPs change significantly the expression of key proteins involved in nitrogen metabolism and electron transfer during the denitrification process, in particular nitrate/nitrite reductases [66]. However, [67] found that AgNPs impacted adversely on the expression of ribosomes and other metabolic enzymes of *E. coli*, as revealed by two-dimensional electrophoresis (2-DE), energy-filtering TEM (EFTEM), and matrix-assisted laser desorption ionization time-of-flight MS (MALDI-TOF MS).

However, in the genetic level, upon binding of NPs with the phosphorus group of DNA, DNA compression, shearing, denaturation, fragmentation, losing its replication ability, inhibiting DNA unwinding, and ultimately interruption in cell division were displayed. As pointed out by [68], TiO_2 NPs showed a significant affinity to G–C reach DNA, as predicted by molecular docking.

9.7.2 Oxidative Stress

ROSs are species of oxygen that have strong positive redox potential and are generated during metabolism. The intracellular defense mechanisms, under ordinary conditions, are originated to adapt with undesired stress to prevent biomolecule damage, whereases in the presence of unfavorable stress circumstances, oxygen passes through undesired reduction states and converts to free radicals rather than water. Such types of ROS include singlet oxygen (O_2), the hydroxyl radical (OH), hydrogen peroxide (H_2O_2), and the superoxide radical (O_2^-), which show various degrees of dynamics and activity [69]. NPs have been shown to produce free radicals that induce massive oxidative stress. However, different types of NPs can generate various ROSs, where zinc oxide NPs can produce H_2O_2 and OH that cause less-acute stress and can be easily neutralized by antioxidants (e.g., catalase and superoxide enzymes) that occur endogenously. On the other hand, MgO-NPs and CaO-NPs produce O_2^-, while CuO-NPs generate four types of ROSs, as referred by [61]. Remarkably, the hydroxyl and superoxide radicals can be retained extracellularly on the cell without penetration inside the cells. On the other hand, peroxide-free radical can cross through the cell membrane. ROS has been acted extracellularly and intracellularly. Wherever their location, ROS triggers lethal impacts on microbial biomolecules via Fenton reaction, which eventually prevents microbial proliferation [70].

9.7.3 Antibiofilm, Antialgal, Antiviral, and Antiparasitic Effects of NPs

The architecture of biofilms makes microbes highly tolerant to foreign chemicals. Recent reports declare the interfering influence of MNPs with biofilm integrity upon interacting with EPSs [71]. MNPs constrain the production of the slimy extracellular polymeric matrix (EPS), which is composed of DNA, proteins, lipoproteins, humic acid, and polysaccharides. Interestingly, [72] stated that the bacterial cells located in the center can survive and withstand stress factors than that in the edges, by virtue of nutrient flow. Nevertheless, several evidences show that MNPs influence not only the rates of microbial metabolic activity but also their adhesion to the inert surfaces and the EPS formation [73]. Meanwhile, MNPs exert significant impact on deteriorating ion channels and long-distance electrical signaling among microbial cells within the biofilm (i.e., quorum sensing). As referred previously, MNPs adhere to and diffuse easily into the biofilm, leading to disordering of membrane potential, improving lipid peroxidation, protein denaturation, ROS generation, osmotic imbalance, and DNA binding, then finally stopping the normal functioning of main physiological processes and biofilm inhibition/eradication.

However, numerous studies have recorded the negative impact of MNPs on algal proliferation [74]. The preponderant toxicity pathway of NPs against algae is through ROS generation, which subsequently causes lipid peroxidation and affects membrane selectivity, permeability, fluidity, and integrity, as referred by [75]. Besides, ROS accumulated in chloroplast results in diminishing of chlorophyll content through changing the lipid–protein ratio of the pigment–protein complexes. In addition, the energy transduction in light reactions was interrupted, hence impairing algal photosynthesis. Regarding virucidal capacity, [76] declared that MNPs inhibit the binding of virus and its fusion to the cells, besides interacting with disulfide bonds located in the glycoproteins, which consequently inhibit viral transmission, as supported by high-angle annular dark field (HAADF) and scanning transmission electron microscopy (STEM) analysis. Moreover, NPs also can restrain viral growth at an early stage of viral replication.

In the same sense, MNPs were studied as antiparasitic agents and proved their efficiency. The weakening in the structure of lipophosphoglycan and glycoprotein molecules that are presented on the parasite surface was mentioned. Additionally, NPs can diffuse into the cell via pores located in cell membrane and ion channels/transporter proteins or through endocytosis, causing exact deterioration features, as recorded previously [77].

9.8 PARAMETERS INFLUENCING BIOCIDAL ACTIVITY OF NPS

The effectiveness of NPs as antimicrobial agents is regulated by several factors, either based on NP properties, the target microbe, or related to environmental conditions surrounding them. The physicochemical traits of NPs encompass surface size, particle shape, chemical structure, surface charge, roughness, and particles hydrophobicity. However, the microbial strain type, the microbe concentration, the microbe defense mechanism, and the time of exposure are the substantial parameters that limit biocidal efficiency. Notably, a large surface area with high surface energy and deficient capping properties can result in NP aggregation and lower activity [61]. Hence, it is plausible to discuss briefly such factors and summarize them graphically, as demonstrated in Figure 9.2.

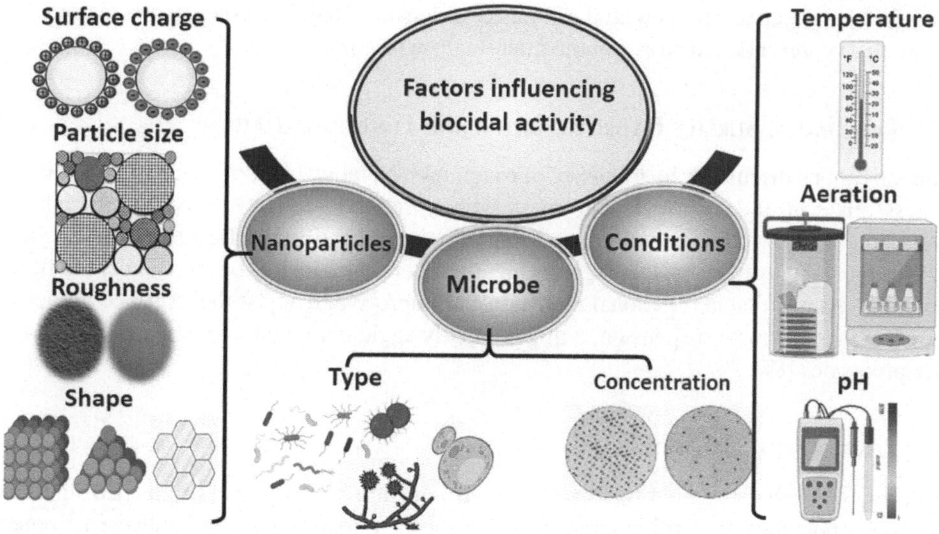

FIGURE 9.2 Graphical illustration declaring the effect of different factors on biocidal efficiency of NPs.

9.8.1 Surface Size and Particle Shape Characteristics

The concept that small NPs seem to be more toxic than large NPs is owed to the fact that small NPs have relatively larger surface-area-to-volume ratio compared to larger NPs. This consequently led to a higher opportunity of direct contact with microbial cell membrane and to easily penetrate it, which could elevate the generation of ROS and, eventually, cellular damage [78]. However, some studies reflect that the smallest Mg $(OH)_2$ NPs exhibited the weakest antibacterial impact [79]. The size is not merely the conclusive factor in antagonistic activity. The association of other properties is collectively influential. Generally, spherical NPs are the most common shape, despite other morphologies, such as rods, cubes, tubes, wires, plates, triangles, polygons, and sheets having also been reported in biocidal activity, but with varied degrees [61]. Remarkably, [80] examined the biocidal influence of spherical, triangular shapes, and rod-like NPs against *E. coli*, in a comparative manner. Probably, the triangular shape gives a higher positive charge to the nanoparticles, which ensures greater activity. Besides, [81] revealed that rods and nanocubes of CeO_2 NPs displayed higher efficiency than other examined shapes. The authors attributed that to the effect of the exposed planes and also the oxidation state of the metals and confirmed that hypothesis through the analysis of the exposed crystal facets, which proposed that less-stable planes require less energy to form oxygen vacancies. Via such manner, the stability of the planes exerts an effect in biocidal activity. Additionally, [82] considered NP surface defects (increased abrasiveness) and the presence of corners and edges in increasing microbial toxicity.

9.8.2 Surface Charge Characteristics

The distribution, adhesion, and metabolism of nanoparticles are also impacted by their surface charge characteristics and zeta potential values. Electrostatic forces established when nanoparticles carrying a positive ζ-potential encounter microbial cell surface with a negative surface charge enhance a deep attraction and closer interaction between both entities and possibly the penetration in their membranes. Actually, cationic NPs are advantageous in preventing microbial growth by limiting their attachment and, thereafter, ROS generation, compared to negatively charged or neutral NPs. As mentioned by [83], NPs can reach key structures via ion exchange, whereases at higher doses, NPs with negative zeta potential offer a certain level of biocidal potency owing to molecular crowding that represents as a stress on the microbial surface [84].

9.8.3 Roughness, Surface Hydrophobicity, and Functionalization

Few investigations discussed the influence of roughness. Notably, [85, 86] stated that the higher NP roughness raises up the surface-area-to-mass ratio, which consequently enhances the adsorption and adhesion on cell wall, eventually increasing microbial inactivation. Besides, surface hydrophilicity/hydrophobicity of NPs display a prominent effect on its kinetics *in vivo*. Thus, the functionalization of NPs in suitable functional molecule can improve and adjust their surface properties. If NPs present in inadequate suspension, it might quickly agglomerate into larger particles and lose its surface properties [87].

9.8.4 Environmental Conditions

A vast array of studies revealed that various environmental circumstances influence significantly antimicrobial potential. Regarding temperature, it shows a great impact on antibacterial potentiality, owing to its implications on the ROS production rate. As an example, [88] found that the exposure of ZnONPs to temperature promoted capturing of electrons at the active sites, which would interact with oxygen, generating ROS, thus improving the antimicrobial potentiality. On the other hand, pH exerts a noticeable effect against microbial growth, as elucidated by [89]. The acidic conditions (3.5–5.8) promote the dissolution rate of ZnONPs, which leads to greater adhesion to the bacterial surface through higher electrostatic interactions, thereafter higher biocidal properties [90]. However, [91] found that the higher solubility of AgNPs occurred in acetic acid (pH 4) than in neutral water (pH 7), which enhanced the oxidative dissolution mechanism and consequently higher interaction of Ag^+ with dissolved oxygen and protons. In general, the characteristics of the solution, like osmotic pressure and pH, exert a significant impact on surface charge, NP solubility, and their aggregation. Further, the mechanical stirring could control ZnONPs physicochemical properties that lead to enhance their antimicrobial activity against *C. albicans*, gram-positive, and gram-negative bacteria, as stated by [92], assigning that to the enhanced release of free Zn ions and zinc complexes. Nonetheless, the abundance of considerable concentrations of NPs on the microbial surface stimulates efficient interaction and facilitates antagonistic activity [65]. That could be explained by higher release of metals ions. The concomitance of longer incubation time in NP treatment also contributes to longer release of metal ions and so decreases microbial viability [93].

9.9 FUTURE PERSPECTIVES OF METAL-BASED NANOMATERIALS

Considerable experimentation and noticeable progress have been accomplished in the recent decades in fields relevant to metallic-based nanomaterials. Overall properties inherent in metallic-based nanomaterials, biological properties in particular, trigger these types of nanomaterials, promising candidates to maximize their effectiveness in a vast array of applications. Although there are huge literature and research articles citing the employment of metallic-based nanomaterials in various applications, there are comparatively fewer studies for their biological applications, especially in the beating parasites and viruses. Also, the biological route of synthesis suffers from shortage in research, particularly those related to polymer/carbon/enzyme-based metallic nanomaterials. Eventually, this shortcoming, coupled with emerging novel biomolecules and new microbial/plant species from extreme or unique conditions, will fill the gap in metallic-based nanocomposite synthesis and technological advances in a significant manner. Upon the synthesis and utilization of such distinctive dual/multicomponent nanomaterials, the near future will definitely endure to present innovative ennoblement. Currently, the recruitment of green synthesis approaches is highly appreciated, owing to their eco-friendly nature, biocompatibility, safety, and low generation of hazards, along with their proficiency. In the future, the technical advances in green synthesis methods boost using new biological moieties, either plant or microbial entities, to open an adequate path for scientists, researchers, and technologists to develop and explore novel nanocomposites of polymers, carbon-based materials, and enzymatic matrices in conjugation with BMNPs or even multimetallic NPs in a more stable, functional, and acceptable formulas.

9.10 CONCLUSION

Metal-based nanomaterials are characterized by possessing a commanding role in biocidal applications. Soaring the antagonistic efficiency of benign biosafe nanomaterials could be implemented via different methods. Preparation of bimetallic/trimetallic/multimetallic nanocomposites, impregnation with organic–inorganic framework, polymeric matrix, biocidal enzyme conjugation, wrapping with carbon-based nanomaterials are among several methods that can fulfill the intended target. However, the achievement of all such options via phyto-synthesis or microbial synthesis methods is still rare. We believe that the detailed perceiving of the basic theory involved in the synthesis methods, optimizing synthesis conditions and better selection of reductants/capping bioagents, facilitates the production of desired nanocomposites with controlled and enhanced properties and effective applicability. Generally, this chapter aims to furnish a comprehensive schematization of the biocidal potency of metals/oxide-based nanocomposites, which are currently fabricated by biogenic methods. Besides, it gives insinuations about various modifications that could be synchronized with them to attain maximum performance. Such modifications are recommended to be implemented by biogenic methods to gain all benefits of novel nanostructure formulas and eco-friendliness, safety, easiness, and biocompatibile features. The amalgamation of both targets concurrently would trigger the researchers' endeavors more easily to reduce the gap between the lab scale and commercialization of new technologies.

ACKNOWLEDGMENTS

The graphical illustrations were performed via biorender.com and without permission from biorender.com.

CONFLICT OF INTEREST

The authors declare that they have no conflicts of interest.

REFERENCES

1. Rodríguez, H.S.; Hinestroza, J.P.; Ochoa-Puentes, C.; Sierra, C.A.; Soto, C.Y. **2014**. Antibacterial activity against *Escherichia coli* of Cu-BTC (MOF-199) metal-organic framework immobilized onto cellulosic fibers. *Journal of Applied Polymer Science* 131: 40815–40820.
2. Tanwar, J.; Das, S.; Fatima, Z.; Hameed, S. **2014**. Multidrug resistance: An emerging crisis. *Interdisciplinary Perspectives on Infectious Diseases* 2014.
3. Owonubi, S.J.; Malima, N.M.; Revaprasadu, N. **2020**. Metal oxide–based nanocomposites as antimicrobial and biomedical agents. In *Antibiotic Materials in Healthcare*. Elsevier, USA: Academic Press, 287–323.
4. Yaqoob, A.A.; Ahmad, H.; Parveen, T. et al. **2020**. Recent advances in metal decorated nanomaterials and their various biological applications: A review. *Frontiers in Chemistry* 19: 341.
5. Hidalgo, T.; Cooper, L.; Gorman, M. et al. **2017**. Crystal structure dependent *in vitro* antioxidant activity of biocompatible calcium gallate MOFs. *Journal of Materials Chemistry B* 5: 2813–2822.
6. Arenas-Vivo, A.; Amariei, G.; Aguado, S.; Rosal, R.; Horcajada, P. **2019**. An Ag-loaded photoactive nano-metal organic framework as a promising biofilm treatment. *Acta Biomaterialia* 97: 490–500.
7. Eltarahony, M.; Abu-Serie, M.; Hamad, H.; Zaki, S.; Abd-El-Haleem, D. **2021**. Unveiling the role of novel biogenic functionalized CuFe hybrid nanocomposites in boosting anticancer, antimicrobial and biosorption activities. *Scientific Reports* 11: 1–22.

8. Perdikaki, A.; Galeou, A.; Pilatos, G.; Prombona, A.; Karanikolos, G. **2018**. Ion-based metal/graphene antibacterial agents comprising monoionic and bi-ionic silver and copper species. *Langmuir* 34: 11156–11166.
9. Belenov, S.; Volochaev, V.; Pryadchenko, V. et al. **2017**. Phase behavior of Pt-Cu nanoparticles with different architecture upon their thermal treatment. *Nanotechnology Russia* 12: 147–155.
10. Arora, N.; Thangavelu, K.; Karanikolos, G. **2020**. Bimetallic nanoparticles for antimicrobial applications. *Frontiers in Chemistry* 28: 412.
11. Omajali, J.; Gomez-Bolivar, J.; Mikheenko, I. et al. **2019**. Novel catalytically active Pd/Ru bimetallic nanoparticles synthesized by Bacillus benzeovorans. *Scientific Reports* 9: 1–20.
12. López-Ubaldo, F.; Sánchez-Mendieta, V.; Olea-Meja, O.; María, S.; Luckie, R. **2020**. Biosynthesis of Ag/Cu bimetallic nanoparticles using Ricinus communis and their antibacterial and antifungical activity. *Advances in Natural Sciences: Nanoscience and Nanotechnology* 11: 025018.
13. Singh, I.; Mazhar, T.; Shrivastava, V.; Tomar, R.S. **2022**. Bio-assisted synthesis of bi-metallic (Ag-Zn) nanoparticles by leaf extract of Azadirachta indica and its antimicrobial properties. *International Journal of Nano Dimensions* 13: 168–178.
14. Bhardwaj, D.; Singh, R. **2021**. Green biomimetic synthesis of Ag–TiO_2 nanocomposite using Origanum majorana leaf extract under sonication and their biological activities. *Bioresources and Bioprocessing* 8: 1–20.
15. Ahmad, M.; Kotb, H.; Mushtaq, S.; Waheed-Ur-Rehman, M.; Maghanga, C.M.; Alam, W. **2022**. Green synthesis of Mn+ Cu bimetallic nanoparticles using vinca rosea extract and their antioxidant, antibacterial, and catalytic activities. *Crystals* 12: 72.
16. Merugu, R.; Gothalwal, R.; Deshpande, P.; De Mandal, S.; Padala, G.; Chitturi, K. **2021**. Synthesis of Ag/Cu and Cu/Zn bimetallic nanoparticles using toddy palm: Investigations of their antitumor, antioxidant and antibacterial activities. *Materials Today: Proceedings* 1: 99–105.
17. Khatak, S.; Wadhwa, N.; Jain, P. **2021**. Monometallic zinc and bimetallic Cu-Zn nanoparticles synthesis using stem extracts of Cissusquadrangularis (Haddjod) and proneness as alternative antimicrobial agents. *Biosciences Biotechnology Research Asia* 17: 763–774.
18. Mittal, A.; Kumar, S.; Banerjee, U.C. **2014**. Quercetin and gallic acid mediated synthesis of bimetallic (silver and selenium) nanoparticles and their antitumor and antimicrobial potential. *Journal of Colloid and Interface Science* 1: 194–209.
19. Sathya, K.; Saravanathamizhan, R.; Baskar, G. **2018**. Ultrasonic assisted green synthesis of Fe and Fe/Zn bimetallic nanoparticles for invitro cytotoxicity study against HeLa cancer cell line. *Molecular Biology Reports* 45: 1397–1404.
20. Merugu, R.; Nayak, B.; Chitturi, K.; Misra, P. **2021**. Bimetallic silver and copper nanoparticles synthesis, characterization and biological evaluation using aqueous leaf extracts of Majorana hortensis. *Materials Today: Proceedings* 44: 2454–2458.
21. Padilla-Cruz, A.L.; Garza-Cervantes, J.A.; Vasto-Anzaldo, X.G. **2021**. Synthesis and design of Ag–Fe bimetallic nanoparticles as antimicrobial synergistic combination therapies against clinically relevant pathogens. *Scientific Reports* 11: 1–10.
22. Kamli, M.; Srivastava, V.; Hajrah, N. **2021**. Phytogenic fabrication of Ag–Fe bimetallic nanoparticles for cell cycle arrest and apoptosis signaling pathways in Candida auris by generating oxidative stress. *Antioxidants* 10: 182.
23. Folarin, O.; Sadiku, E.; Maity A. **2011**. Polymer-noble metal nanocomposites. *International Journal of the Physical Sciences* 6: 4869–4882.
24. Palza, H. **2015**. Antimicrobial polymers with metal nanoparticles. *International Journal of Molecular Sciences* 16: 2099–3116.
25. Razak, N.; Mohamed, R. **2021**. Antimicrobial sustainable biopolymers for biomedical plastics applications–an overview. *Polimery* 66: 574–583.
26. Zahran, M.; Marei, A. **2019**. Innovative natural polymer metal nanocomposites and their antimicrobial activity. *International Journal of Biological Macromolecules* 1: 586–596.
27. Mahira, S.; Jain, A.; Khan, W.; Domb, J. **2019**. Antimicrobial materials – an overview: In antimicrobial materials for biomedical applications. *Royal Society of Chemistry* 1–37.
28. Noori, F.; Megoura, M.; Labelle, M.; Mateescu, M.; Azzouz A. **2022**. Synthesis of metal-loaded carboxylated biopolymers with antibacterial activity through metal subnanoparticle incorporation. *Antibiotics* 11: 439.
29. Lin, B.; Luo, Y.; Teng, Z.; Zhang, B.; Zhou, B.; Wang, Q. **2015**. Development of silver/titanium dioxide/chitosan adipate nanocomposite as an antibacterial coating for fruit storage. *LWT-Food Science and Technology* 63: 1206–1213.

30. Peighambardoust, J.; Peighambardoust, S.; Pournasir, N.; Pakdel, P.M. **2019**. Properties of active starch-based films incorporating a combination of Ag, ZnO and CuO nanoparticles for potential use in food packaging applications. *Food Package and Shelf Life* 22: 100420.
31. Vivekanandhan, S.; Christensen, L.; Misra, M.; Mohanty, A. **2012**. Green process for impregnation of silver nanoparticles into microcrystalline cellulose and their antimicrobial bionanocomposite films. *Journal of Biomaterials and Nanobiotechnology* 3: 371–376.
32. Kumar, S.; Mitra, A.; Halder D. **2017**. Centella asiatica leaf mediated synthesis of silver nanocolloid and its application as filler in gelatin based antimicrobial nanocomposite film. *LWT – Food and Science Technology* 75: 293–300.
33. Percival, S.; Slone, W.; Linton, S.; Okel, T.; Corum, L.; Thomas, J. **2011**. The antimicrobial efficacy of a silver alginate dressing against a broad spectrum of clinically relevant wound isolates. *International Wound Journal* 8: 237–243.
34. Rathnayake, I.; Ismail, H.; Azahari, B.; Darsanasiri, N.; Rajapakse, S. **2012**. Synthesis and characterization of nano-silver incorporated natural rubber latex foam. *Polymer-Plastics Technology and Engineering* 51: 605–611.
35. James, C.; Pugh, T.; Johnson, A.; Jenkins, A. **2011**. An antimicrobial zinc-based molecule for cross linking poly-acrylic acid. *European Polymers Journal* 47: 1338–1345.
36. Lesnichaya, M.; Aleksandrova, G.; Feoktistova, L.; Sapozhnikov, A.; Fadeeva, T.; Sukhov, B.; Trofimov, B. **2010**. Silver-containing nanocomposites based on galactomannan and carrageenan: Synthesis, structure, and antimicrobial properties. *Russian Chemical Bulletin* 59: 2323–2328.
37. Majinyari, D. **2021**. Synthesis, characterization and antimicrobial evaluation of carrageenan/TiO_2 nanocomposite materials. PhD dissertation, The University of Dodma.
38. Goel, A.; Meher, M.; Gupta, P.; Gulati, K.; Pruthi, V.; Poluri, K. **2019**. Microwave assisted κ-carrageenan capped silver nanocomposites for eradication of bacterial biofilms. *Carbohydrate Polymers* 15: 854–862.
39. Ebrahimi, Y.; Peighambardoust, S.; Peighambardoust, S.; Karkaj, S. **2019**. Development of antibacterial carboxymethyl cellulose-based nanobiocomposite films containing various metallic nanoparticles for food packaging applications. *Journal of Food Science* 84: 2537–2548.
40. Azizi-Lalabadi, M.; Ehsani, A.; Divband, B.; Alizadeh-Sani, M. **2019**. Antimicrobial activity of titanium dioxide and zinc oxide nanoparticles supported in 4A zeolite and evaluation the morphological characteristic. *Scientific Reports* 9: 1–10.
41. Xing, Y.; Yi, R.; Yang, H. et al. **2021**. Antifungal effect of chitosan/nano-TiO_2 composite coatings against *Colletotrichum gloeosporioides*, *Cladosporium oxysporum* and *Penicillium steckii*. *Molecules* 26: 4401.
42. Xiong, J.; Cai, X.; Ge, J. **2022**. Enzyme–metal nanocomposites for antibacterial applications. *Particuology* 1: 134–139.
43. Adlerova, L.; Bartoskova, A.; Faldyna, M. **2008**. Lactoferrin: A review. *Veterinarni Medicina* 53: 457–468.
44. Eby, D.; Schaeublin, N.; Farrington, K.; Hussain, S.; Johnson, G. **2009**. Lysozyme catalyzes the formation of antimicrobial silver nanoparticles. *ACS Nano* 3: 984–994.
45. Li, L.-L.; Wang, H. **2013**. Enzyme-coated mesoporous silica nanoparticles as efficient antibacterial agents *in vivo*. *Advanced Healthcare Materials* 2: 1351–1360.
46. Habimana, O.; Zanoni, M.; Vitale, S. **2018**. One particle, two targets: A combined action of functionalised gold nanoparticles, against Pseudomonas fluorescens biofilms. *Journal of Colloid and Interface Science* 15: 419–428.
47. Pomastowski, P.; Sprynskyy, M.; Zuvela, P. et al. **2016**. Silver-lactoferrin nanocomplexes as a potent antimicrobial agent. *Journal of the American Chemical Society* 138: 7899–7909.
48. Díez-Pascual, A.; Luceño-Sánchez, J. **2021**. Antibacterial activity of polymer nanocomposites incorporating graphene and its derivatives: A state of art. *Polymers* 13: 2105.
49. Muthoosamy, K.; Bai, R.; Manickam, S. 2017. Graphene metal nanoclusters in cutting-edge theranostics nanomedicine applications. In *Advances in Biomaterials for Biomedical Applications*, Vol. 66. A. Öchsner, L. Silva, H. Altenbach (eds.). Springer, 429–477.
50. Patel, K.; Singh, R.; Kim, H. **2019**. Carbon-based nanomaterials as an emerging platform for theranostics. *Materials Horizons* 6: 434–469.
51. Han, R.; Song, X.; Wang, Q. et al. **2019**. Microbial synthesis of graphene-supported highly-dispersed Pd-Ag bimetallic nanoparticles and its catalytic activity. *Journal of Chemical Technology & Biotechnology* 94: 3375–3783.
52. Chatzimitakos, T.; Stalikas, C. **2016**. Qualitative alterations of bacterial metabolome after exposure to metal nanoparticles with bactericidal properties: A comprehensive workflow based on 1H NMR, UHPLC-HRMS and metabolic databases. *Journal of Proteome Research* 15: 3322–3330.

53. Madivoli, E.; Kareru, P.; Maina, E.; Nyabola, A.; Wanakai, S.; Nyang'au, J. **2019**. Biosynthesis of iron nanoparticles using Ageratum conyzoides extracts, their antimicrobial and photocatalytic activity. *SN Applied Sciences* 1: 1–11.
54. Li, H.; Chen, Q.; Zhao, J.; Urmila, K. **2015**. Enhancing the antimicrobial activity of natural extraction using the synthetic ultrasmall metal nanoparticles. *Scientific Reports* 5: 11033.
55. Luan, B.; Huynh, T.; Zhou, R. **2016**. Complete wetting of graphene by biological lipids. *Nanoscale* 8: 5750–5754.
56. Gao, W.; Thamphiwatana, S.; Angsantiku, P.; Zhang, L. **2014**. Nanoparticle approaches against bacterial infections. *Wires Nanomedicine and Nanobiotechnology* 6: 532–547.
57. Armentano, I.; Arciola, C.; Fortunati, E. et al. **2014**. The interaction of bacteria with engineered nanostructured polymeric materials: A review. *Scientific World Journal* 2014: 410423.
58. Jung, W.; Koo, H.; Kim, K.; Shin, S.; Kim, S.; Park, Y. **2008**. Antibacterial activity and mechanism of action of the silver ion in *Staphylococcus aureus* and *Escherichia coli*. *Applied Environmental Microbiology* 74: 2171–2178.
59. Shionoiri, N.; Sato, T.; Fujimori, Y. et al. **2012**. Investigation of the antiviral properties of copper iodide nanoparticles against feline calicivirus. *Journal of Bioscience and Bioengineering* 113: 580–586.
60. Wang, L.; Hu, C.; Shao, L. **2017**. The antimicrobial activity of nanoparticles: Present situation and prospects for the future. *International Journal of Nanomedicine* 12: 1227.
61. Eltarahony, M.; Zaki, S.; Kheiralla, Z.; Abd-El-Haleem, D. **2019**. Study on the antagonistic potential of biosynthesized hematite nanoparticles during water and wastewater treatment. *Clean – Soil, Air, Water* 47: 1800418.
62. Nataraj, N.; Anjusree, G.; Madhavan, A. et al. **2014**. Synthesis and anti-Staphylococcal activity of TiO_2 nanoparticles and nanowires in ex vivo porcine skin model. *Journal of Biomedicine and Nanotechnology* 10: 864–870.
63. Ansari, M.; Khan, H.; Khan, A.; Cameotra, S.; Saquib, Q.; Musarrat, J. **2014**. Interaction of Al_2O_3 nanoparticles with *Escherichia coli* and their cell envelope biomolecules. *Journal of Applied Microbiology* 116: 772–783.
64. Franci, G.; Falanga, A.; Galdiero, S.; Rai, L.; Morelli, G.; Galdiero, M. **2015**. Silver nanoparticles as potential antibacterial agents. *Molecules* 20: 8856–8874.
65. Su, Y.; Zheng, X.; Chen, Y.; Li, M.; Liu, K. **2015**. Alteration of intracellular protein expressions as a key mechanism of the deterioration of bacterial denitrification caused by copper oxide nanoparticles. *Scientific Reports* 5: 15824.
66. Shrivastava, S.; Bera, T.; Roy, A.; Dash, D. **2007**. Characterization of enhanced anti-bacterial effects of novel silver nanoparticles. *Nanotechnology* 18: 225103.
67. Cui, Y.; Zhao, Y.; Tian, Y.; Zhang, W.; Lü, X.; Jiang, X. **2011**. The molecular mechanism of action of bactericidal gold nanoparticles on *Escherichia coli*. *Biomaterials* 33: 2327–2333.
68. Su, G.; Zhang, X.; Giesy, J. et al. **2015**. Comparison on the molecular response profiles between nano zinc oxide (ZnO) particles and free zinc ion using a genome-wide toxicogenomic approach. *Environmental Science and Pollution Research International* 22: 17434–17442.
69. Wu, B.; Zhuang, W.; Sahu, M.; Biswas, P.; Tang, Y. **2011**. Cu-doped TiO_2 nanoparticles enhance survival of Shewanella oneidensis MR-1 under ultraviolet (UV) light exposure. *Science Total Environment* 409: 4635–4639.
70. Padmavathy, N.; Vijayaraghavan, R. **2011**. Interaction of ZnO nanoparticles with microbes – a physio and biochemical assay. *Journal of Biomedicine and Nanotechnology* 7: 813–822.
71. Wilking, J.; Zaburdaev, V.; De Volder, M.; Losick, R.; Brenner, M.; Weitz, D. **2013**. Liquid transport facilitated by channels in bacillus subtilis biofilms. *Proceedings of the National Academy of Sciences of the United States of America* 110: 848–852.
72. Eltarahony, M.; Zaki, S.; Abd-El-Haleem, D. **2018**. Concurrent synthesis of zero-and one-dimensional, spherical, rod-, needle-, and wire-shaped CuO nanoparticles by Proteus mirabilis 10B. *Journal of Nanomaterials* 2018.
73. Ramanathan, R.; O'Mullane, A.; Parikh, R.; Smooker, P.; Bhargava, S.; Bansal V. **2011**. Bacterial kinetics-controlled shape-directed biosynthesis of silver nanoplates using Morganella psychrotolerans. *Langmuir* 27: 714–719.
74. Su, H.; Chou, C.; Hung, D. et al. **2009**. The disruption of bacterial membrane integrity through ROS generation induced by nanohybrids of silver and clay. *Biomaterials* 30: 5979–5987.
75. Karageorgou, D.; Thomou, E.; Vourvou, N. **2019**. Antibacterial and algicidal effects of porous carbon cuboid nanoparticles. *ACS Omega* 4: 4991–5001.
76. Huang, J.; Cheng, J.; Yi, J. **2016**. Impact of silver nanoparticles on marine diatom Skeletonema costatum. *Journal of Applied Toxicology* 36: 1343–1354.

77. Salleh, A.; Naomi, R.; Utami, N.; Mohammad, A.; Mahmoudi, E.; Mustafa, N.; Fauzi, M. **2020**. The potential of silver nanoparticles for antiviral and antibacterial applications: A mechanism of action. *Nanomaterials* 10: 1566.
78. Awad, M.; Al Olayan, E.; Siddiqui, M.; Merghani, N.; Alsaif, S.; Aloufi, A. **2021**. Antileishmanial effect of silver nanoparticles: Green synthesis, characterization, *in vivo* and *in vitro* assessment. *Biomedicine & Pharmacotherapy* 1: 111294.
79. Dakal, T.; Kumar, A.; Majumdar, R.; Yadav, V. **2016**. Mechanistic basis of antimicrobial actions of silver nanoparticles. *Frontiers in Microbiology* 1831.
80. Pan, X.; Wang, Y.; Chen, Z. et al. **2013**. Investigation of antibacterial activity and related mechanism of a series of nano-Mg $(OH)_2$. *ACS Applied Mater Interfaces* 5: 1137–1142.
81. Pal, S.; Tak, Y.; Song, J. **2007**. Does the antibacterial activity of silver nanoparticles depend on the shape of the nanoparticle? A study of the Gram-negative bacterium *Escherichia coli*. *Applied Environmental Microbiology* 73: 1712–1720.
82. Morones, J.; Elechiguerra, J.; Camacho, A. et al. **2005**. The bactericidal effect of silver nanoparticles. *Nanotechnology* 16: 2346.
83. Stoimenov, P.; Klinger, R.; Marchin, G.; Klabunde, K. **2002**. Metal oxide nanoparticles as bactericidal agents. *Langmuir* 18: 6679–6686.
84. Fang, B.; Jiang, Y.; Nüsslein, K.; Rotello, V.; Santore, M. **2015**. Antimicrobial surfaces containing cationic nanoparticles: How immobilized, clustered, and protruding cationic charge presentation affects killing activity and kinetics. *Colloid Surface B* 125: 255–263.
85. Gordon, T.; Perlstein, B.; Houbara, O.; Felner, I.; Banin, E.; Margel, S. **2011**. Synthesis and characterization of zinc/iron oxide composite nanoparticles and their antibacterial properties. *Colloids and Surfaces A: Physicochemical and Engineering Aspects* 374: 1–8.
86. Ben-Sasson, M.; Zodrow, K.; Genggeng, Q.; Kang, Y.; Giannelis, E.; Elimelech, M. **2014**. Surface functionalization of thin-film composite membranes with copper nanoparticles for antimicrobial surface properties. *Environmental Science and Technology* 48: 384–393.
87. Sukhorukova, I.; Sheveyko, A.; Kiryukhantsev-Korneev P. et al. **2015**. Toward bioactive yet antibacterial surfaces. *Colloid Surface B* 135: 158–165.
88. Gong, N.; Shao, K.; Feng, W.; Lin, Z.; Liang, C., Sun, Y. **2011**. Biotoxicity of nickel oxide nanoparticles and bio-remediation by microalgae *Chlorella vulgaris*. *Chemistry* 83: 510–516.
89. Su, D.; Li, P.; Ning, M.; Li, G.; Shan, Y. **2019**. Microwave assisted green synthesis of pectin-based silver nanoparticles and their antibacterial and antifungal activities. *Materials Letter* 244: 35–38.
90. Saliani, M.; Jalal, R.; Kafshdare Goharshadi, E. **2015**. Effects of pH and temperature on antibacterial activity of zinc oxide nanofluid against *Escherichia coli* O157:H7 and *Staphylococcus aureus*. *Jundishapur Journal of Microbiology* 8: e17115.
91. Mu, H.; Tang, J.; Liu, Q.; Sun, C.; Wang, T.; Duan, J. **2016**. Potent antibacterial nanoparticles against biofilm and intracellular bacteria. *Scientific Reports* 6: 18877.
92. Peretyazhko, T.; Zhang, Q.; Colvin, V. **2014**. Size-controlled dissolution of silver nanoparticles at neutral and acidic pH conditions: Kinetics and size changes. *Environmental Science and Technology* 48: 11954–11961.
93. Khan, M.; Ansari, A.; Hameedullah, M. et al. **2016**. Sol-gel synthesis of thorn-like ZnO nanoparticles endorsing mechanical stirring effect and their antimicrobial activities: Potential role as nano-antibiotics. *Scientific Reports* 6: 27689.

10 Algal-Mediated Nanotechnology for Environmental and Biochemical Applications

Hania A. Guirguis, Hebatullah H. Farghal, Hu Yang, Tamer Shoeib, and Mayyada M. H. El-Sayed

10.1 INTRODUCTION

Nanotechnology is a multidisciplinary research area that has garnered increased attention partly because of its versatile adaptability in the pharmaceutical and environmental fields. Recent trends in research have been directed toward adopting green approaches that exploit natural resources in the synthesis of nanomaterials [1]. Among the most popular and abundant marine species that have been widely used are algae. There are different classes of algae that occur naturally and abundantly as a multitude of species in various habitats [1, 2]. Algae can be brown (phaeophytes), red (rhodophytes), green (chlorophytes), or blue-green microalgae (cyanobacteria). The broad functionality of algae arises from their rich content of biologically active compounds, such as polysaccharides, proteins, peptides, minerals, and vitamins [3]. By virtue of these compounds, algae and their extracts exhibited various biological activities, including antioxidant, antibacterial, and antitumor activities, among others [4–7]. Hence, they were incorporated into nanomembranes and nanocomposites to produce bioactive nanoforms or successfully deployed as reducing agents for the green synthesis of metal nanoparticles [8–10].

This chapter discusses the synthesis of algal-based metal nanoparticles, nanomembranes, and nanocomposites and their application in environmental bioremediation, water treatment, and catalytic degradation.

10.2 METAL NANOPARTICLES

Metallic nanoparticles, with a diameter range of 1 to 100 nm, have been utilized in medical, electrical, cosmetic, and environmental applications due to their small size and large surface area-to-volume ratio. However, their toxicity and possible persistence in the environment hinder their utility. Accordingly, green synthesized nanoparticles have been gaining traction as an eco-friendly and less-toxic alternative. These green nanoparticles are produced utilizing a natural reducing agent, like bacteria, viruses, and plants of either marine or terrestrial origin [1]. Capping agents, like surfactants and polymers, are often employed to decrease the high surface energy of nanoparticles and prevent their agglomeration. These agents are costly and possibly toxic to humans and the environment. However, green synthesized nanoparticles already possess a natural capping layer that is made up of metabolites, like flavonoids, tannins, alkaloids, and polyphenols, which are present in the natural reducing agent used [11].

One explored class of natural reducing agents is comprised of algal species and their extracts. Green synthesis of nanoparticles using algal species or their extracts can follow the bottom-up approach ending up with the desired nanoparticle [12].

 DOI: 10.1201/9781003430087-10

Synthesis of metallic nanoparticles by algae typically involves incubation of a molar solution of the ionic metal along with algal cell cultures, a certain mass of dried algal powder, or an aqueous or organic solvent extract of the algae. When algal biomass, aqueous, or solvent extracts are used, extracellular nanoparticle synthesis takes place due to naturally present reducing agents, like polysaccharides, proteins, pigments, and others [1]. When viable algal cells are used, mostly in the case of cyanobacteria, synthesis could be extracellular or intracellular [13].

In extracellular synthesis, metal ions are bound to the surface of the cell and reduced by metabolites found outside the cell, like terpenoids, flavonoids, pigments, and alkaloids. These metabolites usually possess bioactive molecules, like amino acids, thiol, and carboxyl groups. In addition to metal reduction, the metabolites also form a stabilizing layer around the particles, usually made up of alcohols and water-soluble compounds [12, 14]. Biomasses of red algae *Chrondrus crispus* and *Kappophycus alvarez*ii were utilized in the synthesis of gold and silver nanoparticles. Fourier transform infrared (FTIR) spectroscopy analysis indicated the presence of thiocarbonyl, hydroxyl, alkane, amine, carboxyl, disulfide, and amide groups, thus confirming that active groups play a role in the reduction process of the nanoparticles and formation of a stabilizing layer [15, 16]. Gold nanoparticles synthesized using *Turbinaria conoides* brown algal aqueous extract and silver nanoparticles synthesized using *Cystophora moniliformis* brown algal aqueous extract had similar FTIR spectra attributed to carboxyl, amine, alkane, phosphate, and hydroxyl groups [17, 18]. Many of the same functional groups were also observed for silver nanoparticles synthesized using organic extracts like *Sargassum polycystum* methanolic extract, representing different active molecules involved in nanoparticle synthesis [19].

For intracellular synthesis taking place inside viable algal cells, which is more common for cyanobacteria, the reduction process of metal ions into their zero-valent form is a coping mechanism of the cell to reduce its exposure to the toxic metal ions [13]. Synthesis is hypothesized to be carried out by algal metabolic activities, like photosynthesis and respiration, the electron transport systems of these activities generate enzymes that act as reducing agents, like nitrogenase. Reduction could take place in thylakoid membranes, cytoplasms, or cell membranes [1]. The cells are vegetative at first but form heterocysts if there is not sufficient fixed nitrogen. Vegetative cells are long and green, surrounded by exopolysaccharides (EPS), while heterocysts have little to no chlorophyll and are surrounded by a polysaccharide layer (HEP). Organisms that form heterocysts are more capable of reducing the metal ions, since heterocysts can transform nitrogen to ammonium *via* the nitrogenase enzyme. Nucleation or initial particle synthesis occurs at different rates in different parts of the cell, depending on synthesis parameters. Particles could form faster in the EPS layer, and as time proceeds, most particles can be found in the vegetative cells, inside the thylakoid membranes, where photosynthesis occurs. Generally, nucleation occurs first in the outer layers, whether EPS or HEP, as was the case with the formation of gold nanoparticles from cyanobacterium *Anabaena* cylindrical cell cultures [13].

Biosynthesized nanoparticles (NPs) are considered non-toxic, or at least less toxic than their chemically synthesized counterparts, and can therefore be used in several applications, like cancer treatment, antibacterial agents, and catalysis [14]. There is a multitude of studies that explore nanoparticles synthesized using algae as reducing agents, especially silver and gold nanoparticles. Silver nanoparticles (AgNPs) have been recognized to exhibit antimicrobial and/or antibacterial activity, most likely due to the silver binding to active groups in the cells rendering them inactive [19]. Gold nanoparticles (AuNPs) are also known for their antibacterial and anticancer activities [20]. Other types of particles have also been synthesized using algae and tested for their bioactivities or environmental remediation capabilities, like dye degradation.

10.2.1 Algal-Synthesized Nanoparticles with Biological Activity

Several algal species have been utilized to synthesize nanoparticles that have been tested for biological activity. The most commonly investigated biological activity is the antibacterial effect, mostly using AgNPs, since silver is already known for being a potent agent against microbes. Other activities exhibited are antifungal, antioxidant, antidiabetic, and anticancer activities. Antibacterial and antifungal activities are nearly exclusively examined using the disk diffusion method, in which the nanoparticles are added to plates containing microbes and the inhibition zone is measured, in millimeter, after some elapsed time. The nanoparticles still contain at least traces of the toxic metal in the solution and are capped by the algal metabolites, which contain groups like thiols and amines. When the nanoparticles bind to the microbial cells, they disrupt the cell membrane and enter the cell, yielding reactive oxygen species (ROS), which could, in turn, lead to cell damage by hindering DNA replication and affecting metabolism. The particles could also bind to active groups in proteins and also lead to cell damage and death [21]. Antioxidant activity is commonly measured using the 1,1-diphenyl-2-picrylhydrazyl (DPPH) antioxidant assay. Upon interaction with a reducing or scavenging agent, DPPH, a stable free radical, turns from violet to yellow [22]. Lastly, anticancer activity is measured by examining the cytotoxicity of the particles against cancer cell lines.

10.2.1.1 Silver Nanoparticles

These nanoparticles were synthesized using either the algal biomass or its extracts, and they were primarily utilized as antibacterial agents. *Chlorella ellipsoidea* green biomass was added to $AgNO_3$ while stirring for 12 h at room temperature. The resulting AgNPs were polydisperse and had different shapes, including spherical and triangular, with a mean diameter of 17.7 nm. Their antibacterial activity was evaluated by measuring inhibition zones when incubated with microbes. They inhibited gram-negative bacteria *Escherichia coli* (9 mm), *Klebsiella pneumonia* (8 mm), *Pseudomonas aeruginosa* (16 mm), and *Staphyloccocus aureus* (15 mm). It was hypothesized that the AgNPs were able to permeate the cells due to the absence of a cell wall in gram-negative bacteria [23]. Eight cyanobacteria strains, eight algal strains, external polysaccharides, and C-phycocyanin pigments were also utilized for the synthesis of AgNPs using $AgNO_3$ precursor, both in light and dark conditions. The algal strains were grown in agar plates, then harvested *via* centrifugation and separated into solid biomass and cell-free culture liquid, both of which were used for synthesis. The algal biomass, liquid, or ethanolic extract was suspended with $AgNO_3$ at 25°C for 48–72 h. None of the strains used was able to synthesize nanoparticles in dark culture-free liquid, but most strains, whether biomass or cell-free culture, were able to facilitate AgNP synthesis in light conditions. Synthesis of AgNPs in cell-free culture liquid indicates that nanoparticle synthesis is mediated by extracellular compounds, possibly polysaccharides. Not all nanoparticles were tested for their antibacterial activity against six bacterial strains, only those synthesized using six cyanobacteria strains and two green algae strains. Cyanobacteria *Anabaena sp. 66–2* AgNPs (average 24.13 nm diameter), *Cylindrospermopsis sp. USC-CRB3* AgNPs, *Synechocystis sp. 48–3* AgNPs (14.67 nm), and green algae *Botryococcus braunii* AgNPs (15.67 nm) exhibited antibacterial activity against all bacterial strains examined (*Bacillus megatarium*, *E. coli*, *B. subtilis*, *Micrococcus luteus*, *P. aeruginosa*, and *S. aureus*). *Syenochocystis sp. 48–3* AgNPs and *Anabaena sp. 66–2* AgNPs exhibited the highest antibacterial activities, possibly due to their smaller size. AgNPs synthesized using *Limnothrix sp. 37–2-1* had the greatest diameter of 31.86 nm and did not exhibit antibacterial activity against any of the bacterial strains, which led to the conclusion that the larger size of the AgNPs hinders their ability to permeate the bacterial cell walls [24].

In addition to algal biomass, aqueous algal extracts were employed for the synthesis of AgNPs. The antibacterial activity of AgNPs synthesized using the aqueous extract of *Enteromorpha flexuosa* green algae was examined against 11 bacterial strains. The extract was incubated with $AgNO_3$ for 1 h to produce AgNPs of circular morphology and 15 nm average diameter. The prepared AgNPs exhibited antimicrobial activity against five gram-positive bacterial strains (*B. subtilis*, *B. pumulis*, *Entercoccus faecalis*, *S. aureus*, and *S. epidermidis*), two gram-negative strains (*E. coli* and *K. pneumonia*), and two fungal strains (*Candida cubicans* and *Saccharomyces cerevisiae*). When compared to algal extract, chemically synthesized AgNPs, or antibiotics, the algal-synthesized AgNPs exhibited higher antimicrobial activity [25]. AgNPs prepared using the aqueous extract of the green algae *Chlorella pyrenoidosa* also exhibited antibacterial activity. After incubation of the extract with $AgNO_3$ for 24 h, the resulting particles had an average diameter of 12 nm and were able to inhibit the growth of *K. pneumonia*, *Aeromonas hydrophilia*, *Acinetobacter* sp., and *S. aureus* [21]. Similarly, *Caulerpa serrulata* green algal extract was also incubated with $AgNO_3$ for 24 h, and the resulting particles had an average diameter of 10 nm. As the synthesis temperature increased from 27° C to 95° C, the UV absorbance intensity of the particles increased, indicating a higher rate of reactant consumption and subsequent nanoparticle formation. The particles exhibited antibacterial activity against *E. coli*, *S. aureus*, *Shigella* sp., *Salmonella typhi*, and *P. aeruginosa* [2]. In a deviation from ordinary metal nanoparticles, *Ulva flexuosa* green algal aqueous extract was used to synthesize hybrid bead AgNPs. $AgNO_3$ was first shaken for 16 h with methacrylic acid copolymer beads, which resulted in the production of silver ions adhered on macroporous methacrylic acid copolymer beads. Then, the aqueous algal extract was poured, and the solution was sonicated then stirred at 120 rpm for 24 h at 25° C. The produced AgNPs appeared to be distributed on the surface of the beads and had an average diameter of 4.9–6.7 nm. The antibacterial activity of the hybrid AgNPs was examined by measuring inhibition zones against *P. aeruginosa* (24 mm), *E. coli* (15 mm), *B. subtilis* (18 mm), and *S. aureus* (12 mm). Atomic force microscopy revealed that the cell walls of the bacterial strains were ruptured or disturbed, forming a hole, and that the cells changed upon addition of the nanoparticles relative to addition of the copolymer beads. These green hybrid AgNPs were synthesized for water disinfection from bacteria, since they are stable and not chemically synthesized [26]. Using an aqueous extract, Kingslin et al. employed the green algae *E. prolifera* to synthesize AgNPs. The aqueous extract was incubated with 1 mM $AgNO_3$ in a water bath at 60 °C for 1 h, followed by another hour at room temperature. The FTIR spectrum of the AgNPs showed alcohol, thiol, and alkene functional groups, suggesting that these groups formed a layer around the AgNPs. Scanning electron microscopy (SEM) images displayed spherical particles with a size range of 169.5–232.6 nm. The AgNPs were tested for their antioxidant activity using the DPPH radical scavenging assay and displayed a maximum inhibition of 60.2%. They were also applied for seed germination and appeared to increase the germination percentage of *Vigna unguiculata* and *Cicer arietinum* L. by 66.66% and 60%, respectively [27].

Organic extracts were also examined for AgNPs synthesis. *Sargassum polycystum* brown algal methanolic extract was mixed with $AgNO_3$ and left at room temperature for 24 h. The resulting nanoparticles had a spherical crystalline morphology with diameters of 50–100 nm, and they inhibited *P. aeruginosa* (32 mm), *K. pneumoniae* (33 mm), *E. coli* (37 mm), and *S. aureus* (38 mm). The AgNPs exhibited higher inhibition zone values than the algal crude extract and the precursor $AgNO_3$ alone [19]. In another study that used methanol extracts of the algae to synthesize AgNPs, the extracts of *Chrococcus disperses* and *Chlorella vulgaris* were incubated with $AgNO_3$ for two weeks. The pellets were then centrifuged and dried to obtain *Chlorella vulgaris* AgNPs with diameters of 27–47 nm and *Chrococcus* disperses AgNPs of 45–55 nm. The antimicrobial activity was measured *via* inhibition zones against plant pathogenic bacteria *Pseudomonas flavescens*, *Agrobacteriu tumefactions*, and *Erwinia amylovora*, as well as fungi *Fusarium solani*, *F. oxysporium*, *Helminthospriom* sp., and *Sclerotinia sclerotiorum*. Generally, *C. disperses* AgNPs exhibited higher antimicrobial activity than *C. vulgaris* AgNPs despite their larger size [28].

An interesting antibacterial therapy strategy is to interfere with quorum sensing, being a means by which bacteria communicate with each other; producing auto-inducers, which are molecules that induce survival factors for bacteria. Brown algal aqueous extract of *Sargassum polyphylum* was mixed with $AgNO_3$ and subjected to agitation at temperatures ranging from 25 °C to 47 °C for 25 h. The resulting nanoparticles were spherical, acquiring sizes of 37–43 nm. The quorum-quenching assay was conducted by measuring the pigment inhibition ability of the AgNPs against *Chromobacterium violaceum* strain. After incubation, the lack of purple violacein pigment indicated that the AgNPs were successful in inhibiting quorum sensing. Additionally, *Sargassum polyphylum* AgNPs exhibited antibacterial activity by inhibiting *E. coli* (18 mm), *P. aeruginosa* (17 mm), *B. subtilis* (14 mm), and *S. aureus* (13 mm) [29].

Since silver is already known for its antibacterial property, most of the studies conducted address the antibacterial activity of green synthesized AgNPs. However, green AgNPs could possess other biological activities. For example, AgNPs synthesized using the brown algae *Lobophora variegate* aqueous extract showed antifungal activity, indicated by inhibition zones, against *C. albicans* (10–16 mm), *C. tropicalis* (10–14 mm), *T. mentagrophytes* (8–16 mm), and *A. flavus* (7–12 mm). While novel solutions like green nanoparticles are important due to pathogen resistance, the AgNPs in the study, which were spherical and had a mean diameter of 40 nm, resulted in lower inhibition zone values than the commercial antifungal fluconazole [30].

El-Sheekh et al. explored the antibacterial and antifungal activities of silver oxide nanoparticles AgO and Ag_2O NPs synthesized using cyanobacteria strains *Spirulina platensis* and *Oscillatoria* sp. The authors used three forms of the algae for the synthesis: an aqueous extract, culture-free cells, and *in vivo* cell cultivation using biomass. All three methods required 200 mM $AgNO_3$, which were stirred with the algae for 30 min at 50 °C while the *in vivo* cultivation required 24 h. Transmission electron microscopy (TEM) images revealed the spherical NPs had a range of size 10.49–45.81 nm. They exhibited antifungal activity against *Candida albicans* and *C. tropicalis*, with an MIC_{90} (minimum inhibitory concentration) of 7.9 and 4.6 µg/mL, respectively. They also showed antibacterial activity against gram-positive *B. subtilis* and gram-negative *S. typhi*, with MIC_{90} of 7.81 and 15.63 µg/mL, respectively [31].

Antibiabetic activity was also explored. Diabetes mellitus is the increase in blood glucose level due to impaired insulin secretion. This could lead to further damage, like renal failure and cardiovascular diseases. Enzymes such as α-amylase and α-glucosidase degrade starch and oligosaccharides into glucose, so their suppression delays glucose intestinal absorption. Researchers rationalized that biosynthesized AgNPs exhibit bioactivities, like antibacterial and antiviral activities, so they can possibly induce antidiabetic effects. Brown algal aqueous extract of *Colpmenia sinuosa* was incubated with $AgNO_3$ for 24 h at room temperature with stirring. The formed nanoparticles were mainly cubic and of 54–65 nm size. Antidiabetic activity of the green synthesized AgNPs was determined by measuring the inhibition of α-amylase and α-glucosidase activity using a colorimetric reaction. As concentration of the AgNPs increased from 0.2 mg/mL to 1 mg/mL, the inhibition activity increased from 28.7% to 94.3% for α-amylase and for α-glucosidase from 32.4% to 90.5%. The study concluded that AgNPs synthesized using *Colpmenia sinuosa* algal extract could have effective *in vitro* antidiabetic activity [32].

10.2.1.2 Gold Nanoparticles

Algae have also been used to synthesize gold nanoparticles (AuNPs) that exhibit biological activity. In the same study that produced AgO and Ag_2O NPs, discussed previously, El-Sheekh et al. synthesized AuNPs using three methods: using aqueous extract, cell culture, and *in vivo* cultivation using biomass of cyanobacteria *S. platensis* and *Oscillatoria* sp. and $HAuCl_4 \cdot 3H_2O$ gold precursor. The resulting AuNPs had octahedral, pentagonal, and triangular shapes and a size range of 15.49–55.08 nm. They showed antifungal activity against *C. tropicalis*, *C. albicans*, and *A glavis*, with an MIC_{90} of 1.1, 2.9, and 6.2 µg/mL, respectively. They also showed antibacterial activity against *B. subtilis* and *S. typhi* with MIC_{90} of 1.95 and 3.9 µg/mL, respectively [31].

Lemanea fluviatilis red algal biomass was added to chloroauric acid ($HAuCl_4$) precursor and stirred for 12 h at room temperature. After centrifugation, the supernatant was collected and dried to obtain polydisperse AuNPs of spherical morphology and 5–15 nm diameter. The AuNPs were tested for their antioxidant activity, which was measured using the DPPH antioxidant assay. The synthesized AuNPs showed antioxidant activity of 25% at 10 mg dose and 90% at 100 mg dose [33]. AuNPs were also prepared using *Hypnea musciformis* red algal biomass, which was incubated with $HAuCl_4$ at room temperature. The particles which were polydisperse and ranged between 6.25 nm and 33.33 nm formed extracellularly due to precipitation of gold on the algal cell wall. AuNPs exhibited antifungal activity against *Aspergillus niger* and *Mucor spp*. It was concluded that growth was inhibited due to the presence of toxic gold ions, which coordinate to protein thiol groups in the fungal cell, thus inhibiting normal functions [34].

The extract of another red algae was used to synthesize bioactive AuNPs. *Gracilaria corticata* aqueous extract was stirred with $HAuCl_4$ solution for 4 h at 40 °C. The resulting particles were monodisperse and had sizes of 45–57 nm. AuNPs were conjugated with ciprofloxacin antibiotic, and their antibacterial and antioxidant activities were assessed. AuNPs inhibited *Enterococcus faecalis* (14 mm), *S. aureus* (19 mm), *Enterbacter aerogenes* (21 mm), and *E. coli* (24 mm). These values were higher than the inhibition zones exhibited by ciprofloxacin alone, and it was inferred that the particles were slightly more effective against gram-negative bacteria. Antioxidant activity was measured using the DPPH assay, and the particles exhibited 75.2% radical scavenging at the 100 μL/mL concentration and 80.6% for 200 μL/mL [35].

10.2.1.3 Copper Oxide Nanoparticles

In an effort to remove microbes from drinking water, chemically synthesized nanoparticles were employed. However, these particles tend to agglomerate, rendering the process counterproductive. Green synthesized nanoparticles are less toxic and already have a green capping layer made up of the extract. As such, algal-synthesized copper oxide nanoparticles were prepared using an algal extract and were utilized to remove microbes. Cyanobacteria *Anabaena cylindrica* aqueous extract was stirred with $CuSO_4$ precursor at 60 °C, then dried. As stirring speed increased from 500 to 900 rpm, the size decreased from 42.54 to 3.6 nm due to restricting crystal growth. Nucleation and particle growth occurred outside the algal cell and resulted in rod-like structures with minimum agglomeration. The antimicrobial activity of the copper oxide nanoparticles was measured against gram-negative *E. coli* versus the antibiotic ampicillin. Water treatment analysis showed complete removal of the 16,000 colonies per 100 mL present in the sample within 30 min. To assess the toxicity of the particles, they were incubated with *Channa punctatus* fish, and after 72 h, there was no mortality. The 14th day indicated 16% mortality when compared with control. Fish blood analysis throughout the 14 days showed that no micronucleus was formed due to CuO presence, so they were deemed not toxic to fish at 40 μg/mL concentration. The particles were also deemed non-toxic to human cells at a concentration of 2,500 μg/mL, as indicated by the MTT cytotoxicity assay [36].

Further work investigated other potential biological activities for algal-synthesized copper oxide nanoparticles. Brown algae *Sargassum polycystum* aqueous extract was incubated with $CuSO_4$ for 24 h at room temperature. The prepared particles were tested for their antibacterial and antifungal activity *via* agar diffusion method as well as their anticancer activity using cytotoxicity assay on human breast cell line. Results indicated that the highest antibacterial activity was reported against *P. aeruginosa* (15 mm), while the lowest belonged to *Shigella dysenteriae* (5 mm). As for the antifungal activity, maximum inhibition was reported against *Aspergillus niger* (20 mm), while the lowest inhibition was recorded against *A. oryzae* (12 mm). In addition, copper oxide nanoparticles inhibited cancer cell growth by more than 93% at a dose of 100 μg/mL [37]. Another brown algal extract was also used to synthesize copper oxide NPs. *Bifurcaria bifurcate* aqueous extract was added dropwise to $CuSO_4$ at 100 °C under stirring conditions for 24 h. The particles produced were spherical for the most part, with some elongated particles of size range 5–45 nm and an average diameter of 20.66 nm. There were two crystalline phases present, monoclinic CuO and cubic Cu_2O. Antibacterial activity of the particles was measured *via* inhibition zone diameters for *Enterobacter aerogenes* (14 mm) and *S. aureus* (16 mm), whereas the extract alone did not exhibit antibacterial activity when incubated with the strains [38].

10.2.1.4 Miscellaneous Nanoparticles

Aside from commonly used metal and metal oxide nanoparticles, zirconia is known to be a catalyst and to exhibit antimicrobial activity. To prepare zirconia nanoparticles (ZrO_2 NPs), zirconium oxynitrate was mixed with *Sargassum whightii* brown aqueous extract and kept inside a furnace at 400°C to release metal nitrates and gases. The mixture was left in the oven for 5 h at the same temperature for calcination, and the formed ZrO_2 NPs were spherical and had an average diameter of 5 nm. Antibacterial activity was measured against gram-positive *B. subtilis* (21 m) as well as gram-negative *E. coli* (19 mm) and *Salmonella typhi* (19 mm). These inhibition zone values were higher than those resulting from the extract alone. Thus, the nanoparticles exhibited strong antibacterial activity due to their small size, and as concentration of the NPs increased, so did their antibacterial activity [39].

Bedoya et al. fabricated zinc oxide (ZnO) and zinc hydroxide ($Zn(OH)_2$) nanoparticles for the application of azo dye photocatalysis in water. They asserted that dyes are considered carcinogenic and that ZnO and $Zn(OH)_2$ nanoparticles could precipitate or photocatalyze these dyes. The authors used macroalgae *Calerpa sertularioides* methanol extract and 1 M zinc nitrate hexahydrate for the synthesis, which required 2 h and stirring at 90 °C, then liquid evaporation at 30 °C. The precipitate was then placed in a furnace for 4 h at 300 °C to produce the NPs. These NPs were encapsulated in a sol gel matrix using tetraethyl orthosilicate, $Si(OC_2H_5)_4$, in order to render them non-toxic for aquatic life by not exposing the water to the NPs directly. TEM images revealed a size range of 45–69 nm. The dye experiment was conducted on naphthol blue-black dye, and the composite achieved 96% transparency of the water containing the dye after 8 h of exposure, which is one day faster than the transparency achieved by NPs that were not encapsulated. This study concluded that tetraethyl orthosilicate–encapsulated ZnO and $Zn(OH)_2$ NPs can be used to treat wastewater from the fabric industry, which contains toxic dyes [9].

All the aforementioned studies dealt with single-component metal nanoparticles; however, one study reported the synthesis of binary metal nanoparticles using green *Chlorella vulgaris* aqueous extract. The extract was added to the precursors along with urea at 160 °C. Cobalt nitrate was used as a precursor for cobalt ferrite nanoparticles ($CoFe_2O_4$), while silver nitrate was added to obtain silver-doped ferrite nanoparticles (Ag-Fe-O). $CoFe_2O_4$ NPs have an average diameter of 20 nm, and their antioxidant activity, measured using DPPH assay, was 3.65%. Additionally, their anticancer activity was measured on human neuroblastoma cells, revealing they were 25% cytotoxic at 100 µg/mL concentration. As for Ag-Fe-O NPs, they had an average size of 15 nm, antioxidant activity of 3.59%, and were 90% cytotoxic at 100 µg/mL, indicating that silver is more potent at penetrating the cell and causing its death [40].

10.3 ALGAL-SYNTHESIZED NANOPARTICLES WITH ENVIRONMENTAL ACTIVITY

Organic dyes, like methylene blue and malachite green, are azo dyes used as staining agents in medical applications. Their persistence in the environment is potentially toxic to humans because they are not biologically degraded. Catalytic degradation of these dyes using algal-synthesized nanoparticles could be an eco-friendly method to remove these dyes from aqueous environments [11, 21]. Catalytic activity of the nanoparticles is measured using UV-visible spectrometry. The absorbance intensity of the dye is measured before and after the addition of the nanoparticles. Decrease in intensity indicates the ability of the nanoparticles to degrade the toxic dyes. The nanoparticles could act as electron transfer agents between a reducing agent like $NaBH_4$ and the dyes or could mediate electron/hydrogen transfer when excited by irradiation.

Methylene blue was among the dyes that were effectively degraded by algal-synthesized AgNPs. *C. pyrenoidosa* green algal aqueous extract was added to $AgNO_3$ and left for 24 h. The obtained AgNPs, which had an average diameter of 12 nm, were added to methylene blue dye solution for 1 h then irradiated with a xenon lamp. After centrifugation, UV-vis spectrophotometry revealed that methylene blue was degraded, and the AgNPs were regenerated after heating at 100 °C for 1 h. The particles retained their stability and their photocatalytic activity with 90% efficiency. As for the photocatalysis mechanism, it was proposed that an electron moved from the valence band of the AgNPs to the conduction band upon irradiation, leaving a positive hole, which, in turn, led to either oxidation or reduction. Water on the surface of the AgNPs consequently got oxidized and resulted in the formation of hydroxyl radicals. The remaining oxygen atoms took up electrons in the conduction band, producing the superoxide radicals, which prompted further oxidation. The superoxide radical then combined to a proton (or H atom), producing hydrogen peroxide, which degraded into hydroxyl radicals that subsequently degraded methylene blue dye by opening the nitrogen bond and yielding smaller compounds like carbon dioxide, water, nitrate, and ammonia [21]. On the other hand, AgNPs, synthesized using the aqueous extract of *Caulerpa racemosa*, acted as electron transfer mediators between $NaBH_4$ and methylene blue. The extract was added to $AgNO_3$ with different concentrations at 25 °C for 24 h. Increase in extract concentration led to increase in absorbance wavelength (due to particle size increase) and intensity (due to particle number increase). AgNPs had an average diameter of 25 nm and had distorted sphere morphology showing no evidence for agglomeration, possibly due to the presence of secondary metabolites forming a capping layer around the particles. This was evidenced by FTIR analysis, which indicated the presence of hydroxyl, carbonyl, methyl, amine, and alkene groups. The catalytic activity of the AgNPs was examined by adding the particles to $NaBH_4$ and methylene blue, where disappearance of the blue color and decrease in absorbance intensity of the dye within 30 min reflected the catalytic activity of the AgNPs, especially since the reducing agent, $NaBH_4$, alone could not degrade the dye. Degradation of methylene blue followed pseudo-first-order kinetics, and equilibrium was explained by the Langmuir–Hinshelwood model. In the degradation mechanism, $NaBH_4$ acted as an electron donor and hydrogen provider, while AgNPs mediated the electron transfer between BH_4^- and the dye. Both BH_4^- and methylene blue adsorbed to the surface of the nanoparticles as hydrogen transfer occurred. Due to the large surface area-to-volume ratio of the AgNPs, they provided many catalytic sites for the degradation of methylene blue [11].

Algal-synthesized AgNPs were also efficient in removing other dyes, like methylene orange and other azo dyes. AgNPs synthesized by *C. ellipsoidea* biomass, which exhibited antibacterial activity, were also examined for their photocatalytic activity by degradation of methylene blue and methylene orange dyes. It was found that as the concentration of AgNPs increased, so did the rate of dye degradation, which corresponded to pseudo-first-order kinetics. This study combined the use of both UV radiation and $NaBH_4$ as precursors. The exact mechanism of electron transfer in the presence of a radiation source and a reducing agent was not mentioned; however, the dye molecules adsorbed onto the surface of the AgNPs, and there was a slight decrease in degradation rate when the AgNPs were absent. The catalytic efficiency of the particles did not decrease after three cycles of experiments [23]. AgNPs synthesized using *Caulerpa serrulata* aqueous extract, which also exhibited antibacterial activity, was found to exhibit catalytic activity by degrading Congo Red azo dye that is utilized in the textile, printing, and cosmetic industries. When $NaBH_4$ and the dye were left for 3 h, there was no change in color, indicating insignificant degradation. However, when AgNPs were added to the mixture, the dye disappeared within 5 min. As the concentration of the catalyst AgNPs increased, the number of active sites increased, leading to a higher degradation rate. Both the dye and $NaBH_4$ were adsorbed on the nanoparticle surface, which acted as a mediator when $NaBH_4$ donated electrons to the dye. While $NaBH_4$ did not degrade the dye on its own, it was observed that in the presence of the AgNPs, the degradation percentage increased when $NaBH_4$ concentration increased. Another parameter that affected dye degradation was synthesis temperature. The AgNPs produced at a temperature of 95 °C displayed greater efficacy compared to the ones synthesized at ambient temperature. A possible explanation is that the higher temperature led to the formation of more nanoparticles, which increased the catalytic activity. Additionally, the catalytic activity increased with increasing the algal extract concentration. It was thus concluded that the AgNPs synthesized from the algal extract are efficient, stable, and cost-effective catalysts for azo dye degradation [2].

In addition to silver nanoparticles, cadmium was used in the green synthesis of catalytic nanoparticles. CdS NPs were synthesized using blue-green *Spirulina (Arthrospira) platensis* protein extract along with $Cd(NO_3)_2{\bullet}4H_2O$ precursor. No external sulfur was added; the sources of sulfur were cysteine amino acid and the photosynthesis pigment phycobiliprotein C-phycocyanin, which are metabolites produced by the algae. The cyanobacteria were cultured in a medium under sunlight at room temperature, while the dried pellets were suspended in a protein extraction buffer and then centrifuged to obtain the protein extract supernatant. Cadmium nitrate solution was added to the protein extract and left to stir overnight at room temperature. Results showed that the protein extract UV-vis absorbance peak was more intense than that of the algal extract due to its main composition of C-phycocyanins, which contain the chromophore phycocyanobilin PCB and apoprotein. *Spirulina* got fragmented upon the synthesis of CdS NPs, but the presence of yellow deposits indicated that 8–12 nm NPs formed within the cell. It was hypothesized that cysteine and glutathione, which both contain sulfur, could have formed Cd-S complexes upon addition of Cd(II) to the cells. The hydrolytic enzymes could hence play a role in the formation of CdS NPs. The toxicity and stress produced by Cd(II) led to the release of sulfur-containing compounds, and so the resulting CdS chelates were less toxic. In addition to that, these CdS NPs were utilized for their photocatalytic activity in the degradation of malachite green dye, which is present in industrial wastewater effluents. CdS is a semiconductor with a narrow bandgap, and this makes it a photocatalysis candidate. Upon irradiation of the particles with a photon of equal energy to the bandgap of the particles, electron hole pairs form. These electron hole pairs could either recombine to emit light or participate in oxidation–reduction reactions with water, producing free radicals which initiate the degradation of malachite green. The high surface area-to-volume ratio means higher surface energy, so most atoms which lie on the surface can play a role in electron transfer to dye molecules. This makes the nanoparticles more efficient photocatalysis agents than bulk CdS. Thus, the recombination of electron hole pairs is less likely, and the participation of the particles in the oxidation–reduction reactions that initiate dye degradation is the more plausible option [41].

Table 10.1 summarizes the main findings of the studies performed on algal-mediated synthesis of metal nanoparticles, while Figure 10.1 illustrates the different types of these nanoparticles along with their biological or environmental activities.

10.4 ALGAL NANOCOMPOSITES

Novel studies have been performed to prepare various algal nanocomposites for water bioremediation via adsorption processes.

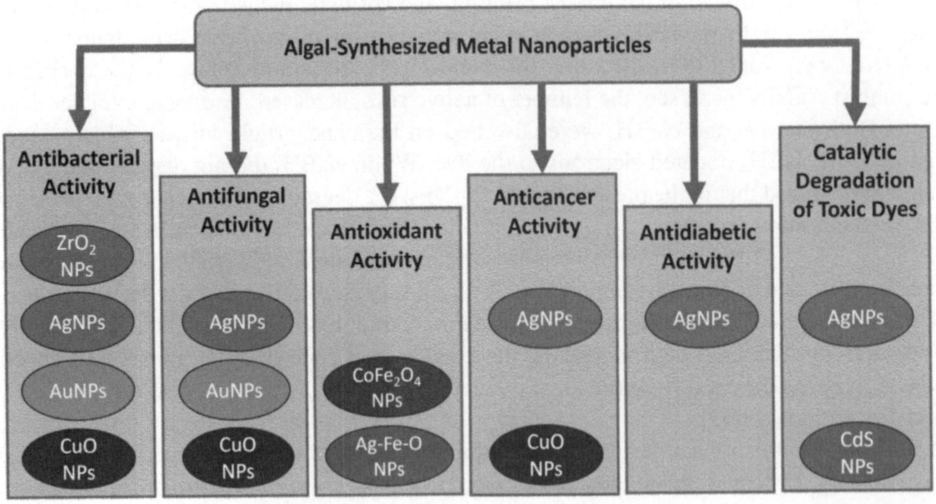

FIGURE 10.1 Algal-synthesized metal nanoparticles and their biological or environmental functionalities.

TABLE 10.1
Summary of the Studies Reported on Algal-Synthesized Nanoparticles

Algal Species	NP Type	Synthesis Conditions	Size (nm)/Morphology	Activity	References
Chlorella ellipsoidea	AgNPs	Biomass + $AgNO_3$, 24 h at room temp.	17.7/spherical	Antibacterial; degradation of methylene blue and methylene orange	[23]
Anabaena sp. 66–2; *Cylindrosper mopsis* sp. USC-CRB3; *Lingbya* sp. 15–2; *Synechococcus* sp. 145–6; *Synechocystis* sp. 48–3; *Butyrococcus braunii*; *Coelastrum* sp. 143–1	AgNPs	Biomass or cell-free culture liquid + $AgNO_3$, 72 h at 25°C	24.1/irregular, spherical; 14.6/irregular; 15.7/spherical; 19.3/ spherical	Antibacterial	[24]
Enteromorpha flexuosa	AgNPs	Aqueous extract + $AgNO_3$, 1 h, in the dark	15/spherical	Antibacterial	[25]
Chlorella pyrenoidosa	AgNPs	Aqueous extract + $AgNO_3$ incubation, 24h	12/–	Antibacterial; methylene blue degradation	[21]
Caulerpa serrulata	AgNPs	Aqueous extract + $AgNO_3$, 25h at 27°C	10/–	Antibacterial; Congo red degradation	[2]
Ulva flexuosa	Hybrid bead AgNPs	Aqueous extract + $AgNO_3$ in polymer beads sonication, 120 rpm for 10 min at room temp.	4.9–6.7/–	Antibacterial	[26]
Enteromorpha prolifera	AgNPs	Aqueous extract + $AgNO_3$, water bath, 60°C, 1 h, then room temp., 1 h	169.5–232.6/spherical	Seed germination; antioxidant	[27]
Chlorella vulgaris; *Chrococcus disperses*	AgNPs	Methanol extract + $AgNO_3$, 2-week incubation	27–47/spherical; 45–55/spherical	Antibacterial; antifungal	[28]
Sargassum polycystum	AgNPs	Methanol extract + $AgNO_3$, 24 h at room temp.	50–100/spherical	Antibacterial; anticancer	[19]
Sargassum polyphylum	AgNPs	Aqueous extract + $AgNO_3$, agitation, 24 h at 24–47°C	37–43/spherical	Antibacterial; quorum-sensing inhibition	[29]
Lobophora variegate	AgNPs	Aqueous extract + $AgNO_3$ stirring at room temp.	40/spherical	Antifungal	[30]

(Continued)

TABLE 10.1 *(Continued)*

Spirulina platensis; *Oscillatoria* sp.	AgO, Ag_2O NPs; AuNPs	Aqueous extract, culture-free cell, or *in vitro* cultivation + $AgNO_3$ or $HAuCl_4•3H_2O$, stirring, 30 min at 50°C	10.5–45.8/spherical; 15.5–55.1/ octahedral, triangular, pentagonal	Antibacterial; antifungal	[31]
Colpomenia sinuosa	AgNPs	Aqueous extract + $AgNO_3$, stirring, 2 h at room temp.	54–65, cubic	Antidiabetic	[32]
Caulerpa racemose	AgNPs	Aqueous extract + $AgNO_3$, 24 h at 25°C	25	Methylene blue degradation	[11]
Lemanea fluviatilis	AuNPs	Biomass + $HAuCl_4$, stirring, 12 h at room temp.	5–15, spherical	Antibacterial	[33]
Gracilaria corticata	AuNPs	Aqueous extract + $HAuCl_4$, stirring, 4 h at 40°C	45–57	Antibacterial	[35]
Hypnea musciformis	AuNPs	Biomass + $HAuCl_4$, 2 h at room temp.	6.25–33.3, spherical	Antifungal	[34]
Anabaena cylindrica	CuO	Aqueous extract + $CuSO_4$, stirring 900 rpm at 60°C	3.6, rod	Antibacterial	[36]
Sargassum polycystum	CuO	Aqueous extract + $CuSO_4$, overnight at room temp.		Antibacterial; antifungal; anticancer	[37]
Bifurcaria bifurcate	CuO	Aqueous extract + $CuSO_4$, 24 h at 100–120°C	20.66, spherical, elongated	Antibacterial	[38]
Caulerpa sertularioides	ZnO, $Zn(OH)_2$	Methanol extract + 1 M $Zn(NO_3)_2$, 2 h, 90°C, furnace 300°C, 4 h	45–69	Naphthol blue-black dye photodegradation	[9]
Sargassum whightii	ZrO_2	Aqueous extract + $ZrO(NO_3)_2 \cdot xH_2O$, 5 h, furnace 400°C	5, spherical	Antibacterial	[39]
Chlorella vulgaris	$CoFe_2O_4$; Ag-Fe-O	Aq. extract + urea + $Fe_3(NO_3)$ + $Co(NO_3)_2$; aq. extract + urea + $Fe_3(NO_3)$ + $AgNO_3$	20, spherical; 15, spherical	Antioxidant; anticancer	[40]
Spirulina (Arthrospira) platensis	CdS	Protein extract + $Cd(NO_3)_2$, overnight, room temp.	8–12	Malachite green dye degradation	[41]

10.4.1 Heat-Treated Algal Nanocomposites

The following studies mainly used heating treatment methods, such as calcination and pyrolysis, to form the nanocomposite. For example, a diatom/ZrO_2 nanocomposite was prepared by calcination for the purpose of de-fluorination of wastewater. The calcination process converted diatom algal biomass of *Phaeodactylum tricornutum* into biosilica that formed the nanocomposite with zirconium dioxide, as was confirmed by TEM and Energy Dispersive X-ray (EDX) spectroscopy measurements. Brunauer-Emmet-Teller (BET) specific surface area for the nanocomposite was 140 m^2/g, with a pore width of 9.7 Å. Percent removal of fluoride reached 90% at acidic pH values due to the positive charge imparted on the composite by the hydronium ions in the solution, which, consequently, enhanced the adsorption of the negative fluoride ions. However, at high pH, the hydroxyl groups competed with fluoride on the composite sites. The effect of addition of nitrates to the medium was negligible; however, silicates and bicarbonates hindered fluoride adsorption on the nanocomposite. Regeneration of the nanocomposite for two cycles was possible by treating it with 0.1 M NaOH, followed by 0.1M H_2SO_4, then distilled water [42]. Similarly, algae were treated by vermiculite solution, then pyrolyzed at 500 °C to form a silicon dioxide–algae biochar nanocomposite to be used for phosphorus bioremediation of wastewater. Gas sorption experiments showed that the nanocomposite had uniform mesopores with surface area of 42.43 m^2/g, pore diameter of 5.17 nm, and a total pore volume of 0.055 cm^3/g. Phosphorus removal was tested on model wastewater while studying the influence of several parameters on adsorption, like pH and ionic strength. The pH range of 3–7 was optimal since higher pHs decreased adsorption due to the competition of hydroxyl ions with the phosphate ones. Low ionic strengths of sodium chloride had a slight effect on phosphate removal, while high ionic strengths hindered the electrostatic attraction and reduced the removal efficiency from 98% to 51% due to the competition on the biochar sites. The nanocomposites were successfully regenerated using NaOH for up to five cycles, maintaining a removal efficiency of not less than 75% [43]. Another pyrolysis experiment was performed on brown algae after being subjected to a current density of 42.46 mA/cm^2 between stainless steel electrodes to form a magnetic biochar capable of removing acid orange 7 from aqueous media. The composite had a heterogeneous surface with a surface area of 337 m^2/g. Adsorption capacities were estimated to be 190, 297, and 382 mg/g at 10, 20, and 30 °C, respectively, indicating endothermic adsorption [44].

10.4.2 Algal-Activated Carbon Nanocomposites

Other studies converted algae into activated carbon to incorporate it with metal nanoparticles into a nanocomposite for the adsorptive removal of dyes and acetylsalicylic acid from water. For example, *Sargassum oligocystum algae* was converted into activated carbon that formed a nanocomposite with magnetite nanoparticles. BET sorption experiments showed that this structure was mesoporous, with 0.1945 cm^3/g, 6.13 nm, and 126.77 m^2/g as pore volume, pore size, and surface area, respectively. The nanocomposite was applied for the removal of methyl violet and methylene blue from wastewater at an optimum pH that lies within the alkaline range. At this pH range, the nanocomposite was negatively charged by virtue of the COOH and Fe-OH groups, which favored adsorption of the cationic dyes. However, low adsorption capacity was observed in the acidic range due to π-π interactions. With increasing the adsorbent dose from 0.05 to 0.15 g/100 mL, the removal efficiency leveled from 75.34% and 75.62% to 98.43% and 99.27% for methylene blue and methyl violet, respectively, and decreased by further increase in the adsorbent concentration. When increasing the dye initial concentration, the efficiency decreased, since all the adsorbent sites became occupied, leaving the excess dye molecules unadsorbed. Moreover, this nanocomposite was regenerated after seven cycles, with a decrease of merely 3% in the removal efficiency after the seventh cycle [45]. In another work concerned with dye removal, cellulose-devoid *Enteromorpha* algae was utilized after its carbonization at 700 °C to form a nanocomposite with Co-based and Mo-based metal nanoparticles for the removal of methylene blue from aqueous solutions. BET surface area was 290.7 m^2/g, where micropores of 2 nm, mesopores of 15.3 nm, and macropores of 62.6 nm were formed. Removal was best achieved at pH 10 due to electrostatic attraction, and maximum adsorption capacity was 1587.3 mg/g. Regeneration of the nanocomposites was conducted for ten cycles using absolute alcohol to obtain removal efficiencies that exceeded 91.6% [46]. In a similar fashion, *Laminaria japonica* was converted to activated carbon, which formed a nanocomposite with iron oxide nanoparticles. Response surface methodology was applied to determine the optimum synthesis conditions of activation temperature, activation time, and impregnation ratio of magnetite to algae-derived activated carbon to investigate their effect on the production yield and adsorption capacity. Optimum parameters were reported to be 727 °C, 129 min, and 2.62:1 for the activation temperature, time, and impregnation ratio, respectively. The composite-removed acetylsalicylic acid with a maximum sorption capacity of 127 mg/g at 10 °C and the isotherm fitted SIPS model [47].

10.4.3 Algae/Metal Nanoparticle Nanocomposites

Other studies prepared algae/metal nanoparticle nanocomposites without converting the algae to biochars or activated carbons, and the composites successfully removed dyes and heavy metals by adsorption. Most of the algae/metal nanoparticle nanocomposites were magnetic nanocomposites prepared using iron oxide nanoparticles. Only one study employed hydrous cerium oxide nanoparticles that were impregnated into *Enteromorpha* sp. macroalgae to form a nanocomposite efficient for the removal of Cr(VI). SEM images showed that the nanoparticles were evenly distributed over the algae surface, causing the roughness needed for adsorption. Maximum adsorption capacity of 7.1 mg/g was obtained at pH 2, where electrostatic attraction occurred between the positive nanocomposite and the negative $HCrO_4^-$ ions – the stable form of Cr(VI) at acidic pHs. Increasing the adsorbent dose from 0.1 to 0.3 g for the same concentration of metal led to a significant drop in the adsorption capacity due to the presence of unoccupied sites on the adsorbent surface, while increasing the metal concentration from 20 to 100 mg/L resulted in a faster adsorption rate and better occupation of the adsorbent sites with maximum adsorption capacity of 16.5 mg/g. In addition, adsorption of Cr(VI) ions significantly declined with the addition of phosphate and nitrate anions due to their interference with the electrostatic attraction between the nanocomposite and the metal ions [48]. Other studies focused on magnetic algae/metal nanoparticle nanocomposites due to their facile handling and regeneration. For example, *Sargassum horneri* biomass was modified with iron oxide nanoparticles to remove five different dyes from wastewater. Equilibrium time amounted to less than 2 h for all tested dyes, where the highest adsorption capacity of 193.8 mg/g belonged to acridine orange and the lowest one of 110.4 mg/g was exhibited by malachite green [49]. Another algal species, *Sargassum swartzii*, was incorporated in a nanocomposite along with nanoscale zero-valent iron for the removal of crystal violet from aqueous solutions. An adsorption capacity of 200 mg/g was obtained at pH 8, with an initial concentration range of 20–100 mg/L. SEM images showed a smoothed-out surface after adsorption due to filling of the pores. FTIR measurements confirmed the role of OH and NH groups in the adsorption of crystal violet [50]. For the removal of another dye, safranin O, *Cymbolia barbata* macroalgae were magnetically modified by iron oxide nanoparticles to treat safranin O–containing aqueous solutions. A maximum adsorption capacity of 192.2 mg/g was attained, and the equilibration time was 90 min. Adsorption equilibrium followed Langmuir and Freundlich isotherms, where the reaction was exothermic, while adsorption kinetics were best explained by the pseudo-second-order model [51]. As for the removal of heavy metals, *Chlorella vulgaris* microalga was attached to the precursor iron oxide nanoparticles, and the nanocomposite was applied for the removal of Cd(II) and Pb(II) from aqueous solutions. The removal of Cd(II) took place through electrostatic attraction, while Pb(II) was chemically bound to the amine group of the nanocomposite. At pH 4, removal efficiencies reached 99.75 and

95.72% for Cd(II) and Pb(II), respectively, whereas it slightly declined upon decreasing the pH to 2. This was attributed to the competition of hydrogen ions with the metals over the adsorbent sites. Adsorption of Cd(II) was influenced by adding 0.001–0.1 mol/L of $NaNO_3$ electrolyte to the solution due to the weak adsorption forces, while Pb(II) adsorption was unaffected due to the formation of aggregation sphere [52].

Photocatalytic degradation of pollutants was also achieved by algal nanocomposites incorporating metal nanoparticles. Brown algae *Padina tetrastromatica* formed a nanocomposite with ZnO nanoparticles for removing turquoise blue and methylene blue from aqueous solutions under direct sunlight. Decolorization efficiency was 90.4% for turquoise blue, and 98.5% for methylene blue. Removal took place by virtue of the holes and electrons generated upon irradiation of the nanocomposite, which, in turn, caused the generation of hydroxyl radicals and superoxide radical anions [53]. Other researchers prepared the same nanocomposites but used the brown macroalgae *Sargassum species* along with ZnO and Co-ZnO. The nanocomposites were used in the photocatalytic degradation – under visible light – of malachite green, where complete removal of the dye was achieved upon using 5 mg/L dye after 10 min at pH 7 for 0.06 g ZnO/algae and 0.05 g Co-ZnO/algae. The percent removal of the formed nanocomposites exceeded that of the plain ZnO nanoparticles by 58% and 75% for ZnO/algae and Co-ZnO/algae, respectively [54]. In addition to dye photodegradation, microalgae–ZnO nanocomposites were also examined for the treatment of hydrocarbon-contaminated wastewater. Two species of microalgae (*Dunaliella tertiolecta* and *Chlorella vulgaris*) were used and exposed to a solution of ZnO nanoparticles, then incubated for seven days. Data obtained by FTIR spectroscopy suggest the attachment of ZnO to microalgae, where a peak at 489 cm^{-1} ascribed to ZnO appeared. Different hydrocarbon concentrations were applied to test the efficiency of the hybrid system as opposed to the individual microalgae and ZnO nanoparticles systems. As a result, the hybrid system turned out to be much more efficient than the other two systems [55].

Table 10.2 summarizes the findings of the studies performed on algal nanocomposites used in water remediation.

10.5 ALGAL NANOMEMBRANES

The idea of preparing nanomembranes containing algae for water treatment is based on the ability of the algal cells to grow and capture pollutants from wastewater. Algae are difficult to retrieve from water-based solutions, so they are affixed to a substrate that has specific properties, such as a high surface area, large pore volume, and uniform porosity, to improve bioremediation [56]. In this regard, *Chlorella vulgaris* immobilized on a porous chitosan nanofiber mat was prepared to treat nitrate-contaminated wastewater. The positively charged chitosan at acidic to neutral pHs favored binding to the negatively charged algal cells as well as the removal of nitrates. Percent removal of nitrate was 87% for the algal-immobilized chitosan, and 32% for the bare chitosan mat, after 15 h [8]. Immobilized algae were also employed for dye removal. The microalgae *C. reinhardtii* was immobilized on a polysulfone nanofibrous web to remove the reactive dyes of Remazol black 5 and reactive blue 221 from wastewater. Algae were left to grow on the nanofibrous web for ten days, and Raman spectroscopy demonstrated the presence of polysulfone compounds, polysaccharides, and glycoproteins on the web surface. At a dye concentration of 10 mg/L and pH 6, removal efficiencies of 73% and 32% were obtained for Remazol black 5 and reactive blue 221, respectively, after 14 days. These efficiencies were higher than those obtained by microalgae alone or the pristine polysulfone web. It was observed that the pH decreased by increasing dye concentration, and this negatively affected algal growth and hindered adsorption due to the high concentration of H^+ ions present. The nanomembrane was regenerated with phosphate-buffered saline (PBS) buffer and reused for three cycles [57]. A more recent study focused on the photocatalytic ability of microalgae in removing heavy metals. *Chlorella vulgaris* were immobilized on TiO_2/Ag chitosan hybrid nanofiber membrane and tested for the visible light-induced photocatalysis of Cr (VI). TiO_2 and

TABLE 10.2
Summary of Algal Nanocomposites Used to Remove Contaminants from Water

Algal Species	Nanocomposite Type	BET Surface Area (m^2/g)	Contaminant Removed from Water	Removal (%)	Maximum Adsorption Capacity (mg/g)	References
Phaeodactylum tricornutum	Diatom/ZrO_2	140.0	Fluorine	90.0	–	[42]
–	SiO_2 algae biochar	42.4	Phosphorous	98.0	159.4	[43]
Brown algae	Fe_3O_4 magnetic biochar	337.0	Acid orange 7	–	382.0	[44]
Sargassum oligocystum	Algal-activated carbon/Fe_3O_4 magnetite	126.8	Methyl violet	99.5	59.9	[45]
			Methyl blue	97.9	60.6	
Enteromorpha sp.	CoO_x/MoO_x multiwrinkled biomass carbon	290.7	Methyl blue	91.6	1587.3	[46]
Laminaria japonica	Activated carbon Fe_3O_4 magnetic composites	420.8	Acetylsalicylic acid	–	127.0	[47]
Enteromorpha sp.	Hydrous cerium oxide NPs	16.5	Cr(VI)	–	–	[48]
Sargassum horneri	Fe_3O_4 NPs	–	Acridine orange Malachite green	–	193.8 110.4	[49]
Sargassum swartzii	Zero-valent iron/algae biocomposite	–	Crystal violet	–	200.0	[50]

silver were incorporated into the nanofiber membrane to improve its photocatalytic power. Algae produced reactive oxygen species, such as $^1O_2$ and O_2^-, along with some hydroxyl radicals; however, they were destroyed under light, and this destruction increased upon immobilization. The prepared immobilized algae system reduced 50% of Cr (VI) ions. By increasing the initial concentration of the metal ions, percent removal and reduction rate decreased due to blockage of algal sites by the excess metal ions. The system was also efficiently reused for five cycles with only 10% reduction in its photocatalytic power [56].

In addition to water treatment applications, algal-incorporated nanomembranes were utilized as biomedical scaffolds. *Spirulina platensis* microalgae were immobilized on a polycaprolactone nanofibrous mat to induce the production of astrocytes necessary for treating brain injuries without causing cytotoxicity as the immobilized microalgae enhanced astrocytes metabolism [58]. The same algal species were immobilized on a silk fibroin nanofiber mat to form scaffolds for blood vessels. The scaffold was tested for platelet aggregation, and they inactivated the platelets that adhered to its surface. Therefore, they were recommended as anticoagulants in intrinsic and common pathways of blood coagulation. Furthermore, these biocompatible mats were found to enhance cell proliferation and growth [58].

10.6 ALGAL EXTRACTS IN NANOCOMPOSITES

For environmental applications, cellulose nanofibers were extracted from *Cladophora rupestris* and were coated with polyaniline nanolayer to form a composite which successfully removed chromium (VI) with an efficiency of 46%. The mechanism of removal involved redox reaction and surface adsorption via coordinate bonds between the nitrogen in polyaniline and the hexavalent chromium [59]. Cellulose nanofibers were also produced from *Cladophora* algae and were dried either by heating or freeze-drying, then pyrolyzed to obtain a char for oil absorption and methylene blue dye adsorption. Performance of the char was compared to that of wood and bacterial-derived cellulose treated in the same way. The char produced from algae-derived cellulose using thermal drying had a BET surface area of 361 m^2/g and a pore volume of 0.64 cm^3/g, which was much higher than that of wood nanofibers. However, the freeze-dried one had less pore volume than the char produced from bacteria or wood and consequently lower capacity for absorption of oil. These properties were also reflected on the capability of the nanofibers to adsorb dyes. Chars from algae-derived cellulose treated by heating showed best removal efficiency of 78% at a 40 mg/L initial concentration of methylene blue, which is greater than that of chars from bacterial-derived fibers, noting that chars from wood-derived fibers gave negligible removal at the same initial concentration [60].

10.7 CONCLUSION

This chapter reported on various nanoparticles biosynthesized utilizing algae for potential use in different environmental and biochemical applications (Figure 10.2). The algal-synthesized nanoparticles, including silver, gold, and copper oxide, are considered safer and more biocompatible relative to their chemically synthesized counterparts and hence have potentials in biomedicines due to their versatile biological activities, as well as in water treatment, owing to their efficiency in catalytic degradation. Algal species and their extracts were also involved in the biosynthesis of nanocomposites and nanomembranes for use in water bioremediation and treatment. Algal nanocomposites, which are prepared using heating treatment methods involving calcination and pyrolysis, have shown enhanced adsorption of impurities from water. Algal nanomembranes were also shown to find use in water treatment and bioremediation based on the sorption capacity of algae immobilized on a substrate of high surface area and uniform porosity.

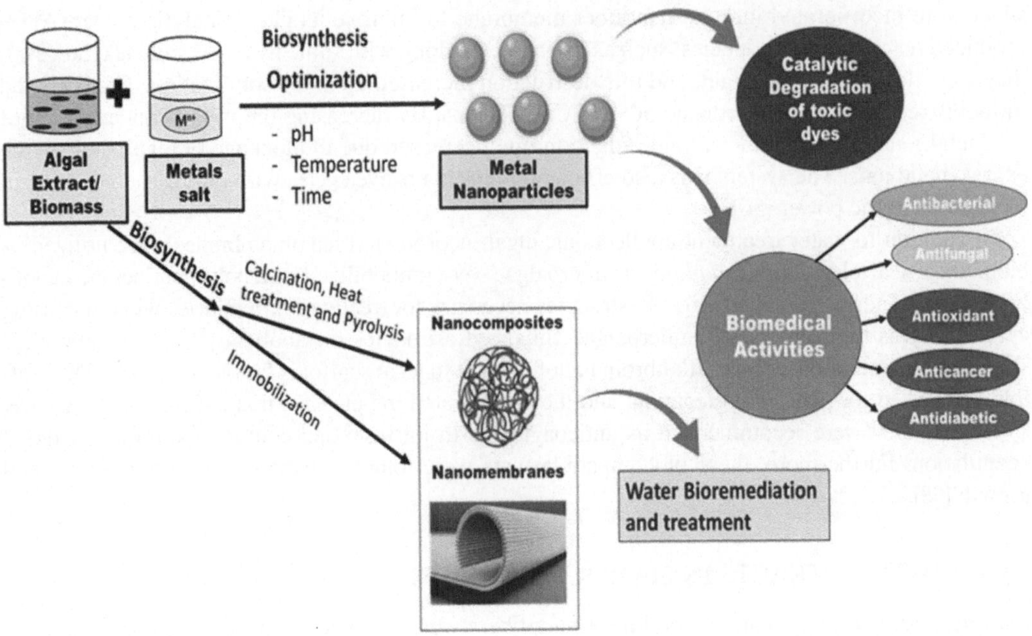

FIGURE 10.2 Algal-mediated synthesis of nanoforms and their applications in the biochemical and environmental fields.

REFERENCES

1. Sharma, A.; Sharma, S.; Sharma, K.; Chetri, S. P. K.; Vashishtha, A.; Singh, P. et al. **2016**. Algae as crucial organisms in advancing nanotechnology: A systematic review. *Journal of Applied Phycology* 28:1759–1774.
2. Aboelfetoh, E. F.; El-Shenody, R. A.; Ghobara, M. M. **2017**. Eco-friendly synthesis of silver nanoparticles using green algae (Caulerpa serrulata): Reaction optimization, catalytic and antibacterial activities. *Environmental Monitoring and Assessment* 189:349.
3. El-Sayed, M. M. H.; Fleita, D.; Rifaat, D.; Essa, H. **2018**. Assessment of the state-of-the-art developments in the extraction of antioxidants from marine algal species. In *Ingredients Extraction by Physicochemical Methods in Food*. A. M.Grumezescu and A. M. Holban (eds.). Cambridge, USA: Academic Press, Elsevier, pp. 367–397.
4. AbouElAzm, N.; Fleita, D.; Rifaat, D.; Mpingirika, E. Z.; Amleh, A.; El-Sayed, M. M. H. **2019**. Production of bioactive compounds from the sulfated polysaccharides extracts of Ulva lactuca: Post-extraction enzymatic hydrolysis followed by ion-exchange chromatographic fractionation. *Molecules* 24:2132.
5. Dayuti, S. **2018**. Antibacterial activity of red algae (Gracilaria verrucosa) extract against *Escherichia coli* and *Salmonella typhimurium*. *IOP Conference Series: Earth and Environmental Science* 137. https://doi.org/10.1088/1755-1315/137/1/012074.
6. Chauhan, R.; Chauhan, A.; Tripathi, A.; Ranjan, A.; Chauhan, S. C. **2021**. Pharmaceutical potential of laboratory grown cultures of blue-green algae: A comprehensive review and future possibilities. *Journal of Experimental Biology and Agricultural Sciences* 9:543–571.
7. Essa, H.; Fleita, D.; Rifaat, D.; Samy, D.; El-Sayed, M. M. H. **2018**. Towards optimizing the conventional and ultrasonic-assisted extraction of sulphated polysaccharides from marine algae. *IOP Conference Series: Materials Science and Engineering* 464(1). https://doi.org/10.1088/1757-899X/464/1/012007.
8. Eroglu, E.; Agarwal, V.; Bradshaw, M.; Chen, X.; Smith, S. M.; Raston, C. L. et al. **2012**. Nitrate removal from liquid effluents using microalgae immobilized on chitosan nanofiber mats. *Green Chemistry* 14:2682–2685.

9. Garcia-Bedoya, D.; Ramirez-Rodriguez, L.; Quiroz-Castillo, J.; Esquer-Miranda, E.; Castellanos-Moreno, A. **2021**. Caulerpa sertularioides extract as a complexing agent in the synthesis of ZnO and $Zn(OH)_2$ nanoparticles and its effect in the azo dye's photocatalysis in water. *Bioresources* 16:1548–1560.
10. Essa, H. L.; Abdelfattah, M. S.; Marzouk, A. S.; Shedeed, Z.; Guirguis, H. A.; El-Sayed, M. M. H. **2021**. Biogenic copper nanoparticles from Avicennia marina leaves: Impact on seed germination, detoxification enzymes, chlorophyll content and uptake by wheat seedlings. *PLoS One* 16(4). https://doi.org/10.1371/journal.pone.0249764.
11. Edison, T. N. J. I.; Atchudan, R.; Kamal, C.; Lee, Y. R. **2016**. Caulerpa racemosa: A marine green alga for eco-friendly synthesis of silver nanoparticles and its catalytic degradation of methylene blue. *Bioprocess and Biosystems Engineering* 39:1401–1408.
12. Khanna, P.; Kaur, A.; Goyal, D. **2019**. Algae-based metallic nanoparticles: Synthesis, characterization and applications. *Journal of Microbiological Methods* 163. https://doi.org/10.1016/j.mimet.2019.105656.
13. Rosken, L. M.; Korsten, S.; Fischer, C. B.; Schonleber, A.; Smaalen, S. V.; Geimer, S. et al. **2014**. Time-dependent growth of crystalline Au_0-nanoparticles in cyanobacteria as self-reproducing bioreactors: 1. Anabaena sp. *Journal of Nanoparticle Research* 16. https://doi.org/10.1007/s11051-014-2370-x.
14. Rai, S.; Wenjing, W.; Shrivastava, A. K.; Prashant, S. K. **2019**. Cyanobacteria as a source of nanoparticles and their applications. In *Role of Plant Growth Promoting Microorganisms in Sustainable Agriculture and Nanotechnology*. A. Kumar, A. K. Singh and K. K. Choudhary (eds.). Sawston: Woodhead Publishing, pp. 183–198.
15. Castro, L.; Blázquez, M. L.; Muñoz, J. A.; González, F.; Ballester, A. **2013**. Biological synthesis of metallic nanoparticles using algae. *IET Nanobiotechnology* 7:109–116.
16. Rajasulochana, P.; Dhamotharan, R.; Murugakoothan, P.; Murugesan, S.; Krishnamoorthy, P. **2010**. Biosynthesis and characterization of gold nanoparticles using the alga Kappaphycus alvarezii. *International Journal of Nanoscience* 9:511–516.
17. Prasad, T. N. V. K. V.; Kambala, V. S. R.; Naidu, R. **2013**. Phyconanotechnology: Synthesis of silver nanoparticles using brown marine algae Cystophora moniliformis and their characterisation. *Journal of Applied Phycology* 25:177–182.
18. Rajeshkumar, S.; Malarkodi, C.; Gnanajobitha, G.; Paulkumar, K.; Vanaja, M.; Kannan, C. et al. **2013**. Seaweed-mediated synthesis of gold nanoparticles using Turbinaria conoides and its characterization. *Journal of Nanostructure in Chemistry* 3(44). https://doi.org/10.1186/2193-8865-3-44.
19. Thangaraju, N.; Venkatalakshmi, R. P.; Chinnasamy, A.; Kannaiyan, P. **2012**. Synthesis of silver nanoparticles and the antibacterial and anticancer activities of the crude extract of sargassum polycystum C. Agardh. *Nano Biomedicine and Engineering* 4:89–94.
20. Kalabegishvili, T.; Murusidze, I.; Kirkesali, E.; Rcheulishvili, A.; Ginturi, E.; Kuchava, N. et al. **2013**. Gold and silver nanoparticles in Spirulina platensis biomass for medical application. *Ecological and Chemical Engineering* 20:621–631.
21. Aziz, N.; Faraz, M.; Pandey, R.; Shakir, M.; Fatma, T.; Varma, A. et al. **2015**. Facile algae-derived route to biogenic silver nanoparticles: Synthesis, antibacterial, and photocatalytic properties. *Langmuir* 31:11605–11612.
22. Fleita, D.; El-Sayed, M.; Rifaat, D. **2015**. Evaluation of the antioxidant activity of enzymatically-hydrolyzed sulfated polysaccharides extracted from red algae; Pterocladia capillacea. *LWT – Food Science and Technology* 63:1236–1244.
23. Borah, D.; Das, N.; Das, N.; Bhattacharjee, A.; Sarmah, P.; Ghosh, K. et al. **2020**. Alga-mediated facile green synthesis of silver nanoparticles: Photophysical, catalytic and antibacterial activity. *Applied Organometallic Chemistry* 34(3). https://doi.org/10.1002/aoc.5597.
24. Patel, V.; Berthold, D.; Puranik, P.; Gantar, M. **2015**. Screening of cyanobacteria and microalgae for their ability to synthesize silver nanoparticles with antibacterial activity. *Biotechnology Reports* 5:112–119.
25. Yousefzadi, M.; Rahimi, Z.; Ghafori, V. **2014**. The green synthesis, characterization and antimicrobial activities of silver nanoparticles synthesized from green alga Enteromorpha flexuosa (wulfen) J.Agardh. *Materials Letters* 137:1–4.
26. Dixit, D.; Gangadharan, D.; Popat, K. M.; Reddy, C. R. K.; Trivedi, M.; Gadhavi, D. K. **2018**. Synthesis, characterization and application of green seaweed mediated silver nanoparticles (AgNPs) as antibacterial agents for water disinfection. *Water Science and Technology* 78:235–246.
27. Kingslin, A.; Kalimuthu, K.; Kiruthika, M. L.; Khalifa, A. S.; Nhat, P. T.; Briindhadevi, K. **2022**. Synthesis, characterization and biological potential of silver nanoparticles using Enteromorpha prolifera. *Applied Nanoscience*. https://ui.adsabs.harvard.edu/link_gateway/2023ApNan.13.2165K/doi:10.1007/s13204-021-02105-x.

28. Hafez, E. E.; Kabeil, S. S. **2013**. Antimicrobial activity of nano-silver particles produced by micro algae. *Journal of Pure and Applied Microbiology* 7:35–42.
29. Arunkumar, M.; Suhashini, K.; Mahesh, N.; Ravikumar, R. **2014**. Quorum quenching and antibacterial activity of silver nanoparticles synthesized from Sargassum polyphyllum. *Bangladesh Journal of Pharmacology* 9:54–59.
30. Sathyaseelan, T.; Subbiah, M.; Sivamugugan, V. **2015**. Green synthesis of silver nanoparticles using marine brown algae Lobohora variegata and its efficacy in antifungal activity. *World Journal of Pharmaceutical Research* 4:2137–2145.
31. El-Sheekh, M. M.; Hassan, L. H. S.; Morsi, H. S. **2021**. Evaluation of antimicrobial activities of blue-green algae-mediated silver and gold nanoparticles. *Rendiconti Lincei-Scienze Fisiche e Naturali* 32:747–759.
32. Kiran, M. V.; Murugesan, S. **2014**. Biological synthesis of silver nanoparticles from marine alga Colpomenia sinuosa and its *in vitro* anti-diabetic activity. *American Journal of Bio-Pharmacology Biochemistry and Life Sciences* 3:1–7.
33. Sharma, B.; Purkayastha, D. D.; Hazra, S.; Thajamanbi, M.; Bhattacharjee, C. R.; Ghosh, N. N. et al. **2014**. Biosynthesis of fluorescent gold nanoparticles using an edible freshwater red alga, Lemanea fluviatilis (L.) C.Ag. and antioxidant activity of biomatrix loaded nanoparticles. *Bioprocess and Biosystems Engineering* 37:2559–2565.
34. Murugesan, S.; Bhuvaneswari, S.; Shanthi, N.; Murugakoothan, P.; Sivamurugan, V. **2015**. Red alga Hypnea musciformis (Wulf) Lamour mediated environmentally benign synthesis and antifungal activity of gold nano particles. *International Journal of Nanoscience and Nanotechnology* 6:71–83.
35. Naveena, B. E.; Prakash, S. **2012**. Biological synthesis of gold nanoparticles using marine algae Gracilaria corticata and its application as a potent antimicrobial and antioxidant agent. *Asian Journal of Pharmaceutical and Clinical Research* 6:179–182.
36. Bhattacharya, P.; Swarnakar, S.; Ghosh, S.; Majumdar, S.; Banerjee, S. **2019**. Disinfection of drinking water via algae mediated green synthesized copper oxide nanoparticles and its toxicity evaluation. *Journal of Environmental Chemical Engineering* 7(1). https://doi.org/10.1016/j.jece.2018.102867.
37. Ramaswamy, S. V. P.; Narendhran, S.; Sivaraj, R. **2016**. Potentiating effect of ecofriendly synthesis of copper oxide nanoparticles using brown alga: Antimicrobial and anticancer activities. *Bulletin of Materials Science* 39:361–364.
38. Abboud, Y.; Saffaj, T.; Chagraoui, A.; Bouari, A. E.; Brouzi, K.; Tanane, O. et al. **2014**. Biosynthesis, characterization and antimicrobial activity of copper oxide nanoparticles (CONPs) produced using brown alga extract (Bifurcaria bifurcata). *Applied Nanoscience* 4:571–576.
39. Kumaresan, M.; Anand, K. V.; Govindaraju, K.; Tamilselvan, S.; Kumar, V. G. **2018**. Seaweed Sargassum wightii mediated preparation of zirconia (ZrO_2) nanoparticles and their antibacterial activity against gram positive and gram negative bacteria. *Microbial Pathogenesis* 124:311–315.
40. Satheeshkumar, M. K.; Kumar, E. R.; Indhumathi, P.; Srinivas, C.; Deepty, M.; Sathiyaraj, S. et al. **2020**. Structural, morphological and magnetic properties of algae/$CoFe_2O_4$ and algae/Ag-Fe-O nanocomposites and their biomedical applications. *Inorganic Chemistry Communications* 111. https://doi.org/10.1016/j.inoche.2019.107578.
41. Mandal, R. P.; Sekh, S.; Sarkar, N. S.; Chattopadhyay, D.; De, S. **2016**. Algae mediated synthesis of cadmium sulphide nanoparticles and their application in bioremediation. *Materials Research Express* 3(5).
42. Thakkar, M.; Wu, Z.; Wei, L.; Mitra, S. **2015**. Water defluoridation using a nanostructured diatom–ZrO_2 composite synthesized from algal Biomass. *Journal of Colloid and Interface Science* 450:239–245.
43. Wang, Y. Y.; Lu, H. H.; Liu, Y. X.; Yang, S. M. **2016**. Removal of phosphate from aqueous solution by SiO_2–biochar nanocomposites prepared by pyrolysis of vermiculite treated algal biomass. *RSC Advances* 6:83534–83546.
44. Jung, K. W.; Choi, B. H.; Jeong, T. U.; Ahn, K. H. **2016**. Facile synthesis of magnetic biochar/Fe_3O_4 nanocomposites using electro-magnetization technique and its application on the removal of acid orange 7 from aqueous media. *Bioresource Technology* 220:672–676.
45. Foroutan, R.; Mohammadi, R.; Razeghi, J.; Ramavandi, B. **2019**. Performance of algal activated carbon/Fe_3O_4 magnetic composite for cationic dyes removal from aqueous solutions. *Algal Research* 40. https://doi.org/10.1016/j.algal.2019.101509.
46. Yang, L.;Wang, Y.; Liu, A.; Zhang, Y. **2019**. CoO x/MoO y-anchored multi-wrinkled biomass carbon as a promising material for rapidly selective methyl blue removal. *Journal of Materials Science* 54:11024–11036.

47. Jung, K. W.; Choi, B. H.; Song, K. G.; Choi, J. W. **2019**. Statistical optimization of preparing marine macroalgae derived activated carbon/iron oxide magnetic composites for sequestering acetylsalicylic acid from aqueous media using response surface methodologys. *Chemosphere* 215:432–443.
48. Rangabhashiyam, S.; Selvaraju, N.; Mohan, B. R.; Anzil, P. K. M.; Amith, K. D.; Ushakumary, E. R. **2016**. Hydrous cerium oxide nanoparticles impregnated Enteromorpha sp. for the removal of hexavalent chromium from aqueous solutions. *Journal of Environmental Engineering* 142(9).
49. Angelova, R.; Baldikova, E.; Pospiskova, K.; Maderova, Z.; Safarikova, M.; Safarik, I. **2016**. Magnetically modified Sargassum horneri biomass as an adsorbent for organic dye removal. *Journal of Cleaner Production* 137:189–194.
50. Jerold, M.; Vasantharaj, K.; Joseph, D.; Sivasubramanian, V. **2017**. Fabrication of hybrid biosorbent nanoscale zero-valent iron-Sargassum swartzii biocomposite for the removal of crystal violet from aqueous solution. *International Journal of Phytoremediation* 19:214–224.
51. Mullerova, S.; Baldikova, E.; Prochazkova, J.; Safarik, I. **2019**. Magnetically modified macroalgae Cymopolia barbata biomass as an adsorbent for safranin O removal. *Materials Chemistry and Physics* 225:174–180.
52. Gupta, P. L.; Jung, H.; Tiwari, D.; Kong, S. H.; Lee, S. M. **2017**. Insight into the mechanism of Cd (II) and Pb (II) removal by sustainable magnetic biosorbent precursor to Chlorella vulgaris. *Journal of the Taiwan Institure of Chemical Engineers* 71:206–214.
53. Pandimurugan, R.; Thambidurai, S. **2016**. Novel seaweed capped ZnO nanoparticles for effective dye photodegradation and antibacterial activity. *Advanced Powder Technology* 27:1062–1072.
54. Rabie, A. M.; Abukhadra, M. R.; Rady, A. M.; Ahmed, S. A.; Labena, A.; Mohamed, H. S. et al. **2020**. Instantaneous photocatalytic degradation of malachite green dye under visible light using novel green Co–ZnO/algae composites. *Research on Chemical Intermediates* 46:1955–1973.
55. Salehi, M.; Biria, D.; Shariati, M.; Farhadian, M. **2019**. Treatment of normal hydrocarbons contaminated water by combined microalgae–photocatalytic nanoparticles system. *Journal of Environmental Management* 243:116–126.
56. Wang, L.; Zhang, C.; Gao, F.; Mailhot, G.; Pan, G. **2017**. Algae decorated TiO_2/Ag hybrid nanofiber membrane with enhanced photocatalytic activity for Cr (VI) removal under visible light. *Chemical Engineering Journal* 314:622–630.
57. Keskin, N. O. S.; Celebioglu, A.; Uyar, T.; Tekinay, T. **2015**. Microalgae immobilized by nanofibrous web for removal of reactive dyes from wastewater. *Industrial and Engineering Chemistry Research* 54:5802–5809.
58. Kim, S. H.; Shin, C.; Min, S. K.; Jung, S. M.; Shin, H. S. **2012**. *In vitro* evaluation of the effects of electrospun PCL nanofiber mats containing the microalgae Spirulina (Arthrospira) extract on primary astrocytes. *Colloids and Surfaces B: Biointerfaces* 90:113–118.
59. Camacho, D. H.; Gerongay, S. R. C.; Macalinao, J. P. C. **2013**. Cladophora cellulose-polyaniline composite for remediation of toxic chromium (VI). *Cellulose Chemistry and Technology* 47:125–132.
60. Štefelová, J.; Slovák, V.; Siqueira, G.; Olsson, R. T.; Tingaut, P.; Zimmerman, T. et al. **2017**. Drying and pyrolysis of cellulose nanofibers from wood, bacteria, and algae for char application in oil absorption and dye adsorption. *ACS Sustainable Chemistry and Engineering* 5:2679–2692.

11 Cyanobacteria as Bioagents for Synthesis of Nanoparticles

Komal, Ashutosh Sharma, Yadvinder Singh, and Rahul Badru
Sri Guru Granth Sahib World University, Punjab, India

11.1 INTRODUCTION

Nanotechnology is an important area of research uniting material science and engineering, involving synthesis of inorganic, organic, and composite nature particles in the nanodimensional range (i.e., 1–100 nm), with divergent chemical composition, dispersity, and shapes. The past two decades have seen remarkable advances in the field of nanotechnology in terms of synthetic methodologies, materials, material processing, characterization techniques, and widespread applications. Nanoparticles (NPs) attributed entirely different optical, electronic, chemical, and photoelectrochemical properties from those of bulk materials due to their high surface-area-to-volume ratio and scalable size [1–2]. NPs have remained the center of curiosity for intensive research work. The science of NPs has also become an inseparable part of a multidisciplinary field of research due to their eccentric behavior, properties, and applications. Their large surface-area-to-volume ratio, ability to interact with other particles, and other characteristics make them attractive tools in various fields. They find applications in the fields of environmental sensing and monitoring [3a,b], wastewater treatment [4], shape and size selective catalysis [5], molecular imaging [6], photocatalysis [7], semiconductors [8], and medicine [9].

Much of the literature in the past is devoted to the synthesis of metallic nanoparticles (MNPs) of Au, Ag, Pd, Pt, Zn, Ni, Cu, Rh, and Ir. Besides these, several NPs of metal oxides, hydroxides, and halides have also been synthesized and worked upon.

11.2 SYNTHESIS OF NANOPARTICLES

Nanoparticles are synthesized by different chemical, physical, photochemical, electrochemical, radiolytic, and biological methods. The prime requirements in the synthesis of NPs are metal salt solutions (whose NPs are to be synthesized); appropriate reducing agent, such as $NaBH_4$/NH_2NH_2; and capping agents to avoid self-aggregation of the synthesized NPs and to increase their stability. In the chemical mode of NP synthesis, $NaBH_4$ reduces the metal ion from metal salt solution to metal NPs that are encapsulated by the capping agent, thus stabilizes them. With the advancement in time, several indirect modes of reduction have also been exploited in place of direct usage of reducing agents. Among these, the biogenic mode of synthesis has caught much pace in the last few decades. The biogenic mode of synthesis represents the usage of biomass, extracts, or metabolites (polyphenols, proteins, flavonoids, alkaloids, and polysaccharides) of bioagents like bacteria, algae, fungi, and plants toward the synthesis of NPs. In recent years, numerous reports have been published focused on the biosynthesis of MNPs by using bacteria [10], fungi [11], yeast [12], actinomycetes [13], viruses [14], and many reviews on the same [15–17]. The reviews focused on algae-mediated NP synthesis have also been published previously [18–21]. These articles disseminate information on the biosynthesis of a variety of nanostructures using various strains of algae, and their potential applications in diverse fields. But the hypothetical biochemical mechanism underlying such biosyntheses is only defined in few cases [15, 19].

 DOI: 10.1201/9781003430087-11

Nanobiotechnology aims to make use of all biological models in the production of nanoscaled products for a broad range of applications [18–20]. Various biomolecules of natural origin are identified that act as a driving force for the designing of greener, safer, and environment friendly protocols for the synthesis of NPs of distinct shape and size [16]. Microbes like bacteria, fungi, viruses, and algae are able to grab several of the metal ions from the environment and convert them into nanoscale particles with the help of enzymes, which are involved in intracellular or extracellular activities of cell [[10, 13–14]]. Plant-extracted metabolites have been found to play a dual role as reducing and capping agent for NPs during the synthesis process [21–22]. The morphology and sizes of NPs mainly depend on experimental conditions involved in the extraction and plant part/extract in use [23]. The underlying mechanism might differ from bioagent to bioreagent, but all of them stand better in terms of environmental sustainability over chemical reduction procedures. Cost-effectiveness, feasibility, and scalability are the other fundamental features of these biogenic technologies. Further, purification and isolation were also found to perform an important role in deciding the characteristics of the obtained NPs.

The biogenic mode of synthesis of NPs is sometimes referred to as bioremediation, as in biogenic synthesis, the biological organisms like bacteria, fungi, algae, and plants, as a whole or their products, are utilized for the removal of potentially toxic contaminants, like heavy metal ions, pesticides, and other organic compounds from contaminated environment. As a consequence, it will be fitting to state, bioremediation actually sets the stage for nanomaterials to be biosynthesized using bioagents such as algal tools, prokaryotes, fungi, and plants. It is like killing two birds with one stone, that is, to bioremediate metal cations and to biosynthesize nanomaterials of the same. On one end, it helps bioagents in getting rid of toxic metal ions from the surrounding growth medium of plants, and on the other end, one gets a handful of metal NPs. The transition from bioremediation to biosynthesis can be seen in the headings of a number of publications in recent years on biogenic synthesis of NPs [23–25].

The biosynthesis of NPs offers the following advantages over other methods:

- Carried out in aqueous and ambient conditions of temperature and pressure, thus are lesser toxic than other methods.
- Biological components are used as reducing and capping agent, hence reduces the cost of synthesis process
- Due to slow kinetics, they offer a better control over the crystal growth. Biological systems thus serve better among various green synthetic methodologies for NP synthesis.

11.3 SYNTHESIS OF NPS BY ALGAE

The high metal uptake ability and predominance in all types of environments make algae a low-cost raw material for biogenic synthesis of NPs [21]. The use of algae in the biosynthesis of NPs has been regarded as a green and reliable process. Figure 11.1 represents the share of algae based on a number of publications involving biogenic synthesis of NPs, which is merely 4%, but almost all algal groups have been explored for their potential for the biosynthesis of NPs. The different algal species explored for their metallic NP synthesis potential by various workers are presented in Table 11.1. The autotrophic nature, ease in culturing and handling, higher rate of metal accumulation capability, with other advantages, including utilization for synthesis at low temperature with greater energy efficiency, less toxicity, and less risk to the environment, make algae a better choice for NP synthesis.

In one of the preliminary studies, *Chlorella* has been found to accumulate heavy metals ions of uranium, copper, cadmium, and nickel [84]. *Chlorella* has also been reported to be enriched with pigments like chlorophyll; carotenoids; phenols; xanthines; flavonoids; compounds like proteins, minerals, and vitamins [85]; and functional groups, like hydroxyl, carboxyl, and amines that serve a dual role as metal-reducing and capping agents to provide stabilized metal NPs out of their aqueous

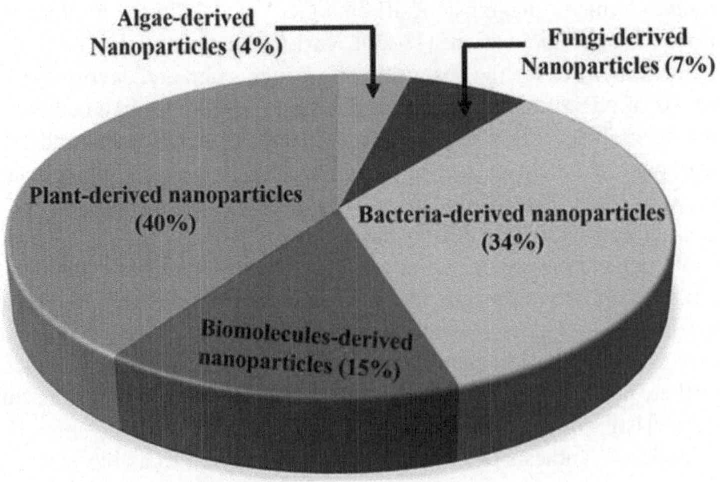

FIGURE 11.1 Share of algae, fungi, bacteria, plants, and biomolecules toward the biogenic synthesis of NPs. Source: With permission from Elsevier [15].

TABLE 11.1
List of Different Species of Red, Brown, and Green Algae Used in Biogenic Synthesis of NPs

Class	Species	Metal Synthesized	References
Chlorophyceae (green algae)	*Acutodesmus* sp.	Ag	[26]
	Botryococcus sp.	Ag	[27]
	Caulerpa sp.	Ag	[28]
	Chaetomorpha sp.	Ag	[25]
	Chlorella sp.	Ag, Au	[27, 29]
	Chlamydomonas sp.	Ag	[27]
	Codium sp.	Ag	[25]
	Coelastrum sp.	Ag	[27]
	Cosmarium sp.	Au	[30]
	Enteromorpha sp.	Ag	[31]
	Klebsormidium sp.	FeOOH	[32]
	Pithophora sp.	Ag	[33]
	Prasiola sp.	Au	[34]
	Rhizoclonium sp.	Au	[35]
	Spirogyra sp.	Ag, Au	[32, 36]
	Scenedesmus sp.	Ag	[27]
	Tetraselmis sp.	Au	[37]
	Ulva sp.	Ag, Au	[31, 38]
	Urospora sp.	Ag	[39]
Cyanophyceae (blue-green algae)	*Anabaena* sp.	Ag, ZnO	[27, 40]
	Limnothrix sp.	Ag	[27]
	Lyngbya sp.	Ag, Au	[27, 41]
	Microchaete sp.	Ag	[42]
	Microcoleus sp.	Au	[38]

(Continued)

TABLE 11.1 *(Continued)*
List of Different Species of Red, Brown, and Green Algae Used in Biogenic Synthesis of NPs

Class	Species	Metal Synthesized	References
	Nostoc sp.	Ag, Au	[43]
	Oscillatoria sp.	Ag, Au	[44a,b, 45]
	Phormidium sp.	Ag, Au, CuO, CdS	[38, 46–48]
	Plectonema sp.	Ag, Au, Pt, Pd	[49]
	Spirulina sp.	Ag, Au	[50]
	Synechocystis sp.	Ag, Au	[27, 51]
Phaeophyceae (brown algae)	*Bifurcaria* sp.	CuO	[52]
	Colpmenia sp.	Ag	[53]
	Cystophora sp.	Ag	[54]
	Cystoseira sp.	Au	[55]
	Dictyota sp.	Au	[56]
	Ecklonia sp.	Au	[57]
	Fucus sp.	Au	[58]
	Gilidie sp.	Ag	[59]
	Laminaria sp.	Au	[60]
	Padina sp.	Ag, Au	[61–62]
	Sargassum sp.	Ag, Au, Pd, ZnO, Fe_3O_4, AgCl	[31, 63–68]
	Stoechospermum sp.	Au	[69]
	Turbinaria sp.	Ag, Au	[70]
Rhodophyceae (red algae)	*Acanthophora* sp.	Ag	[71]
	Amphiroa sp.	Ag	[72]
	Chondrus sp.	Au	[73]
	Galaxaura sp.	Au	[74]
	Gracilaria sp.	Ag, Au, ZnO	[75–77]
	Jania sp.	Ag	[53]
	Kappaphycus sp.	Au, Ag	[78–79]
	Lemanea sp.	Au	[80]
	Porphyra sp.	Ag	[81]
	Pterocladia sp.	Ag	[53]
Bacillariophyceae (diatoms)	*Amphora* sp.	Ag	[82]
	Navicula sp.	Au	[41]
Euglenophyceae (euglenoids)	*Euglena* sp.	Ag	[83]

solutions. An extract containing live culture of *Chlorella vulgaris* was able to reduce Ag^+ ions to silver NPs, even at higher concentrations, within a short interval of time [86]. Similar results were obtained with *Euglena gracilis* [87]. The aqueous solution of gold (III) ions could be reduced to gold NPs even with the dried algal biomass [88]. Approximately 88% of gold was attained in the metallic state, the crystals of which got deposited into the interior as well as the exterior cell surface in various shapes, like tetrahedral, icosahedral, and decahedral structures.

Thus, biogenic synthesis by algae can be broadly classified into intracellular and extracellular modes. The importance of both modes has been explained separately by scientists in their reports. The extracellular biosynthesis process is analogous to biosorption, which means the ability of dead biomass as biosorbent has the capacity to bind heavy metals, thus enabling the removal of these pollutants from aqueous solutions, even when present in very low concentration [45]. On the other hand, the intracellular process corresponds to bioaccumulation, which depends on the heavy metal uptake capability of living cells from polluted environment and mineralize them through a metabolic pathway. When the concentration of metal ions reaches beyond a certain limit, the living cells and biomass start biosynthesizing the metallic nanostructures.

11.3.1 Intracellular Mode of Biogenic Synthesis of MNPs

In the proposed intracellular mode of reduction, initially metal ion uptake by cell takes place, which gets reduced intracellularly and later on ejected into the surrounding medium either directly or dispersed only after lysis of the cell. The whole process of intracellular formation of NPs by algal biomass is understood partially. In the mechanisms proposed by Mandal et al. [89] and Parial et al. [38], due to electrostatic interaction between ions and charged groups, the metal ions are first attached on the cell surface. After that, the cellular enzymes are responsible for the reduction of ions, which leads to the formation of nuclei, and then finally the reduction of metal ions (Figure 11.2).

11.3.2 Extracellular Mode of Biogenic Synthesis of MNPs

The extracellular synthesis of NPs involves the synthesis of NPs using biomass or the cell-free extract. In extracellular mode, different biomolecules, including pigments, proteins, certain hormones, and enzymes, play an important role, like reducing and capping agents, in the synthesis process [90]. Proteins, through amino groups or cysteine residues and sulfated polysaccharides, are found to be responsible for the stabilization and capping of metal NPs in aqueous solutions [91]. The various existing modes of extracellular synthesis of MNPs, including the use of cell biomass extracts, cell-free culture media, and biomolecules, are depicted in Figure 11.3.

In few cases, simultaneous involvement of extracellular as well as intracellular enzymes toward the reduction of metal ion to metal NPs is reported [35, 92]. Sau et al. [92] reported the involvement of both processes in algae-mediated synthesis of AuNPs, which was either intracellular in *Rhizoclonium fontinale* and, in case of *Lyngbya majuscula* and *Spirulina subsalsa*, as extracellular in the growth medium [35].

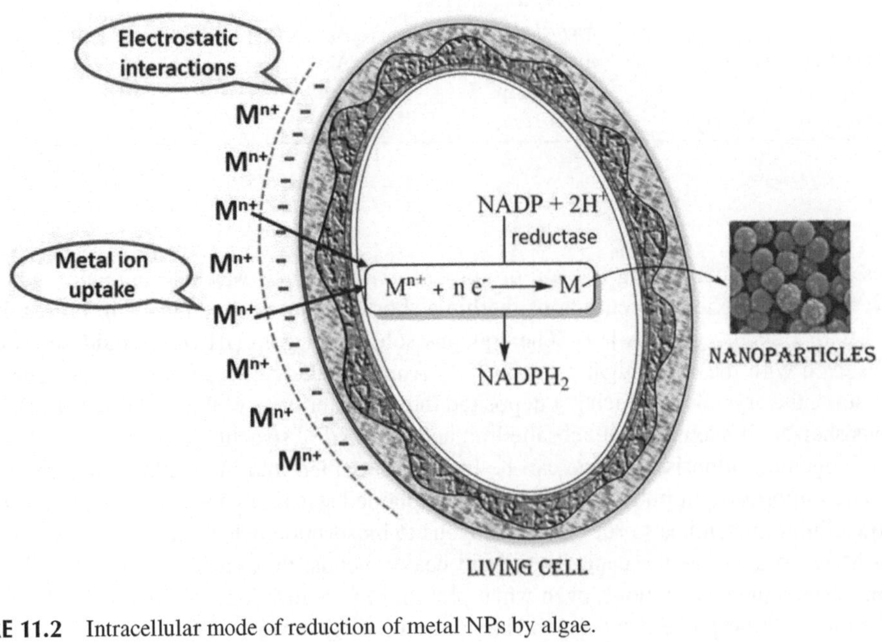

FIGURE 11.2 Intracellular mode of reduction of metal NPs by algae.

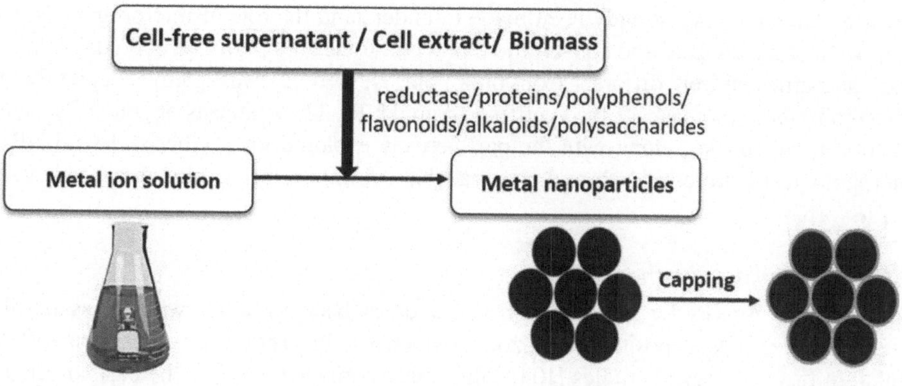

FIGURE 11.3 Extracellular mode of reduction of metal NPs by algae.

11.4 CYANOBACTERIA AS A BIOAGENT FOR NANOPARTICLE SYNTHESIS

Cyanobacteria (blue-green algae) are oxygen-evolving photosynthetic prokaryotic organisms distributed in a wide range of aquatic and terrestrial habitats all over the world. Cyanobacteria have a high growth rate and biomass productivity, with the ability to accumulate elements required for their growth and metabolism from their environment. These organisms are now gaining more attention in the field of nanobiotechnology due to their ability to solubilize metal ions and in reducing them efficiently. Cyanobacteria are considered as one of the most important sources of pigments, protein, vitamins, lipids, and other biomolecules [93], and therefore, they are considered a better biological template for the synthesis of nanoscale particles. Cyanobacteria have been employed as biological template for the synthesis of nanoscale particles via intracellular as well as extracellular modes [27, 94].

Initially, cyanobacteria were employed for the biosynthesis of Ag- and AuNPs [95–96], but later, Pd, Pt, Ni, Zn, Cu, and Cd NPs have also been synthesized using the extract of cyanobacteria. The present review signifies and summarizes the synthesis and application of NPs synthesized by using various cyanobacterial strains. The mechanistic aspects involved in the biosynthesis of NPs by cyanobacteria reported earlier can also be taken into account to depict the probable mechanism of NP synthesis by cyanobacteria. This review is structured with a goal in mind to assemble and disseminate the role of cyanobacteria in nanobiosynthesis up to the maximum extent.

11.4.1 Mechanism of NP Synthesis by Cyanobacteria

Cyanobacteria are well-known for synthesizing metallic nanoparticles by intracellular process. Mubarak-Ali et al. [96] have reported that the intracellular protein molecules of cyanobacterium *Oscillatoria willei* NTDM01 are responsible for the reduction of silver ion to metallic silver, which was released outside after cell death. Moreover, biomolecules, including polysaccharides and different amino acids (such as tyrosine and aspartic acid), have been identified as the bioagents for Ag ion reduction and for directing the anisotropic growth of Ag nanoplates on the basis of different functional groups, such as hydroxyl (-OH) and carboxyl anions (-COOH) [97]. Sharma et al. reported high amounts of uptake of Au (III) ions by *Sesbania drummondii* and its subsequent intracellular reduction from Au (III) ions to Au (0) [98].

11.4.1.1 Synthesis of Gold NPs

A number of studies reported the synthesis of gold NPs by using algal biomass. Cyanobacteria-mediated NP synthesis is a simplistic process in which the cyanobacterial extract is able to reduce the metal ion from its aqueous solution to NPs of variable shape within minutes to hours at room temperature. Attempts have already been made to understand the role of microorganisms, that is, how they accumulate the gold and convert them into non-toxic NPs [99]. The colloidal gold solution possesses properties entirely different from bulk gold. Numerous investigations depict the utilization of cyanobacterial biomass for the synthesis of gold NPs. The synthesis of gold NPs using various cyanobacterial species, along with the part actively explored for synthesis of gold NPs along with morphological features of NPs and pharmacological importance explored, are enlisted in the Table 11.2.

11.4.1.1.1 Extracellular Synthesis

The formation of AuNPs by algae/cyanobacteria dates back to 1984, when Dissanayake and Kritsotakis found rich Au deposits in algal zones (which also incorporate cyanobacteria) of Mannar, Sri Lanka, in their geological studies [104]. This study perhaps served as the basis of thought for interactions among algal biomass and metals. Following which, several reports link up the accumulation of Au with algae biomass [105–106]. But the practical confirmation has been done only in 2006, when Lengke et al. [49b] were able to recover gold out of the aqueous solution of Au (I) and Au (III) using biomass of the cyanobacterium *Plectonema boryanum*. The AuNPs obtained were cubical in shape and sized 10–25 nm, as confirmed by SEM, TEM, and XPS analysis. Gold NPs were found to be synthesized both on the interior and exterior of the cell-studied cyanobacterial species

TABLE 11.2
Cyanobacterial Species Employed for Biosynthesis of Gold NPs, Morphology of As-Synthesized NPs, and Their Applications

S. No.	Organism	Active Part/Mode	Morphology of NPs	Application	References
1	*Anabaena* sp.	Cultured biomass, polysaccharides	Spherical, 10–30nm	–	[100]
2	*Calothrix* sp.	Cultured biomass, polysaccharides	Spherical, 5–30 nm	–	[100]
3	*Gloeocapsa* sp.	Centrifuged, culture	Spherical, < 100 nm	Antitumor	[101]
4	*Leptolyngbya* sp.	Cultured biomass, polysaccharides	Spherical, 5–7 nm	–	[100]
5	*Lyngbya majuscule*	Thallus	Spherical, < 20 nm	–	[41]
6	*Microcoleus chthonoplastes*	Biomass	Hexagonal, triangular, 8–50 nm	–	[38]
7	*Nostoc ellipsosporum*	Biomass	Nanorods, 137–209 length, 33–69 dia	–	[43b]
8	*Oscillatoria limnetica*	Biomass	–	–	[45]
9	*Phormidium valderianum, P. tenue*	Biomass	Spherical and hexagonal, 8–50 nm	–	[38]
10	*Phormidium willei*	Biomass	Triangular, 25 nm	Antioxidant, DNA–Au binding	[44b]
11	*Plectonema boryanum*	CNB culture	Cubic, 10–25 nm	–	[49b]
12	*Rhizoclonium heiroglyphicum, R. Frontinale*	Biomass	Spherical, < 20 nm	–	[38, 41]
13	*Spirulina subsalsa, S. platensis*	Biomass, protein extract	Spherical, < 20 nm; spherical, 2–8 nm	Antibacterial against gram-positive (*Bacillus subtilis* and *Staphylococcus aureus*)	[41, 50b, 102–103]
14	*Synechocystis* sp.	Culture	Spherical, 11–15 nm	–	[51]

(Figure 11.4). Since then, interest arose in the cyanobacterial-mediated synthesis of NPs using bioactive agents or their biomass, and literature is brimming with numerous attempts in this direction.

Later, Chakraborty et al. [35] prepared NPs of Au using other cyanobacterial strains: *Lyngbya majuscula*, *Spirulina subsalsa*, and *Rhizoclonium heiroglyphicum*. The change in color of the bioextract indicated the biochemical reduction to gold NPs, which were characterized using TEM. About 30–60% of Au was accumulated in the biomass within just 10 min under slightly basic conditions and reduced slowly using cellular metabolism. Almost the whole of the Au was eluted from the algae with nitric acid.

11.4.1.1.2 Intracellular Synthesis

Parial et al. reported that cyanobacterial strains *Phormidium valderianum*, *Phormidium tenue*, *Microcoleus chthonoplastes* [38], and *Nostoc ellipsosporum* [43b] are able to synthesize gold NPs by intracellular mode. The gold was isolated as nanorods, nanospheres, and few as triangular-shaped, as confirmed by TEM analysis. However, AuNPs were not obtained by treatment of aqueous Au (III) solutions with dried algal biomass, indicating the biosynthesis to be linked to cellular metabolism. Geetha et al. studied and explained the synthesis of gold NPs intracellularly by marine CNB *Gloeocapsa* species [101]. They established that NP size can be tuned by controlling the cultural parameters. Further, they discovered gold NPs to be active against human cervical cancer. Govindaraju et al. have emphasized the synthesis of gold and silver NPs by *Spirulina platensis* due to its rich protein content and nutrient [107]. *Spirulina* is a filamentous cyanobacterium and is considered as one of the richest and most natural sources of nutrition all over the world. This alga is known to contain 60–70% vegetable protein, β-carotene, good amount of amino acids, vitamins, and other biologically active essential substances. In a similar way, other researchers carried out the biosynthesis of AuNPs using different cyanobacterial strains, including *Synechocystis* sp. PCC 6803 [51], *Phormidium willei* [44b], *Leptolyngbya* sp., *Anabaena* sp., and *Calothrix* sp. [100].

11.4.1.2 Synthesis of Silver NPs

Different shapes and sizes of silver NPs, such as spheres, rods, wires, and plates, can be synthesized by using different methods. Silver NPs have proven to be less toxic to human health, and they have a higher rate of toxicity to various microorganisms [108]. They are used in less quantity as an antimicrobial agent in comparison to the standard antimicrobial agents. The number of reports showed the important properties possessed by silver NPs and found a suitable application in different areas

FIGURE 11.4 SEM and TEM images of a cyanobacterial cell encrusted with gold NPs.

Source: With permission from ACS [49b].

and how they act as antifungal and anticancer drugs, very good antioxidants, and in the treatment of complications related to diabetes [35]. The few common cyanobacterial strains employed by various workers for synthesis of AgNPs are listed in Table 11.3.

Lengke et al. [49a] showed that the cyanobacterium *Plectonema boryanum* produce silver NPs by extracellular mode (Figure 11.5). Govindaraju et al. [31] have reported the extracellular synthesis of gold, silver, and Au/Ag bimetallic NPs using an edible and protein-rich alga, *Spirulina platensis*. Ali et al. [32] demonstrated that the protein molecules secreted outside in the medium from cells of *Oscillatory willei* NTDMO1 help in extracellular reduction of silver ions into silver NPs and their further stabilization. Sudha et al. [30] have reported the production of silver NPs by seaweeds (*Sargassum plagiophyllum*, *Ulva reticulate*, and *Enteromorpha compressa*). Ahmed et al. [39] used an eco-friendly approach by using *Spirulina platensis* and *Nostoc* sp. for the synthesis of silver NPs. The production of silver NPs has contributed one more step toward medicines concerning wound dressing and artificial implantation and in the prevention of contamination caused by microbes.

TABLE 11.3
Cyanobacterial Strains Employed for the Synthesis of AgNPs, Morphology of As-Synthesized NPs, and Their Applications

S. No.	Organism	Active Part	Morphology of NPs	Pharmacological Importance	References
1	*Anabaena ambigua* and *A. variabilis*	Biomass	Spherical, 50 nm	–	[109]
2	*Anabeana oryzae*	Cell extract	Spherical, 8–16 nm	Antitumor	[110]
3	*Aulosira fertilissma*	Biomass	Spherical, 50 nm	–	[109]
4	*Arthrospira indica*	Biomass	Spherical, 50 nm	–	[109]
5	*Calothrix marchica*	Cell extract	Spherical, 8–16 nm	Antitumor	[110]
6	*Chroococcus* sp.	Biomass	Spherical	–	[109]
7	*Cylindrospermospsis* sp.	Biomass	Variable, 13 and 31 nm	Antibacterial	[27]
8	*Gloeocapsa gelatinosa*	Biomass	Spherical	–	[109]
9	*Haloleptolyngbya alcalis*	Cell extract	Spherical, <50 nm	**Ammonia sensing**	[111]
10	*Hapalosiphon fontinalis*	Biomass	Spherical	–	[109]
11	*Limnothrix* sp.	Biomass	Variable, 13 and 31 nm	**Antibacterial**	[27]
12	*Lyngbya majuscula*	Thallus	Spherical, 5–25 nm	–	[112]
13	*Microchaete* sp.	Biomass	Spherical	–	[109]
14	Multiple species	Cell culture	Spherical, 40–80 nm	**Antibacterial**	[113]
15	*Nostoc linckia*	Biomass	Spherical, 20–50 nm	–	[114]
16	*Nostoc muscorum*	Cell extract	Spherical, 8–16 nm	**Antitumor**	[115]
17	*Oscillatory willei*	Cell extract	Spherical, 100–200 nm	–	[44a]
18	*Oscillatoria limnetica*	Biomass	Quasi-spherical, 3.30–17.97 nm	Cytotoxic effects against both human breast and human colon cancer cell line	[116]
19	*Phormidium fragile*	Cell culture	Spherical, 100–200 nm	–	[117]
20	*Plectonema boryanum*	Cell culture	Spherical and octahedral, 1–200 nm	–	[49a]
21	*Scytonema* sp.	Biomass	Spherical	–	[109]
22	*Spirulina platensis*	Single-cell protein	Spherical, 6–16 nm	–	[107]
23	*Synechocystis* sp.	Biomass	Variable, 13 and 31 nm	Antibacterial	[27]
24	*Tolypothrix tenuis*	Biomass	Spherical	–	[109]
25	*Westiellopsis prolifica*	Biomass	Spherical	–	[109]

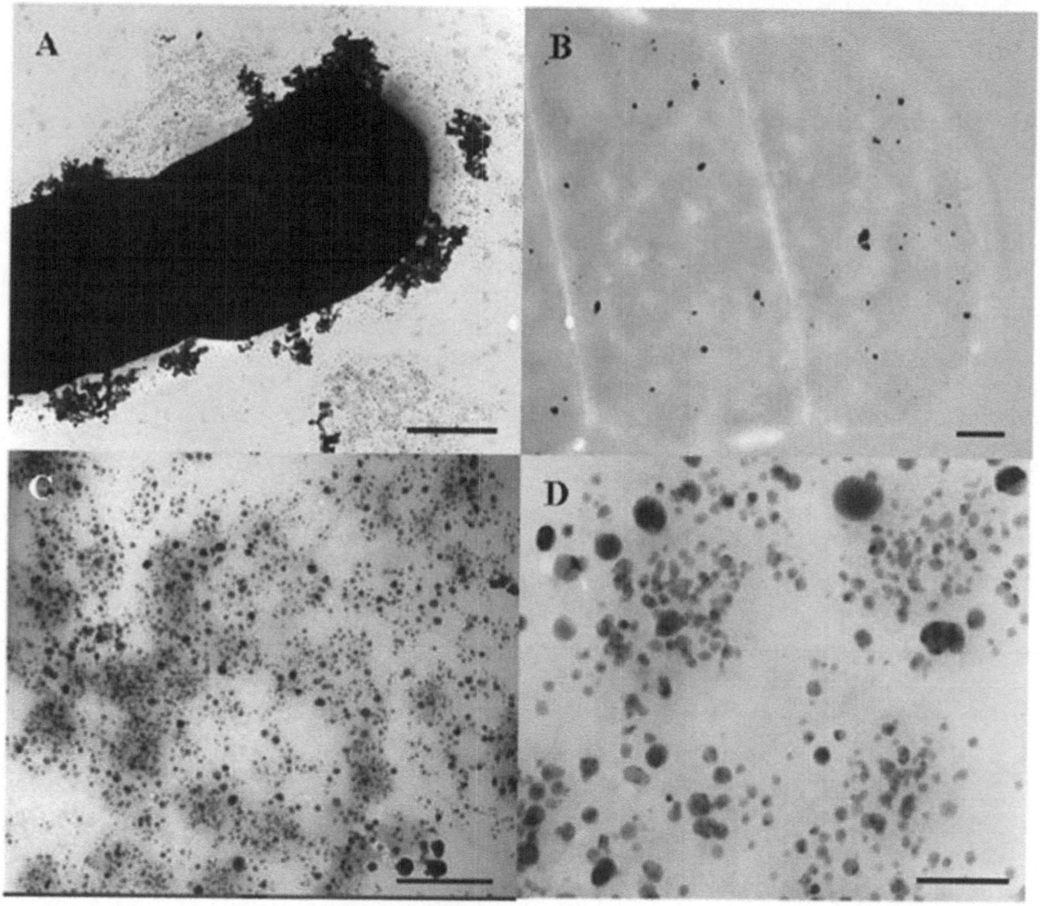

FIGURE 11.5 SEM Images Showing Extracellular AgNP Synthesis by Cyanobacteria.

Source: With permission from ACS [49a].

Xie et al. [97] carried out the process of biosynthesis of silver nanoparticle by using the raw extract made up of the supernatant of lyophilized cells of *Chlorella vulgaris.*

11.4.1.3 Synthesis of Other NPs

Apart from Ag and Au (NPs that are mostly worked upon in context to their wider applications), there are fewer reports on biogenic synthesis of NPs using other metals, like zinc, copper, platinum, palladium, cadmium, and selenium (Table 11.4).

There are very few reports shown by the review of literature about the biosynthesis of copper nanoparticles using microorganisms or plant extracts. Rahman et al. [47] employed the cyanobacterium *Phormidium* sp. for the synthesis copper oxide nanoparticles extracellularly and reported that the NP synthesis process was carried out by hydrolysis of the cationic copper by secreted proteins or reductase by the studied organism under normal conditions. The synthesis of CuO NPs by using cell-free extract of cyanobacterium *Spirulina platensis* has been reported by Saran et al. [118] and Doamn et al. [119]. Asif et al. [120] reported synthesis of CuO NPs by using the cell extract of *Phormidium* sp. and hypothesized that the phytochemicals present in cell extract of the studied organism act as bioreductant of Cu^{2+} ions and stabilizing agent of synthesized CuO NPs.

Moskvina et al. [121] studied the role of heterotrophic bacteria associated with cyanobacterium *Nostoc muscorum* for the synthesis of cadmium sulfide NPs. The biologically synthesized

TABLE 11.4
Cyanobacterial Strains Employed for Synthesis of Other NPs, Morphology of As-Synthesized NPs, and Their Applications

S. No.	Organism	Active Part	Morphology of NPs	Application	References
1	*Anabaena* sp.	Cell culture	β-FeOOH tetragonal	–	[118]
2	*Anabaena* strain L31	Cell extract	ZnO, 80 nm	Antioxidant	[40]
3	*Anabaena* sp., *Calothrix* sp., and *Leptolyngbya* sp.	*In vitro* enzymatic reduction	Pt, Pd	–	[100]
4	*Anabaena* sp. and *Calothrix* sp.	Cell extract	FeOOH tetragonal structure, 100 nm	–	[32]
5	*Limnothrix planctonica*, *Synechoccus leopoldiensis*, and *Pseudoanabaena limnetica*	Cell culture	Hg	–	[119]
6	*Nostoc* sp. *EA03*	Cell extract	ZnO, 50–80 nm	Cytotoxic, antibacterial	[120]
7	*Phormidium* sp.	Biomass	CuO, 10–40 nm, quasi-spherical	–	[47]
8	*Phormidium tenue*	Pigment (C-phycoerythrin)	CdS, 5 nm	–	[48]
9	*Plectonema boryanum* UTEX 485	Cell culture	Pt, Pd; 30 nm–0.3 μm, spherical	–	[49c–d]
10	*Spirulina platensis*	Cell-free extract	CuO, 30–40nm	**Antibacterial**	[118]

cadmium sulfide nanoparticles are photoluminescent and may act as highly efficient photocatalysts, due to which they are used for the degradation of the dye pollutant malachite green. Thus, cadmium sulfide is one of the desired criteria of wastewater treatment, that is, biosorption of pollutants or contaminants and their subsequent degradation. So they are known for their potential use in bioremediation by conversion of toxic cadmium^{+2} ion into less-toxic cadmium sulfide nanoparticles [122]. MubarakAli et al. [48] used c-phycoerythrin pigment derived from marine cyanobacterium *Phormidium tenue* NTDM05 for the synthesis of CdS nanoparticles and reported that the fluorescent protein present in pigment is responsible for the formation of the CdS NPs and their stabilization.

Zinc oxide nanoparticles have found potential applications, such as antibacterial, antidiabetic, anti-inflammatory, wound healing, and antioxidant properties. Singh et al. [40] used the cell extract of cyanobacterial strain *Anabaena* L31 for the synthesis of ZnO NPs and attempted to conjugate biosynthesized ZnO NPs with shinorine, a UV-absorbing water-soluble compound, and suggested that this framework can be used for developing environment-friendly sunscreen filters originated from a biological source. The synthesis of ZnO NPs using common *Anabaena flos-aquae* cyanobacteria and *Euglena gracilis* microalgae was also studied by Brayner et al. [123]. Ebadi et al. [124] used the cell extract of the cyanobacterium *Nostoc* sp. EA03 for the synthesis of ZnO NPs and suggested that the palmitic acid present in the cell extract plays a key role as a reducing agent during the green synthesis of ZnO NPs. Asif et al. [125] screened 30 cyanobacterial strains for their potential to synthesize nanosized ZnO NPs and found that the minimum-sized (~40 nm) spherical-shaped biosynthesized ZnO NPs by extract of *Oscillatoria* sp. Further, they optimized reaction condition like pH, temperature, reaction time, concentration of extract, and metal precursor for maximum yield of NPs and reported that 0.02 M zinc nitrate, 10 ml of extract volume, pH 8, at 80°C for 3 h, respectively, are the best conditions for synthesis of ZnO NPs.

The synthesis of selenium NPs (SeNPs) by using biogenic catalyst has received increasing attention due to their wide biomedical applications. Pandey et al. [126] evaluated selenite toxicity and extracellular synthesis of SeNPs by using cyanobacterium *Anabaena* sp. PCC 7120 and reported that the polysaccharide content may act as the capping agent for synthesized SeNPs. Afzal et al. [127] used the aqueous extract of cyanobacterium *Anabaena variabilis* NCCU-441 to synthesize SeNPs and demonstrated that the various functional groups and biochemical compounds present in the extract of this organism are responsible for synthesis of SeNPs.

Siddiqui et al. [128] screened 37 cyanobacterial strains for fabrication of titanium dioxide NPs (TiO2 NPs) and reported *Synechocystis* NCCU-370 as the suitable organism for synthesis among all screened cyanobacterial strains in terms of size of the synthesized NPs and time of synthesis.

Lengke et al. [49c–d] screened the ability of *Plectonema boryanum* UTEX 485 to form palladium and platinum NPs under variable temperatures. The *Plectonema boryanum* UTEX 485 biomass after incubation at 25–100°C for up to 28 days biosynthesizes spherical and elongated-shaped PdNPs of ≤ 30 nm size and amorphous spherical-shaped PtNPs of sized ≤ 0.3 μm. Brayner et al. [100] intracellularly synthesized Au, Ag, Pt, and Pd NPs utilizing the biomass of cyanobacterial strains *Calothrix pulvinata*, *Anabaena flos-aquae*, and *Leptolyngbya foveolarum* and suggested that the intracellular process of biosynthesis of NPs by studied organism is enzyme-mediated, and possibly, nitrogenase enzyme is related to efficiency of the reduction process.

Brayner et al. [100] studied the biosynthesis process of beta-iron oxyhydroxide (β-FeOOH) NPs using cyanobacterial species *Anabaena flos-aquae* and *Calothrix pulvinata* and green algae *Klebsormidium* sp. and proposed the mechanism of iron oxide NP formation by cyanobacteria.

Dahmoune et al. [123] reported the intracellular biosynthesis of akaganeite β-FeOOH nanorods by using planktonic cyanobacterium *Anabaena flos-aquae* strain ALCP B24.

11.5 BIOMEDICAL APPLICATIONS OF NANOPARTICLES SYNTHESIZED BY USING CYANOBACTERIA

The nanoparticles synthesized through green route are non-toxic and biocompatible and have potential to be utilized in wide biomedical applications [123]. Some of the important applications of cyanobacteria-mediated biogenic NPs are listed here.

11.5.1 Antimicrobial

Microalgae, including various strains of cyanobacteria, possess some biologically active compounds and, extracted from them, are known to have high antimicrobial activity. The production of these active compounds by cyanobacteria totally depends upon numerous factors, like temperature, incubation period, pH of culture medium, and light intensity. The biologically synthesized metal and metal oxide NPs (silver, gold, zinc oxide, and copper oxide) have found significant activity against pathogenic bacteria, virus, fungi, and yeasts in comparison to the natural compounds and NPs being synthesized by the non-biological route. The various cyanobacterial strains used for synthesis of different metal and metal oxide NPs having antimicrobial activity are presented in Table 11.5. Small-sized NPs have higher antibacterial activity than big nanoparticles; that's why antibacterial activity depends upon size and shape. Different types of NPs are known to show antibacterial activities, but out of all, silver NPs exhibit a broad inhibitory nature toward microbes even more significantly against antibiotic-resistant bacterial strains. The marine cyanobacterial–mediated synthesis of silver NPs showed high tendency of antibacterial activity in human pathogens such as *Escherichia coli*, *Klebsiella* sp., *Proteus vulgaris*, and *Pseudomonas aeruginosa* [38a]. The AgNPs are used in various fields to reduce infections in case of burn treatment (Klaus et al. 1999) and to destroy microorganisms on textile fabrics [81, 82]. AgNPs synthesized by using unicellular cyanobacterium *Synechococcus* sp. Are reported to have antibacterial properties against *Bacillus subtilis*, *Staphylococcus aureus*, and *Escherichia coli*, out of which *B. subtilis* was found to be

TABLE 11.5
Antibacterial Applications of CNB-Mediated Biogenic NPs

Cyanobacteria	Activity against Pathogen	Metal Nanoparticles	Concentration	Inhibition (in mm)	Method	References
Anabaena sp. *66–2*	*B. megaterium, E. coli, B. subtilis, M. luteus, P. aeruginosa, S. aureus*	Ag	70 μL/100 μL	17 ± 0.05, 16 ± 0.10, 16 ± 0.11, 20 ± 0.10, 22 ± 0.05, 17 ± 0.15	Agar well diffusion	[27]
Chroococcus sp. *MBKG08*	*E. coli, S. aureus, S. paratyphi, K. pneumonia, M. luteus*	Ag	40 μL, 60 μL, 80 μL	1, 1, 3, 1, 1	Well diffusion	[133]
Cylindrospermopsis sp. *USC CRB3*	*B. megaterium, E. coli, B. subtilis, M. luteus, P. aeruginosa, S. aureus*	Ag	70 μL/100 μL	15 ± 0.11, 12 ± 0.10, 13 ± 0.17, 11 ± 0.10, 15 ± 0.05, 10 ± 0.11	Agar well diffusion	[27]
Cylindrospermospsis sp. *121–1*	*S. mutans*	Ag	1.5 mg/mL	11.4 ± 0.12	Agar well diffusion	[131, 134]
Desertifilum sp. *IPPAS B-1220*	*S. mutans, E. coli, K. pneumoniae, S. typhimurium, MRSA*	Ag	1.54 mg/100 μL	13.3 ± 0.03, 14.8 ± 0.07, 16.5 ± 0.09, 18.5 ± 0.06, 23.7 ± 0.08	Agar well diffusion	[134]
	P. aeruginosa, S. flexneri, B. subtilis, B. cereus, S. enterica		–	15.0 ± 0.00, 22.67 ±0.33, 16.33±0.33, 17.33±0.33, 16.67±0.33	Agar well diffusion	[131]
Lyngbya majuscule CUH/Al/ MW-150	*B. megaterium, E. coli, M. luteus, P. aeruginosa*	Ag	0.1 mg/mL, 0.5, 1	4 ± 0.15, 8 ± 0.1, 6 ± 0.11	Agar well diffusion	[32]
Leptolyngbya valderianum	*P. aeruginosa*	Ag	0.5 mg/mL	–	Agar well diffusion	[135]
Leptolyngbya JSC-1	*E. coli, S. aureus*	Ag	25 μL, 50 μL, 100 μL, 25 μL, 50 μL, 100 μL	9 ± 2 mm, 15 ± 2 mm, 21 ± 2 mm, 7± 2 mm, 12 ± 2 mm, 17 ± 2 mm	Agar well diffusion	[136]
Microcoleus sp.	*E. coli, P. vulgaris, S. typhi, V. cholerae, B. subtilis, S. aureus, Streptococcus* sp., *Corynebacterium* sp.	Ag	10 μL, 20 μL, 50 μL	11, 12, 9, 9, 12, 11, 10, 12	Agar well diffusion	[113]
Microcystis sp. *MBKG16*	*E. coli, S. aureus, S. paratyphi, K. pneumonia, M. luteus*	Ag	–	–	Well diffusion	[133]
Nostoc sp. *Phormidium* sp., *Scytonema* sp.	*E. coli, S. aureus, P. aeruginosa, MRSA ATCC*	Ag	70 μL	9, 10, 9, 2 10, 9, 8, 11.5 9.5, 7, 9, 12	Agar well diffusion	[137]
Nostoc Bahar M	*K. pneumoniae*	Ag	–	15.33 ± 0.1	Agar well diffusion	[138]

Cyanobacteria	Activity against Pathogen	Metal Nanoparticles	Concentration	Inhibition (in mm)	Method	References
Nostoc carneum	*S. aureus*, *Streptococcus* sp., *E. coli*, *E. aerogenes*	Ag	20 mM	16, 18, 21, 18	Disc-diffusion method	[94]
Nostoc commune	*E. coli*	Ag	–	–	Agar well diffusion	[139]
Nostoc Linckia	*S. aureus*, *P. aeruginosa*, *E. coli*, *K. pneumoniae*	Ag	5 mM	9, 10, 10, 9	Disc-diffusion method	[132]
Nostoc sp. *EA03*	*P. aeruginosa PAO1*, *E. coli*, *S. aureus*	ZnO	65 µg/mL	–	Flow cytometry	[124]
Nostoc sp. *Strain HKAR-2*	*R. solanacearum*, *X. campestris*	Ag	5 µg/mL, 10, 15, 5 µg/mL, 10, 15	15 ± 4.5, 25 ± 4.0, 25 ± 5.0, 18 ± 1.5, 23 ± 4.0, 23 ± 10.0	Agar well diffusion	[140]
Nostoc sp., *Spirulina platensis*	*S. aureus*, *S. epidedymis*, *P. aeruginosa*, *E. coli*, *K. pneumoniae*	Ag	40 µL	13.6±0.5, 11.6±0.57, 4, 16, 15.6±0.57	Agar well diffusion	[141]
Oscillatoria limnetica	*E. coli*, *B. cereus*	Ag	–	22, 20	Disc-diffusion method	[116]
Oscillatoria sp.	*E. coli*, *Klebsiella* sp., *Salmonella* sp., *Pseudomonas* sp.	Ag	1 µg/ml	10, 13, 14, 15	Disc-diffusion method	[134]
Spirulina platensis	*E. coli*, *P. vulgaris*, *K. pneumonia*, *S. epidermidis*, *S. aureus*, *B. cereus*	CuO	50 µg/100 µL	21.3 ± 0.85, 28.0 ± 0.41, 23.3 ± 1.11, 24.0 ± 0.91, 23.8 ± 0.48, 27.0 ± 1.47	Disc diffusion	[118]
Synechococcus sp. *145–6*	*B. megaterium*, *E. coli*, *M. luteus*, *P. aeruginosa*	Ag	5–10 mg/mL	09 ± 0.10, 10 ± 0.11, 11 ± 0.05	Agar well diffusion	[27]
Synechocystis sp. *48–3*	*B. megaterium*, *E. coli*, *B. subtilis*, *M. luteus*, *P. aeruginosa*, *S. aureus*	Ag	5–10 mg/mL	16 ± 0.10, 15 ± 0.10, 14 ± 0.15, 18 ± 0.11, 20 ± 0.10, 17 ± 0.15	Agar well diffusion	[27]

more sensitive to the biogenic AgNPs in comparison to other studied bacterial strains [129, 130]. Hamida et al. [131] reported that the silver NPs synthesized by using *Desertifilum* sp. have significant antibacterial effect at dosage 1.54 mg/mL against five bacterial strains (*Pseudomonas aeruginosa*, *Bacillus cereus*, *Bacillus subtilis*, *Shigella flexneri*, and *Salmonella enterica*), including both gram-negative and gram-positive bacteria. The significant inhibitory activity against various pathogenic bacterial strains (*S. aureus*, *P. aeruginosa*, *E. coli*, and *K. pneumonia*) has been reported by AgNPs synthesized by using phycocyanin, photosynthetic accessory pigment, extracted from cyanobacterium *N. linckia* [132]. AgNPs are known to significantly block the pathways of respiratory enzyme, alter the DNA of microbes and the disruption of cell wall, and finally, result in the death of microbes [84–87].

Copper NPs are considered as an antimicrobial agent, and it was approved by the United States Environmental Protection Agency in 2008. Copper oxide NPs have potential antimicrobial activities against infectious organisms like *Escherichia coli*, *Bacillus subtilis*, *Vibrio cholera*, *Pseudomonas aeruginosa*, *Syphilis typhus*, and *Staphylococcus aureus* [90–91]. Inside the bacterial cells, copper ions are known to disrupt all biochemical processes. The antibacterial activity of nickel NPs and bimetallic copper–nickel tested very effectively against human pathogens like *Staphylococcus aureus*, *Escherichia coli*, etc. [90–92]. Zinc oxide NPs are also known to have antibacterial activity and inhibit the growth of microorganisms like *Bacillus subtilis*, *E. coli* [93]. Recently, the zinc oxide NPs have also been investigated to show a high rate of antibacterial activities at both microscale and nanoscale formulation. Zinc oxide NPs show their antibacterial activities against bacterial strains like *Enterococcus faecalis*, *Sallmonella*, *Campylobacter jejuni*, and *Clostridium perfringens*. Various mechanisms for the antibacterial activity of zinc oxide NPs have been investigated [94]. For instance, firstly, zinc oxide NPs actively interact with membrane-bounded lipids; after this interaction, they disorganize the membrane structure, which finally leads to the disintegrity of membrane, malfunctioning, and finally, bacterial death [95–96].

11.5.2 Antioxidant

Many different kinds of nanomaterials have inherent antioxidant capacity that arises from the surface properties of the material rather than from their functionalization with antioxidants. These materials mostly consist of inorganic metal nanoparticles. In the literature, there are several examples of metal and metal oxide NPs synthesized from cyanobacteria displaying antioxidant activities. Mubarak Ali et al. [44b] demonstrated the synthesis of gold nanoparticles by using cyanobacterial strain *Phormidium* sp. and with different concentrations of AuNPs in methanol, analyzing them against citric acid using DPPH (1,1-diphenyl-2-picrylhydrazyl) radical scavenging assay, and these AuNPs have the capability to neutralize hydroxyl radicals generated by H_2O_2–ascorbic acid redox system. Singh et al. [40] have emphasized the synthesis of ZnO NPs formed by using cell-free extract of *Anabaena* strain L31, which was tested in vivo for their antioxidative potential. After 12 h incubation, only ZnONPs exhibited maximum ROS, whereas cells treated with ZnONPs–shinorine conjugate revealed diminished ROS production. Strong antioxidant potential was exhibited by AgCl-Ag_3PO_4 nanoparticles synthesized by using cell-free extract of two cyanobacterial strains *Nostoc* sp. IBRC-M5064 and *Nostoc pruniforme*. Siddiqui et al. [142] demonstrated the formation of TiO_2 NPs by using cell-free extracts of cyanobacterial strain *Synechocystis* sp. NCCU-370 and indicated antioxidant potential by DPPH, ABTS, and FRAP assay. DPPH assay manifests 90% free radical scavenging activity at 25 µg/mL concentrations. ABTS assay revealed 71.79% scavenging activity at 500 µg/mL, and FRAP assay showed its maximum FRAP values at 500 µg/mL. Pandey et al. [126] reported the synthesis of selenium nanoparticles (SeNPs) using cyanobacterial strain *Anabaena* sp. PCC 7120, and the synthesized SeNPs exhibited significant antioxidant activity at minimum concentration of 50 µg/mL. Husain et al. [42] biosynthesized silver nanoparticles (AgNPs) using cyanobacterial strain *Nostoc muscorum* NCCU 442, which showed the highest free radical scavenging activity. Afzal et al. [127] studied the biomedical applications of selenium nanoparticles synthesized by using extracts of *Anabaena variabilis* NCCU-441, and the as-synthesized SeNPs showed strong free radical scavenging activity.

11.5.3 Anticancer and Antitumor

Presently, the production of nanoproducts based on biogenically synthesized metallic nanoparticles having potent anticancer and antibacterial activity is an emerging discipline. Hamida et al. [131] used extracts of a novel cyanobacterium *Desertifilum* sp. IPPAS B-1220 for the synthesis of silver nanoparticles. The biofabricated AgNPs showed significant cytotoxic effects against MCF-7, HepG2, and Caco-2 cancer cell lines. The new cyanobacterial strain *Nostoc* sp. Bahar M was used for the synthesis of AgNPs, and the results of cytotoxicity assays revealed that AgNPs showed dose-dependent significant cytotoxic activity against human colon cancer (Caco-2) cells with an IC_{50} of 150 μg/mL [138]. Roychoudhury et al. [135] demonstrated the biofabrication of AgNPs by using *Lyngbya majuscule* and checked its antiproliferative activity against K562, MOLT-3, and REH leukemic cell lines. Among these three, REH cell line was most sensitive, with IC50 dose of 620 ± 3.73 μg/mL, K562 (IC50 1000 ± 2.79 μg/mL), and MOLT-3 (IC50 1000 ± 4.38 μg/mL) of AgNPs. El-Naggar et al. [132] had tested the AgNPs synthesized using cyanobacterial strain *Nostoc linckia* that manifest cytotoxic activity against breast cancer MCF-7 cell line. AgNPs were more effective against MCF-7 (IC50 27.79 ± 2.3 μg/mL) cell line as compared to other cell lines. Pandey et al. [126] reported synthesis of selenium NPs (SeNPs) by using *Anabaena* sp. PCC 7120 and studied the antiproliferative activity of synthesized SeNPs against HeLa cell line, which showed IC50 value of 5.5 μg/mL. Ebadi et al. [124] emphasized the synthesized ZnO NPs using cell extract of *Desertiflum* sp. and revealed effective antiproliferative activity against lung alveolar epithelial (A549) cancer cell line and MTT assay of ZnO NPs results with IC50 value of 100 μg/mL.

11.6 CONCLUSION

It is concluded that due to increased cases of multidrug-resistant (MDR) bacterial infections and cancer, the demand to develop and discover alternative, eco-safe anticancer agents is high. The autotrophic nature, ease in culturing and handling, with other advantages, like capability for NP synthesis at low temperature, less toxicity, and less risk to the environment, are advantageous to cyanobacteria as the better choice for synthesis of NPs at large scale. Various biomolecules, like chlorophyll, carotenoids, phenols, xanthines, flavonoids, proteins, minerals, and vitamins present in cyanobacterial extract are known to be responsible for bioreduction of NPs. Although a number of reports explain the mechanism of intracellular and extracellular synthesis of NPs by cyanobacteria, to understand the precise mechanism, more details studies are required. The NPs synthesized by using different cyanobacterial species have wide applications in the medical field, like as anticancer agents, in drug delivery, as biosensors, and in medical imaging.

ACKNOWLEDGMENTS

The authors are thankful to Sri Guru Granth Sahib World University, Fatehgarh Sahib, for providing the necessary facilities.

CONFLICT OF INTEREST

The authors declare that there is no conflict of interests regarding the publication of this chapter.

REFERENCES

1. Thangadurai, T. D.; Manjubaashini, N.; Thomas S. et al. **2020**. Nanomaterials, properties and applications. In *Nanostructured Materials: Engineering Materials*, Springer. https://doi.org/10.1007/978-3-030-26145-0_2
2. Barhoum, A.; Makhlouf, A. S. H. **2018**. *Fundamentals of Nanoparticles*, 1st edition, Elsevier, Amsterdam, The Netherlands.
3. a) Wang, L.; Ma, W.; Xu, L.; Chen, W.; Zhu, Y.; Xu, C.; Kotov, N. A. **2010**. Nanoparticle-based environmental sensors. *Materials Science and Engineering: Reports* 70(3–6): 265–274.
   b) Zhao, Q. **2019**. *Advanced Nanomaterials for Pollutant Sensing and Environmental Catalysis*, 1st edition, Elsevier, Amsterdam, The Netherlands.
4. Kumari, P.; Alam, M.; Siddiqi, W. A. **2019**. Usage of nanoparticles as adsorbents for waste water treatment: An emerging trend, *Sustainable Materials and Technologies* 22: e00128.
5. a) Liu, L.; Corma, A. **2018**. Metal catalysts for heterogeneous catalysis: From single atoms to nanoclusters and nanoparticles, *Chemical Reviews* 118: 4981–5079.
   b) An, K.; Somorjai, G. A. **2012**. Size and shape control of metal nanoparticles for reaction selectivity in catalysis, *ChemCatChem* 4: 1512–1524.
6. a) Dreaden, E.C.; Alkilany, A. M.; Huang, X.; Murphy, C. J.; El-Sayed, M. A. **2012**. The golden age: Gold nanoparticles for biomedicine, *Chemical Society Reviews* 41: 2740–2779.
   b) Jonghoon, K.; Nohyun, L.; Taeghwan, H. **2017**. Recent development of nanoparticles for molecular imaging, *Philosophical Transactions of the Royal Society A* 375: 0022.
7. Xu, C.; Anusuyadevi, P. R.; Aymonier, C. et al. **2019**. Nanostructured materials for photocatalysis, *Chemical Society Review* 48: 3868–3902.
8. Ong, C. B.; Ng, L. Y.; Mohammad, A. W. **2018**. A review of ZnO nanoparticles as solar photocatalysts: Synthesis, mechanisms and applications, *Renewable and Sustainable Energy Reviews* 81: 536–551.
9. Lee, J. E.; Lee, N.; Kim, T.; Kim, J.; Hyeon, T. **2011**. Multifunctional mesoporous silica nanocomposite nanoparticles for theranostic applications, *Accounts of Chemical Research* 44: 893–902.
10. Saravanan, C.; Rajesh, R.; Kaviarasan, T. et al. **2017**. Synthesis of silver nanoparticles using bacterial exopolysaccharide and its application for degradation of azo-dyes, *Biotechnology Reports* 15: 33–40.
11. a) Molnar, Z.; Bodai, V.; Szakacs, G. et al. **2018**. Green synthesis of gold nanoparticles by thermophilic filamentous fungi, *Scientific Reports* 8: 3943.
    b) Feroze, N.; Arshad, B.; Younas, M. et al. **2019**. Fungal mediated synthesis of silver nanoparticles and evaluation of antibacterial activity, *Microscopy Research and Technique* 83: 72–80.
12. a) Skladanowski, M.; Wypij, M.; Laskowski, D. et al. **2017**. Silver and gold nanoparticles synthesized from Streptomyces sp. isolated from acid forest soil with special reference to its antibacterial activity against pathogens, *Journal of Cluster Science* 28: 59–79.
    b) Eugenio, M.; Muller, N.; Frases, S. **2016**. Yeast-derived biosynthesis of silver/silver chloride nanoparticles and their antiproliferative activity against bacteria, *RSC Advances* 6: 9893–9904.
13. Manimaram, M.; Kannabiran, K. **2017**. Actinomycetes-mediated biogenic synthesis of metal and metal oxide nanoparticles: Progress and challenges, *Letters in Applied Microbiology* 64: 401–408.
14. a) Le, D. H.; Lee, K. L.; Shukla, S. et al. **2017**. Potato virus X, a filamentous plant viral nanoparticle for doxorubicin delivery in cancer therapy, *Nanoscale* 9: 2348–2357.
    b) Slocik, J. M.; Knecht, M. R.; Wright, D. W. **2011**. Biomimetic synthesis of nanoparticles. In *Encyclopedia of Inorganic and Bioinorganic Chemistry*, John Wiley & Sons, New Jersey, USA.
15. Rana, A.; Yadav, K.; Jagadevan, S. **2020**. A comprehensive review on green synthesis of nature-inspired metal nanoparticles: Mechanism, application and toxicity, *Journal of Cleaner Production* 272: 122880.
16. Gahlawat, G.; Choudhury, A. R. **2019**. A review on the biosynthesis of metal and metal salt nanoparticles by microbes, *RSC Advances* 9: 12944–12967.
17. Akhtar, M. S.; Panwar, J.; Yun, Y. S. **2013**. Biogenic synthesis of metallic nanoparticles by plant extracts, *ACS Sustainable Chemistry & Engineering* 1: 591–602.
18. Khanna, P.; Kaur, A.; Goyal, D. **2019**. Algae-based metallic nanoparticles: Synthesis, characterization and applications, *Journal of Microbiological Methods* 163: 105656.
19. Dahoumane, S. A.; Mechouet, M.; Wijesekera, K. et al. **2017**. Algae-mediated biosynthesis of inorganic nanomaterials as a promising route in nanobiotechnology – a review, *Green Chemistry* 19: 552–587.
20. Shankar, P. D.; Shobana, S.; Karuppusamy, I. et al. **2016**. A review on the biosynthesis of metallic nanoparticles (gold and silver) using biocomponents of microalgae: Formation mechanism and applications, *Enzyme and Microbial Technology* 95: 28–44.

21. Sharma, A.; Sharma, S.; Sharma, K. **2016**. Algae as crucial organisms in advancing nanotechnology: A systematic review, *Journal of Applied Phycology* 1759–1774.
22. Jadoun, S.; Arif, R.; Jangid, N. K. et al. **2021**. Green synthesis of nanoparticles using plant extracts: A review. *Environmental Chemistry Letters* 19: 355–374.
23. Ghorbani, H. R.; Safekordi, A. A.; Attar, H. et al. **2011**. Biological and non-biological methods for silver NPs synthesis, *Chemical and Biochemical Engineering* 25: 317–326.
24. Sheng, P. X.; Ting, Y. P.; Chen, J. P.; Hong, L. **2004**. Sorption of lead, copper, cadmium, zinc, and nickel by marine algal biomass: Characterization of biosorptive capacity and investigation of mechanisms, *Journal of Colloid and Interface Science* 275: 131–141.
25. Kannan, R.R.R.; Arumugam, R.; Ramya, D. et al. **2013**. Green synthesis of silver nanoparticles using marine macroalga Chaetomorphalinum, *Applied Nanoscience* 3: 229–233.
26. Chokshi, K.; Pancha, I.; Ghosh, T. et al. **2016**. Green synthesis, characterization and antioxidant potential of silver nanoparticles biosynthesized from de-oiled biomass of thermotolerant oleaginous microalgae Acutodesmusdimorphus, *RSC Advances* 6: 72269–72274.
27. Patel, V.; Berthold, D.; Puranik, P. et al. **2015**. Screening of cyanobacteria and microalgae for their ability to synthesize silver nanoparticles with antibacterial activity, *Biotechnology Reports* 5: 112–119.
28. Karthikeyan, P.; Mohan, D.; Abishek, G. et al. **2015**. Synthesis of silver nanoparticles using Phytoplankton and its characteristics, *International Journal of Fisheries and Aquatic Studies* 2: 398–401.
29. Xie, J.; Lee, J. Y.; Wang, D. I. C. et al. **2007**. Identification of active biomolecules in the high yield synthesis of single crystalline gold nanoplates in algal solutions, *Small* 3: 672–682.
30. Dahoumane, S.A.; Yepremian, C.; Djediat, C. et al. **2014**. A global approach of the mechanism involved in the biosynthesis of gold colloids using micro-algae, *Journal of Nanoparticle Research* 16: 2607.
31. Dhanalakshmi, P. K.; Riyazulla, A.; Rekha, R. et al. **2012**. Synthesis of silver nanoparticles using green and brown seaweeds, *Phykos* 42: 39–45.
32. Roychoudhury, P.; Pal, R. **2014**. Spirogyra submaxima-a green alga for nanogold production, *Journal of Algal Biomass Utilization* 5: 15–19.
33. Sinha, S. N.; Paul, D.; Halder, N. et al. **2015**. Green synthesis of silver nanoparticles using fresh water green alga Pithophoraoedogonia (Mont.) Wittrock and evaluation of their antibacterial activity, *Applied Nanoscience* 5: 703–709.
34. Sharma, B.; Purkayastha, D. D.; Hazra, S. et al. **2014**. Biosynthesis of gold nanoparticles using a freshwater green alga, Prasiola crispa, *Materials Letters* 116: 94–97.
35. Chakraborty, N.; Banerjee, A.; Lahiri, S. et al. **2009**. Biorecovery of gold using cyanobacteria and an eukaryotic alga with special reference to nanogold formation-a novel phenomenon, *Journal of Applied Phycology* 21: 145–152.
36. Pinjarkar, H.; Gaikwad, S.; Ingle, A.P. et al. **2016**. Phycofabrication of silver nanoparticles and their antibacterial activity against human pathogens, *Advanced Materials Letters* 7: 1010–1014.
37. Shakibaie, M.; Forootanfar, H.; Mollazadeh-Moghaddam, K. et al. **2010**. Green synthesis of gold nanoparticles by the marine microalga Tetraselmis suecica, *Biotechnology and Applied Biochemistry* 57: 71–75.
38. Parial, D.; Patra, H. K.; Dasgupta, A. K. R. et al. **2012**. Screening of different algae for green synthesis of gold NPs, *European Journal of Phycology* 47: 22–29.
39. Suriya, J.; Bharathi Raja, S.; Sekar, V. et al. **2012**. Biosynthesis of silver nanoparticles and its antibacterial activity using seaweed Urospora sp. *African Journal of Biotechnology* 11: 12192–12198.
40. Singh, G.; Babele, P. K.; Kumar, A. et al. **2014**. Synthesis of ZnO nanoparticles using the cell extract of the cyanobacterium, Anabaena strain L31 and its conjugation with UV-B absorbing compound shinorine, *Journal of Photochemistry and Photobiology* 138: 55–62.
41. Chakraborty, N.; Banerjee, A.; Lahiri, S. et al. **2009**. Biorecovery of gold using cyanobacteria and an eukaryotic alga with special reference to nanogold formation – a novel phenomenon, *Journal of Applied Phycology* 21: 145–152.
42. Husain, S.; Afreen, S.; Yasin, D. et al. **2019**. Cyanobacteria as a bioreactor for synthesis of silver nanoparticles-an effect of different reaction conditions on the size of nanoparticles and their dye decolorization ability, *Journal of Microbiological Methods* 162: 77–82.
43. a) Cepoi, L.; Rudi, L.; Chiriac, T. et al. **2015**. Biochemical changes in cyanobacteria during the synthesis of silver nanoparticles, *Journal of Microbiology* 61: 13–21.

    b) Parial, D.; Patra, H. K.; Roychoudhury, P. et al. **2012**. Gold nanorod production by cyanobacteria-a green chemistry approach, *Journal of Applied Phycology* 24: 55–60.
44. a) Mubarak Ali, D.; Sasikala, M.; Gunasekaran, M. et al. **2011**. Biosynthesis and characterization of silver nanoparticles using marine cyanobacterium, Oscillatoria willei NTDM01, *Digest Journal of Nanomaterials and Biostructures* 6: 385–390.

b) Mubarak Ali, D.; Kumar, J. A.; Nag, K. H. et al. **2013**. Gold nanoparticles from pro and eukaryotic photosynthetic microorganisms-comparative studies on synthesis and its application on biolabeling, *Colloids and Surfaces B: Biointerfaces* 103: 166–173.

45. Asmathunisha N.; Kathiresan K. **2011**. A review on biosynthesis of nanoparticles by marine organisms. *Colloids and Surfaces B: Biointerfaces* 103: 283–287.
46. Elkomy, R. G. **2020**. Antimicrobial screening of silver nanoparticles synthesized by marine cyanobacterium Phormidiumformosum, *Iran Journal of Microbiology* 12: 242–248.
47. Rahman, A.; Ismail, A.; Jumbianti, D. et al. **2009**. Synthesis of copper oxide nano particles by using Phormidium cyanobacterium, *Indonesia Journal of Chemistry* 9: 355–360.
48. Mubarak Ali, D.; Gopinath, V.; Rameshbabu, N. et al. **2012**. Synthesis and characterization of CdS nanoparticles using C-phycoerythrin from the marine cyanobacteria, *Material Letters* 74: 8–11.
49. a) Lengke, M. F.; Fleet, M. E.; Southam, G. **2007**. Biosynthesis of silver nanoparticles by filamentous cyanobacteria from a silver(I) nitrate complex, *Langmuir* 23: 2694–2699.
    b) Lengke, M. F.; Fleet, M. E.; Southam, G. **2006**. Morphology of gold nanoparticles synthesized by filamentous cyanobacteria from gold(I)-thiosulfate and gold(III)–chloride complexes, *Langmuir* 22: 2780–2787.
    c) Lengke, M. F.; Fleet, M. E.; Southam, G. **2006**. Synthesis of platinum nanoparticles by reaction of filamentous cyanobacteria with platinum (IV)-chloride complex, *Langmuir* 22: 7318–7323.
    d) Lengke, M. F.; Fleet, M. E.; Southam, G. **2007**. Synthesis of palladium nanoparticles by reaction of filamentous cyanobacterial biomass with a palladium(II) chloride complex, *Langmuir* 23: 8982–8987.
50. a) Mahdieh, M.; Zolanvari, A.; Azimee, A. S. et al. **2012**. Green biosynthesis of silver nanoparticles by Spirulina platensis, *Scientia Iranica* 19: 926–929.
    b) Kalabegishvili, T., Kirkesali, E.I., Rcheulishvili, A.N., Ginturi, E., Murusidze, I., Kuchava, N., Bagdavadze, N., Tsertsvadze, G., Gabunia, V., Frontasyeva, M., et al. **2012**. Synthesis of gold nanoparticles by blue-green algae Spirulina platensis. *Advanced Science, Engineering and Medicine*, 4: 1–7.
51. Focsan, M.; Ardelean, I. I.; Craciun, C.; Astilean, S. **2011**. Interplay between gold nanoparticle biosynthesis and metabolic activity of cyanobacterium Synechocystis sp. PCC 6803, *Nanotechnology* 22: 485101.
52. Abboud, Y.; Saffaj, T.; Chagraoui, A. et al. **2014**. Biosynthesis, characterization and antimicrobial activity of copper oxide nanoparticles (CONPs) produced using brown alga extract (Bifurcariabifurcata), *Applied Nanoscience* 4: 571–576.
53. El-Rafie, H. M.; El-Rafie, M.; Zahran, M. K. **2013**. Green synthesis of silver nanoparticles using polysaccharides extracted from marine macro algae, *Carbohydrates Polymers* 96: 403–410.
54. Prasad, T. N. V. K. V.; Kambala, V. S. R.; Naidu, R. **2013**. Phyconanotechnology: Synthesis of silver nanoparticles using brown marine algae Cystophora moniliformis and their characterisation, *Journal of Applied Phycology* 25: 177–182.
55. Gonzalez-Ballesteros, N.; Prado-Lopez, S.; Rodriguez-Gonzalez, J. B. **2017**. Green synthesis of gold nanoparticles using brown algae Cystoseirabaccata: Its activity in colon cancer cells, *Colloids Surfaces B: Biointerfaces* 153: 190–198.
56. Varun, S.; Sudha, S.; Kumar, P. S. **2014**. Biosynthesis of gold nanoparticles from aqueous extract of Dictyota Bartayresiana and their antifungal activity, *Indian Journal of Advances in Chemical Science* 2: 190–193.
57. Venkatesan, J.; Manivasagan, P.; Kim, S. K. et al. **2014**. Marine algae-mediated synthesis of gold nanoparticles using a novel Ecklonia cava, *Bioprocess and Biosystems Engineering* 37: 1591–1597.
58. Mata, Y. N.; Torres, E.; Blazquez, M. L. et al. **2009**. Gold (III) biosorption and bio-reduction with brown alga Fucusvesiculosus, *Journal of Hazardous Material* 166: 612–618.
59. Raut, R. W.; Lakkakula, J. R.; Kolekar N. S. et al. **2009**. Phytosynthesis of silver nanoparticle using Gliricidiasepium (Jacq.), *Current Nanoscience* 5: 117.
60. Ghodake, G.; Lee, D. S. **2011**. Biological synthesis of gold nanoparticles using the aqueous extract of the brown algae Laminaria japonica, *Journal of Nanoelectron: Optoelectron* 6: 268–271.
61. Jegadeeswaran, P.; Rajeshwari, S.; Venckatesh, R. **2012**. Green synthesis of silver nanoparticles from extract of Padina tetrastromatica leaf, *Digest Journal of Nanomaterials and Biostructures* 7: 991–998.
62. Isaac, G.; EmilinRenitta, R. **2015**. Brown algae mediated synthesis, characterization of gold nano particles using *Padina pavonica* and their antibacterial activity against human pathogens, *International Journal of PharmTech Research* 8: 31–40.
63. Govindaraju, K.; Kiruthiga, V.; Kumar, V. G. et al. **2009**. Extracellular synthesis of silver nanoparticles by a marine alga, *Sargassum wightii* Grevilli and their antibacterial effects, *Journal of Nanoscience and Nanotechnology* 9: 5497–5501.

64. Singaravelu, G.; Arockiamary, J. S.; Kumar, V. G. et al. **2007**. A novel extracellular synthesis of monodisperse gold nanoparticles using marine alga, *Sargassum wightii* Greville, *Colloids Surfaces B: Biointerfaces* 57: 97–101.
65. Prasad, B.; Padmesh, T. **2014**. Seaweed (*Sargassum ilicifolium*) assisted green synthesis of palladium nanoparticles, *International Journal of Scientific Engineering and Research* 5: 229–231.
66. Nagarajan, S.; Arumugam Kuppusamy, K. **2013**. Extracellular synthesis of zinc oxide nanoparticle using seaweeds of gulf of Mannar, *Indian Journal of Nanobiotechnology* 11: 39.
67. Mahdavi, M.; Namvar, F.; Ahmad, M. B. et al. **2013**. Green biosynthesis and characterization of magnetic iron oxide (Fe_3O_4) nanoparticles using seaweed (*Sargassum muticum*) aqueous extract, *Molecules* 18: 5954–5964.
68. Dhas, T. S.; Kumar, V. G.; Karthick, V. et al. **2014**. Facile synthesis of silver chloride nanoparticles using marine alga and its antibacterial efficacy. *Spectrochimica Acta Part A: Molecular and Biomolecular* 120: 416–420.
69. Rajathi, F. A. A.; Parthiban, C.; Ganesh Kumar, V. **2012**. Biosynthesis of antibacterial gold nanoparticles using brown algae, *Stoechospermum marginatum* (kutzing), *Spectrochimica Acta Part A* 99: 166–173.
70. a) Rajeshkumar, S.; Kannan, C.; Annadurai, G. **2012**. Green synthesis of silver nanoparticles using marine brown algae Turbinaria conoides and its antibacterial activity, *International Journal of Pharmacy and Biological Sciences* 3: 502–510.
b) Rajeshkumar, S.; Malarkodi, C.; Gnanajobitha, G. **2013**. Seaweed-mediated synthesis of gold nanoparticles using Turbinariaconoides and its characterization, *Journal of Nanostructure in Chemistry* 3: 44.
71. Ibraheem, I. B. M.; Abd-Elaziz, B. E. E.; Saad, W. F. **2016**. Green biosynthesis of silver nanoparticles using marine Red Algae Acanthophora specifera and its antimicrobial activity, *Journal of Nanomedicine & Nanotechnology* 7: 1–4.
72. Sajidha Parveen, K.; Lakshmi, D. **2016**. Biosynthesis of silver nanoparticles using red algae, Amphiroa fragilissima and its antibacterial potential against Gram positive and Gram negative bacteria, *International Journal of Current Science and Research* 19: 93–100.
73. Castro, L.; Blazquez, M. L.; Munoz, J. A. **2013**. Biological synthesis of metallic nanoparticles using algae, *IET Nanobiotechnology* 7: 109–116.
74. Abdel-Raouf, N.; Al-Enazi, N.M.; Ibraheem, I. B. **2017**. Green biosynthesis of gold nanoparticles using Galaxaurae longata and characterization of their antibacterial activity, *Arabian Journal of Chemistry* 10: S3029–S3039.
75. Murugesan, S.; Elumalai, M.; Dhamotharan, R. **2011**. Green synthesis of silver nano particles from marine alga Gracilaria edulis, *Bioscience Biotechnology Research Communications* 4: 105–110.
76. Naveena, B. E.; Prakash, S. **2013**. Biological synthesis of gold nanoparticles using marine algae Gracilariacorticata and its application as a potent antimicrobial and antioxidant agent, *Asian Journal of Pharmaceutical and Clinical Research* 6: 179–182.
77. Priyadharshini, R. I.; Prasannaraj, G.; Geetha, N. **2014**. Microwave-mediated extracellular synthesis of metallic silver and zinc oxide nanoparticles using macro-algae (Gracilaria edulis) extracts and its anticancer activity against human PC3 cell lines, *Applied Biochemistry and Biotechnology* 174: 2777–2790.
78. Ganesan, V.; Aruna Devi, J.; Astalakshmi, A. **2013**. Ecofriendly synthesis of silver nanoparticles using a sea weed, Kappaphycus Alvarezii (Doty) Doty ex P.C. Silva, *International Journal of Engineering and Advanced Technology* 2: 559–563.
79. Rajasulochana, P.; Krishnamoorthy, P.; Dhamotharan, R. **2012**. Potential application of Kappaphycusalvarezii in agricultural and pharmaceutical industry, *Journal of Chemical and Pharmaceutical Research* 4: 33–37.
80. Sharma, B.; Purkayastha, D. D.; Hazra, S. **2014**. Biosynthesis of fluorescent gold nanoparticles using an edible freshwater red alga, Lemaneafluviatilis (L.) C. Ag. and antioxidant activity of biomatrix loaded nanoparticles, *Bioprocess and Biosystems Engineering* 37: 2559–2565.
81. Venkatpurwar, V.; Pokharkar, V. **2011**. Green synthesis of silver nanoparticles using marine polysaccharide: Study of *in vitro* antibacterial activity, *Material Letters* 65: 999–1002.
82. Jena, J.; Pradhan, N.; Dash, B. P. et al. **2015**. Pigment mediated biogenic synthesis of silver nanoparticles using diatom Amphora sp. and its antimicrobial activity, *Journal of Saudi Chemical Society* 19: 661–666.
83. a) Li, X.; Schirmer, K.; Bernard, L. et al. **2015**. Silver nanoparticle toxicity and association with the alga Euglena gracilis, *Environmental Science: Nano* 2: 594–602.
b) Li, Y.; Tang, X.; Song, W. et al. **2015**. Biosynthesis of silver nanoparticles using Euglena gracilis, Euglena intermedia and their extract, *IET Nanobiotechnology* 9: 19–26.

84. Cho, D. Y.; Lee, S. T.; Park, S. W. **2008**. Studies on the biosorption of heavy metals onto *Chlorella vulgaris, Journal of Environmental Science and Health. Part A: Environmental Science and Engineering and Toxicology* 29(2): 389–409.
85. Volesky, B. **2003**. *Sorption and Biosorption*, BV-Sorbex Inc, St. Lambert, Canada.
86. Greene, B.; Hosea, M.; McPherson, R. **1986**. Interaction of gold (I) and gold (III) complexes with algal biomass, *Environmental Science and Technology* 20: 627–632.
87. Dahoumane, S. A.; Yepremian, C.; Djediat, C. et al. **2014**. A global approach of the mechanism involved in the biosynthesis of gold colloids using micro-algae, *Journal of Nanoparticle Research* 16: 2607.
88. Hosea, M.; Greene, B.; Mcpherson, R. et al. **1986**. Accumulation of elemental gold on the alga Chlorella vulgaris, *Inorganica Chimica Acta* 123: 161–165.
89. Mandal, D., Bolander, M. E., Mukhopadhyay, D. et al. **2006**. The use of microorganisms for the formation of metal NPs and their application, *Applied Microbiology and Biotechnology* 69: 485–492.
90. Mata, Y. N.; Torres, E.; Blazquez, M. L. et al. **2009**. Gold(III) biosorption and bioreduction with the brown alga Fucusvesiculosus, *Journal of Hazardous Material* 166: 612–618.
91. Kannan, R. R. R.; Stirk, W. A.; Staden, J. V. **2013**. Synthesis of silver NPs using the seaweed *Codium capitatum* P.C. Silva (Chlorophyceae), *South African Journal of Botany* 86: 1–4.
92. Sau, T. K.; Murphy, C. J. **2004**. Room temperature, high-yield synthesis of multiple shapes of gold nanoparticles in aqueous solution, *Journal of the American Chemical Society* 126: 8648–8649.
93. a) Ahlgren, G.; Gustafsson, I. B.; Boberg, M. **1992**. Fatty acid content and chemical composition of freshwater microalgae, *Journal of Phycology* 28: 37–50.
b) Khan, Z.; Bhadouria, P.; Bisen, P. **2005**. Nutritional and therapeutic potential of *Spirulina. Current Pharmaceutical Biotechnology* 6: 373–379.
94. El-Naggar, N. E.; Hussein, M. H.; El-Sawah, A. A. **2018**. Phycobiliprotein-mediated synthesis of biogenic silver NPs, characterization, *in vitro* and *in vivo* assessment of anticancer activities, *Scientific Reports* 8: 8925.
95. Sharma, A.; Sharma, S.; Sharma, K. et al. **2016**. Algae as crucial organisms in advancing nanotechnology: A systematic review, *Journal of Applied Phycology* 28: 1759–1774.
96. Ali, D. M.; Sasikala, M.; Gunasekaran, M. et al. **2011**. Biosynthesis and characterization of silver NPs using marine cyanobacterium, Oscillatoria willei NTDM01, *Digest Journal of Nanomaterials and Biostructures* 6: 385–390.
97. Xie, J., Lee, J. Y., Wang, D. C. et al. **2007**. Silver nanoplates: From biological to biomimetic synthesis, *ACS Nano* 1: 429–439.
98. Sharma, N. C.; Sahi, S. V., Nath, S. et al. **2007**. Synthesis of plant-mediated gold NPs and catalytic role of biomatrix-embedded nanomaterials, *Environmental Science and Technology* 41: 5137–5142.
99. Mehta, S. K.; Gaur, J. P. **2005**. Use of algae for removing heavy metal ions from freshwater, *Critical Reviews in Biotechnology* 25: 113–152.
100. Brayner, R.; Baberhousee, H.; Hernadi, M. et al. **2007**. Cyanobacteria as bioreactors for the synthesis of Au, Ag, Pd, and Pt nanoparticles via an enzyme-mediated route, *Journal of Nanoscience and Nanotechnology* 7: 2696–2708.
101. Geetha, S.; Sathakkathulzaria, J.; Aarthi, R. et al. **2014**. Green synthesis of gold nanoparticles using cyanobacteria Gloeocapsa sp. and the antitumour potential, *Journal of Chemical and Pharmaceutical Science* 4: 172–174.
102. Suganya, K. U.; Govindaraju, K.; Kumar, V. G. et al. **2015**. Blue green alga mediated synthesis of gold NPs and its antibacterial efficacy against Gram positive organisms, *Materials Science and Engineering* C47: 351–356.
103. Parial, D.; Pal, R. **2014**. Green synthesis of gold nanoparticles using cyanobacteria and their characterization, *Indian Journal of Applied Research* 4: 69–72.
104. Dissanayake, C.B.; Kritsotakis, K. **1984**. The geochemistry of Au and Pt in peat and algal mats – a case study from Sri Lanka, *Chemical Geology* 42: 61–76.
105. Greene, B.; Hosea, M.; McPherson, R. et al. **1986**. Interaction of gold(I) and gold(III) complexes with algal biomass, *Environmental Science and Technology* 20: 627–632.
106. Nayak, D.; Nag, M.; Banerjee, S. et al. **2006**. Preconcentration of 198 Au in a green alga, Rhizoclonium, *Journal of Radioanalytical and Nuclear Chemistry* 268: 337–340.
107. Govindaraju, K.; Basha, S. K.; Kumar, V. G. et al. **2008**. Silver, gold and bimetallic nanoparticles production using single- cell protein (Spirulina platensis) Geitler, *Journal of Materials Science* 43: 5115–5122.
108. Silva, L. P.; Silveira, A. P.; Bonatto, C. C.; Reis, I. G.; Milreu, P. V. **2017**. Chapter 26 – Silver nanoparticles as antimicrobial agents: Past, present, and future. In Anton Ficai, Alexandru Mihai Grumezescu (eds.), *Micro and Nano Technologies, Nanostructures for Antimicrobial Therapy*, Elsevier, Amsterdam, The Netherlands, pp. 577–596.

109. Husain, S.; Sardar, M.; Fatma, T. **2015**. Screening of cyanobacterial extracts for synthesis of silver nanoparticles, *World Journal of Microbiology Biotechnology* 31: 1279–1283.
110. El Bialy, B. E.; Hamouda, R. A.; Khalifa, K. S. et al. **2017**. Cytotoxic effect of biosynthesized silver NPs on Ehrlich ascites tumor cells in mice, *International Journal of Pharmacy* 13: 134–144.
111. Tomer, A. K.; Rahi, T.; Neelam, D. K. et al. **2019**. Cyanobacterial extract-mediated synthesis of silver nanoparticles and their application in ammonia sensing, *International Microbiology* 22: 49–58.
112. Roychoudhury, P.; Ghosh, S.; Pal, R. **2016**. Cyanobacteria mediated green synthesis of gold-silver nanoalloy, *Journal of Plant Biochemical Biotechnology* 25: 73–78.
113. Sudha, S. S.; Rajamanickam, K.; Rengaramanujam, J. **2013**. Microalgae mediated synthesis of silver nanoparticles and their antibacterial activity against pathogenic bacteria, *Indian Journal of Experimental Biology* 51: 393–399.
114. Cepoi, L., Rudi, L., Chiriac, T. et al. 2014. Biochemical changes in cyanobacteria during the synthesis of silver NPs, *Canadian Journal of Microbiology* 61: 13–21.
115. Khalifa, K.; Hamouda, R.; Hanafy, D. et al. 2016. *In vitro* antitumor activity of silver NPs biosynthesized by marine algae, *Digest Journal of Nanomaterials and Biostructures* 11: 213–221.
116. Hamouda, R. A.; Hussein, M. H.; Abo-Elmagd, R. A. et al. **2019**. Synthesis and biological characterization of silver nanoparticles derived from the cyanobacterium Oscillatoria limnetica. *Scientific Reports* 9: 13071.
117. Satapathy, S.; Shukla, S. P. **2017**. Application of a marine cyanobacterium Phormidium fragile for green synthesis of silver NPs, *Indian Journal of Biotechnology* 16: 110–113.
118. Saran, S.; Sharma, G.; Kumar, M. et al. **2017**. Biosynthesis of copper oxide nanoparticles using cyanobacteria spirulina platensis and its antibacterial activity, *International Journal of Pharmaceutical Sciences and Research* 8: 3887–3892.
119. Doman, K. M., Gharieb, M. M., Abd El-Monem, A. M. et al. **2023**. Synthesis of silver and copper nanoparticle using Spirulina platensis and evaluation of their anticancer activity, *International Journal of Environmental Health Research*. https://soi.org/10.1080/09603123.2022.2163987
120. Asif, N.; Ahmad, R.; Fatima, S. et al. 2023. Toxicological assessment of Phormidium sp. derived copper oxide nanoparticles for its biomedical and environmental applications, *Scientific Reports* 13: 6246.
121. Moskvina, M. I.; Brekhovskikh, A. A.; Nikandrov, V. V. **2003**. The role of the heterotrophic bacteria associated with the cyanobacterium *Nostoc muscorum* in the formation of cadmium sulfide, *Microbiology* 72(2): 246–247.
122. Mandal, R. P.; Sekh, S.; Sarkar, N. S. et al. **2016**. Algae mediated synthesis of cadmium sulphide nanoparticles and their application in bioremediation, *Materials Research Express* 3: 11.
123. Brayner, R.; Dahoumane, S. A.; Yéprémian, C. et al. **2010**. ZnO nanoparticles: Synthesis, characterization, and ecotoxicological studies, *Langmuir* 26: 6522–6528.
124. Ebadi, M.; Reza, M.; Seyyed, Z. et al. **2019**. A bio-inspired strategy for the synthesis of zinc oxide nanoparticles (ZnO NPs) using the cell extract of cyanobacterium Nostoc sp. EA03: From biological function to toxicity evaluation, *RSC Advances* 9: 23508–23525.
125. Asif, N.; Fatima, S.; Aziz, M. N. et al. **2021**. Biofabrication and characterization of cyanobacteria derived ZnO NPs for their bioactivity comparison with commercial chemically synthesized nanoparticles, *Bioorganic Chemistry* 113: 104999.
126. Pandey, S.; Awasthee, N.; Shekhar, A. et al. **2021**. Biogenic synthesis and characterization of selenium nanoparticles and their applications with special reference to antibacterial, antioxidant, anticancer and photocatalytic activity, *Bioprocess and Biosystems Engineering* 44: 2679–2696.
127. Afzal, B., Yasin, D., Naaz, H. et al. **2021**. Biomedical potential of Anabaena variabilis NCCU-441 based Selenium nanoparticles and their comparison with commercial nanoparticles. *Scientific Reports* 11: 13507.
128. Asif, N.; Ahmad, R.; Fatima, S. et al. **2023**. Toxicological assessment of *Phormidium* sp. derived copperoxide nanoparticles for its biomedical and environmental applications, *Scientific Reports* 13: 6246.
129. Lefebvre, D. D.; Kelly D.; Budd, K. **2007**. Biotransformation of Hg(II) by cyanobacteria, *Applied and Environmental Microbiology* 73: 243–249.
130. Keskin, N. O. S.; Kocberber, K. N.; Donmez, G. et al. 2016. Green synthesis of silver nanoparticles using cyanobacteria and evaluation of their photocatalytic and antimicrobial activity, *Journal of Nano Research* 40: 120–127.
131. Hamida, R. S.; Abdelmeguid, N. E.; Ali, M. A. **2020**. Synthesis of silver nanoparticles using a novel cyanobacteria Desertifilum sp. extract: Their antibacterial and cytotoxicity effects, *International Journal of Nanomedicine* 15: 49–63.

132. El-Naggar, N. E.; Hussein, M. H.; El-Sawah, A. A. **2017**. Bio-fabrication of silver nanoparticles by phycocyanin, characterization, *in vitro* anticancer activity against breast cancer cell line and *in vivo* cytotoxicity. *Scientific Reports* 7: 10844.
133. Kuraganti, G.; Edla, S.; Dasari, T. **2020**. Cyanobacteria assisted bio-reduction of silver nanoparticle conjugates and study on their characterization and antibacterial activity against pathogenic bacteria, *International Journal of Pharmaceutical Sciences and Research* 11: 163–172.
134. Hamida, R. S.; Ali, M. A.; Goda, D. A. et al. **2020**. Novel biogenic silver nanoparticle-induced reactive oxygen species inhibit the biofilm formation and virulence activities of methicillin-resistant staphylococcus aureus (MRSA) strain, *Frontiers in Bioengineering and Biotechnology* 8: 433.
135. Roychoudhury, P.; Pal, R. **2014**. Synthesis and characterization of nanosilver-a blue green approach, *Indian Journal of Applied Research* 4: 1.
136. Zada, S.; Ahmad, A.; Khan, S. et al. **2018**. Biogenic synthesis of silver nanoparticles using extracts of Leptolyngbya JSC-1 that induce apoptosis in HeLa cell line and exterminate pathogenic bacteria, *Artificial Cells, Nanomedicine, and Biotechnology* 46: S471–S480.
137. Rashed, S. A.; Shehri, S. A.; Moubayed, N. M. **2018**. Extracellular biosynthesis of silver nanoparticles from cyanobacteria, *Biomedical Research* 29: 2859–2862.
138. Hamida, R. S.; Albasher, G.; Bin-Meferij, M. M. **2020**. Oxidative stress and apoptotic responses elicited by nostoc-synthesized silver nanoparticles against different cancer cell lines, *Cancers* 12: 2099.
139. Morsy, F. M.; Nafady, N. A.; Abd-Alla, M. H. et al. **2014**. Green synthesis of silver nanoparticles by water soluble fraction of the extracellular polysaccharides/matrix of the cyanobacterium Nostoc commune and its application as a potent fungal surface sterilizing agent of seed crops, *University Journal of Microbiological Research* 2: 36–43.
140. Sonker, A. S.; Pathak, J.; Kannaujiya, V. K. et al. **2017**. Characterization and *in vitro* antitumor, antibacterial and antifungal activities of green synthesized silver nanoparticles using cell extract of Nostoc sp. strain HKAR-2, *Canadian Journal of Biotechnology* 1(1): 26.
141. Ahmed, E.; Hafez, A.; Ismail, F. et al. **2015**. Biosynthesis of silver nanoparticles by Spirulina platensis and Nostoc sp, *Advanced Journal of Microbiology Research* 4: 36–49.
142. Siddiqui, T.; Khan, N.; Asif, N. et al. **2022**. Screening, characterisation and bioactivities of green fabricated TiO_2 NP via cyanobacterial extract, *Environmental Science and Pollution Research* 29: 39052–39066.

12 Biomedical Applications of Biogenic Carbon-Based Fluorescent Nanoparticles

Kamalpreet Kaur, Gagandeep Singh, Rahul Badru, Navneet Kaur, and Narinder Singh

12.1 INTRODUCTION

Nanotechnology is an active area of research which deals with the synthesis of materials having the structural property of bulk materials and atoms at least one dimension in the nanoscale.[1] It is developing very fast, as at nanoscale the materials exhibit properties such as varying size, shape, and morphology for application in different areas, such as chemistry, electronics, medicine, energy, etc.[2] Nowadays, materials consisting of optical characteristics have drawn attention in imaging analysis and sensing. Among all, the most important fluorescent labelling technique has evolved. On the other hand, conventional luminescent labelling materials exhibit the disadvantage of complex manufacturing process and self-aggregation in water, limiting their application. Relative to conventional fluorescent dyes, fluorescent nanomaterials get easily dispersed in water possessing higher stability. Fluorescent[3] as well as phosphorescent[4] materials possess a broad potential in different areas, which include phototherapy,[5] pollution control,[6] bioimaging,[7] photopolymerization,[8] and so on. On the other hand, the preparation of fluorescent materials needs raw materials extracted from crude oil with complex processing procedures, causing environmental issues. Therefore, it is of utmost importance to develop environmentally friendly techniques for the development of valuable luminescent materials. Green synthesis is a promising approach for biocompatible materials preparation by eradicating the use of toxic and costly chemicals, less energy consumption, and producing harmful by-products in the environment.[9] The biogenic sources give rise to the transformation of no aggregated valued materials to nanomaterials having great importance and applicable in different fields.[10] The use of plant, microbes, and fungi extract is an environment-friendly, low-cost, easily available source, having nitrogen and oxygen, which makes it an excellent carbon source precursor for CQDs synthesis.[11] Carbon dots luminophores are a new shining star in the family of carbon and can be defined as quasi-zero-dimensional (0D) carbon-based material with size below 20 nm and exhibit low toxicity, excellent photostability, high sensitivity, favorable compatibility, large stokes shift, excellent selectivity for target analytes, high quantum yield, and tunable fluorescence emission and excitation.[12] Luminescent carbon nanoparticles were first reported in the year 2004, obtained coincidentally while purifying single-walled carbon nanotubes (SWNTs). Sun et al. (2006) illustrated fluorescent CD synthesis by laser ablation and surface passivation, but the quantum yield was only about 10%. Complicated synthesis procedure and low QY limited the development of CDs.[13] Much development has been made in the synthesis, application, and properties of CDs during the last few years, as reviewed by Lee et al.[14] Lim et al.[15] Zhu et al.[16] Baker et al.[17] and Wang et al.[18] Researchers have designed different technologies to develop CDs with multicolor, deep-red, near-infrared emission (NIR), room temperature phosphorescence, chirality, thermally activated delayed luminescence, narrow full width at half maximum, and various other applications. Carbon dots are categorized into graphene quantum dots (GQDs), carbon quantum

DOI: 10.1201/9781003430087-12

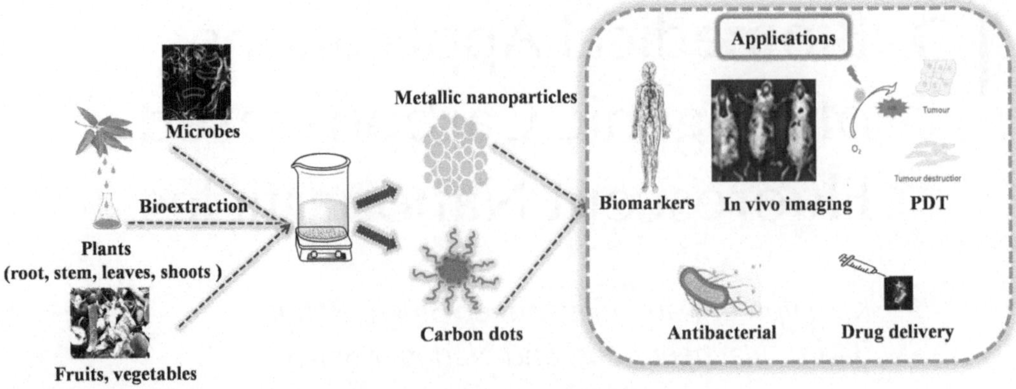

FIGURE 12.0 Various biomedical applications exhibited by carbon dots.

dots (CQDs), and carbonized polymer dots (CPDs), depending on the synthetic mechanism, properties, and nanostructures. GQDs refer to graphene fragments having size below 20 nm, where the electron transport is confined in all three dimensions. The edge effects and quantum confinement of the GQDs induce photoluminescence. The CQDs and CPDs are spherical-shaped, produced from smaller polymers, biomass, or molecules via "bottom-up" approach. The quantum confinement size effect and intrinsic state fluorescence are their PL mechanism. The CPDs contain π-conjugated polymers that in small-sized particles exhibit brightness. Here, we will introduce the basics of fluorescence, a brief history of luminescence from biogenic sources, and a selection of suitable biogenic sources for luminescent material fabrication. Moving further, preparation and optical properties of fluorescent carbon dot materials will be illuminated. Further, application of materials for environmental contamination, anticounterfeiting, and photothermal and photodynamic therapy in antibacterial applications will be discussed and outlined. At last, we will give a discussion on the current challenges and future outline opinions for the development of luminescent materials derived from biogenic sources.

12.2 BRIEF HISTORY AND FUNDAMENTALS OF FLUORESCENCE

Fluorescence was introduced first by Eilhard Wiedemann (a German physicist) as luminescence in 1888 for the light phenomenon not conditioned by temperature increase, that is, incandescence.[19] This word is derived from the Latin root "lumen", which means light. In the middle of the 19th century after fluorescence was introduced by G. G. Stokes, the difference between phosphorescence and fluorescence depends on the emission time scale after ending excitation, that is, in phosphorescence, emitted light exists for a longer time after the end of excitation, whereas in fluorescence, the emission of light disappears immediately after the end of excitation. But such criteria are not clearly understood, as there are short-lived phosphorescence (zinc sulfide) and long-lived fluorescence (europium salts) having comparable durations with hundreds of nanoseconds. The modern definition of *fluorescence* is "a continuous radiation emission from vibrationally and electronically excited molecule which is in non-thermal equilibrium with its own environment". The principal mechanism of luminescence is photon absorption to lift an electron from the ground state (S_0) to an excited state. The excited state contains unstable first- and second-singlet excited state as S_1 and S_2, respectively. The fluorescence arises when excited electron returns back from S_1 or S_2 to S_0. In addition to this, it is probable that the electron from S_1 jumps to T_1 (first-triplet excited state), known as intersystem crossing, which is non-radiative transition occurring from reversal of electron spin, and phosphorescence occurs by the return of electron from T_1 to S_0 (Figure 12.1).

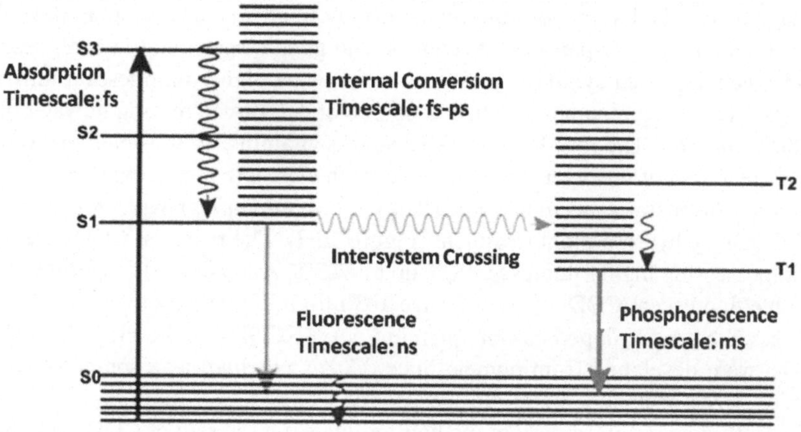

FIGURE 12.1 Schematic illustration of Jablonski diagram.

Source: Reproduced with copyright from the American Chemical Society.

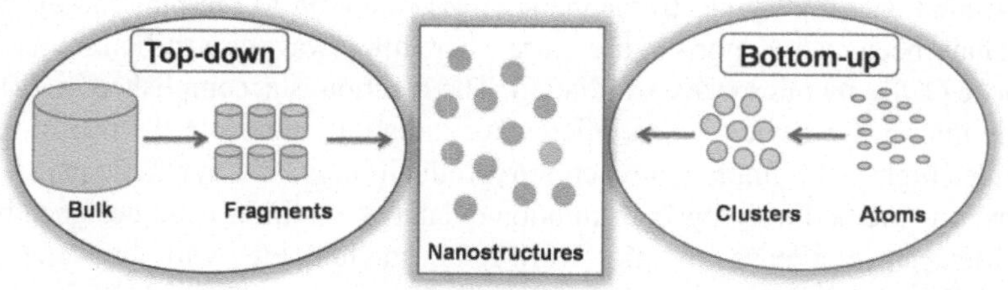

FIGURE 12.2 Schematic illustration of top-down and bottom-up approaches.

12.3 SYNTHESIS OF NANOMATERIALS

The two major approaches used for the preparation of nanomaterials are bottom-up and top-down approaches (Figure 12.2). The top-down method includes ion sputtering, mechanical milling, laser ablation, ball milling, cutting, grinding, and so on. Very small-sized particles cannot be achieved by top-down approaches, though it is easier to carry out. The bottom-up approach is based on the principle of molecular recognition, that is, self-assembly.[20] The bottom-up method includes chemical vapor deposition, hydrothermal method, laser pyrolysis, physical vapor deposition, solgel method, and so on. The high cost and toxic quality of synthesis approach limit its use on a large scale.

12.3.1 Synthesis Strategy of Carbon Dots

There are various methods for the synthesis of CQDs, which are categorized into physical synthesis and chemical synthesis, according to the feasibility of the chemical reaction. The physical synthesis may cause destruction of the morphology, defects on the material surface, and contamination of the material itself in the synthesis process. Thus, chemical synthesis has been mostly used by the researchers, which will be discussed further.

12.3.1.1 Hydrothermal Method

Hydrothermal synthesis of CQDs involves a chemical reaction in aqueous solution under high conditions of pressure 1 MPa ~ 1 GPa and temperature of 100 ~ 1,000°C. The benefit of using hydrothermal synthesis is that the ions are mixed evenly in an aqueous solution; thus, the product produced has high purity, easy particle size control, and good dispersion. Liu et al. described cost-effective and green approach synthesis pathways to fabricate highly luminescent nitrogen-doping carbon dots (N-CDs) using rose-heart radish as source of carbon via mixing in water, followed by heating at 180°C for 3 h.[21] The rose-heart radish contains vitamins, amino acids, and crude protein, resulting in a great deal of nitrogen, oxygen, and carbon elements producing CQDs with high QY of 13.6%. In a similar manner, neem leaf extracts (*Azadirachta indica*) have been employed for the synthesis of CQDs by hydrothermal treatment (Figure 12.3). The resultant CQDs are eco-friendly and low-cost, possessing highly fluorescent QY up to 27.2%. Another study reported the formation of water-soluble fluorescent CQDs (QY 14%, size 1–3 nm) utilizing lemon peel waste with hydrothermal process.[22] Nitrogen-doped carbon quantum dots (NCQDs) and undoped carbon quantum dots (UCQDs) were developed from pomelo juice (*Citrus maxima*) as a source of carbon.[23] The average particle sizes of undoped and nitrogen-doped CQDs were 3 nm and 7 nm, respectively, displaying hydrothermal as an important method for adjusting the size of CQDs. Krishnaiah et al. synthesized N-CDs from Kentucky bluegrass, which shows strong cyan-blue fluorescence under 365 nm UV light. *Poa pratensis* (Kentucky bluegrass, or KB) consists of flavonoids, anthocyanins, phenols, tannins, etc., acting as a source of carbon for N-CDs.[24]

12.3.1.2 Microwave-Assisted Method

Microwave irradiations modify electromagnetic (EM) radiation to thermal energy for producing CQDs. There have been many reports in literature where microwave as a carbon source has been utilized to produce CQDs by microwave irradiation. The reaction is accomplished in different single- or multi-mode microwave device, where microwave radiations can raise the rate of reaction. This method provides high yield, improved reaction reproducibility, safer synthetic paths, and purified by constant heating method with low consumption of energy. Architha et al. designed blue emission CQDs from Mexican mint leaves by microwave reflux methodology with an average diameter of nearly 2.43 ± 0.02 nm and quantum yield of 17% for cytotoxicity impact on MCF-7 cells.[25] Gul et al. utilized banana peels, a greener and renewable source, for the synthesis of CDs.[26] The banana peels were carbonized in an oven and then ground into powder, which was further dissolved in water and heated in a microwave, giving blue fluorescent CDs. Monte-Filho et al. designed blue luminescent carbon dots by the use of lemon and onion juices via one-step carbonization procedure, utilizing a household microwave oven (Figure 12.4).[27] The resultant CDs exhibit high QY of 23.6%, particle size 4.23–8.22 nm, and an excellent water-solubility. Zhao et al. prepared CDs using green natural kelp material which contains sugar, fibers, and protein as carbon source and ethylene diamine as nitrogen dopant via one-pot microwave irradiation. Venugopalan et al. synthesized biogenic, highly fluorescent CQDs by using wild lemon leaves with QY 7.2%.[28]

12.3.1.3 Microwave Hydrothermal Method

In this method, microwave is utilized to heat up the solvent, which rapidly attains high temperature, pressure, and short reaction time due to combined heating hydrothermal and microwave methods.

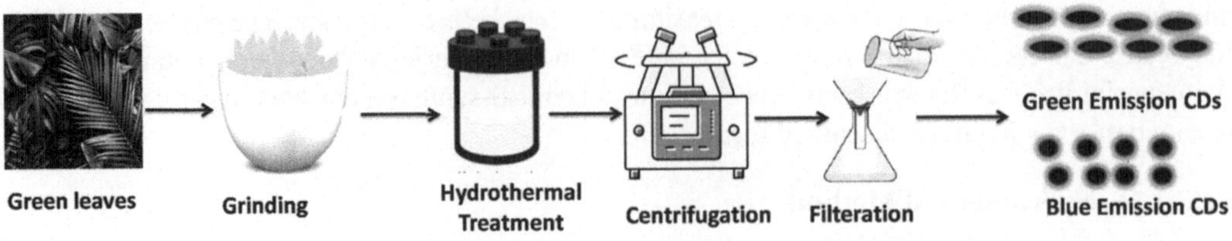

FIGURE 12.3 Schematic illustration of preparation of N-CQDs by hydrothermal method.

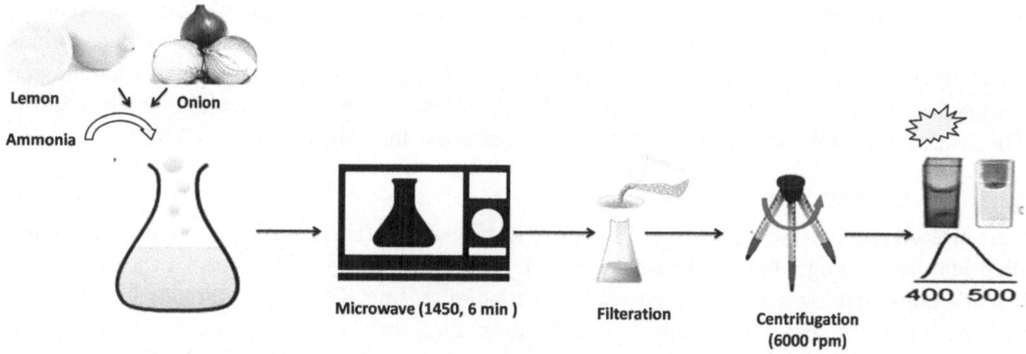

FIGURE 12.4 Schematic illustration of preparation of CQDs by microwave-assisted method.

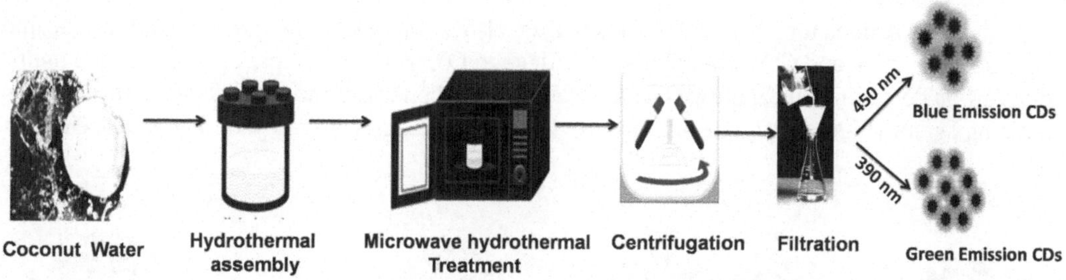

FIGURE 12.5 Schematic illustration of preparation of CQDs by microwave–hydrothermal method.

The advantage of this procedure relative to traditional hydrothermal is that if the sample exhibits definite depth, microwaves can easily pierce and each depth can be warmed equally, avoiding temperature difference and heat conduction, thereby improving the speed of the reaction. However, minimum studies have been reported in literature utilizing microwave–hydrothermal method for the synthesis of CQDs. Purbia et al. reported the development of green and blue luminescent CDs via microwave–hydrothermal technique from coconut water. The size of the particle varies, depending on the heating time and temperature of the microwave, between 1 and 6 nm (Figure 12.5). Liang et al. synthesized blue-emitting CQDs with bamboo tar with good fluorescent quantum yield of 19.3%, exhibiting strong emission when excited at 450 nm wavelength, good solubility in water, and size range of 1–7.5 nm.[29] The CDs prepared using palm kernel shell as carbon precursor exhibit high quantum yield of 44%, with size of 6–7 nm in diameter.[30]

12.3.1.4 Pyrolysis Carbonization Method

Pyrolysis is a very classical method for the synthesis of CDs which involves the decomposition of a substance by heat. The size of particle prepared by pyrolysis is between 0.4 and 6 nm, and quantum yield lies between 3% and 25%. Dehvari et al. developed nitrogen-doped carbon dots utilizing crab shell via one-pot sonochemical approach in an aqueous medium.[31] The N@CDs are spherical-shaped, with an average size of 8 nm and high QY of 14%. O. Sanni et al. fabricated highly luminescent pine cone CDs having an average diameter of 15.2 ± 2.7 nm, with QY of 17%.[32] In another study, the CDs were synthesized from essential Indian oils, basil, clove, cardamom, and turmeric, having particle size lying below 10 nm.[33] Moreover, luminescent CDs were also synthesized using gynostemma as carbon source via pyrolysis method. CDs display mean size of particle about 2.5 nm, emit blue luminescence under 365 nm, and exhibit good water-solubility.[34] Agnol et al. illustrated the synthesis of photoluminescent CDs by pyrolysis of spirulina microalgae without the use of any surface passivation agent.[35] The developed CQDs exhibit an average size of 10 nm and blue luminescence emission at around 450 nm wavelength, with QY of 23.5%. Singh et al. prepared luminescent carbon dots using mango (*Mangifera indica, M. indica*) leaves by pyrolysis method for 3 h at 300°C. The synthesized CQDs had a spherical shape, with size less than 10 nm (Figure 12.6).[36]

12.3.1.5 Chemical Oxidation Method

This method requires the utilization of strong oxidizing agents (H_2SO_4, HNO_3, or H_2O_2) which oxidize a target, causing it to lose electrons. The CDs prepared by this method have a variety of functional groups, making it a more extensive application in sensing. The chemical oxidation method is fast and convenient, but it also exhibits great disadvantages, such as expensive oxidants, causing pollution to the environment and low production capacity. Boruah et al. developed CDs from sugarcane bagasse, taro peels, and garlic peels via biocompatible chemical oxidation method.[37] It has been reported that the CQDs prepared were highly water-soluble, exhibiting strong blue luminescence with good QY of 4–27%. Hu et al. chose green tea leaf as a raw material and sulfuric acid as an oxidant to synthesize CDs.[38] The tea leaf was dried and ground into a fine powder, and then further calcinations were done for 2 h at 350 °C. Thereafter, H_2SO_4 was added into the resultant residue and carbonized for 20 h, and CDs were obtained with good QY of 14.8%. Gupta et al. developed multicolor carbon dots using pineapple extract as a source of carbon using phosphoric and sulfuric acid as oxidizing agents (Figure 12.7).[39] Kailasa et al. designed blue-, green-, and yellow-luminescent CDs

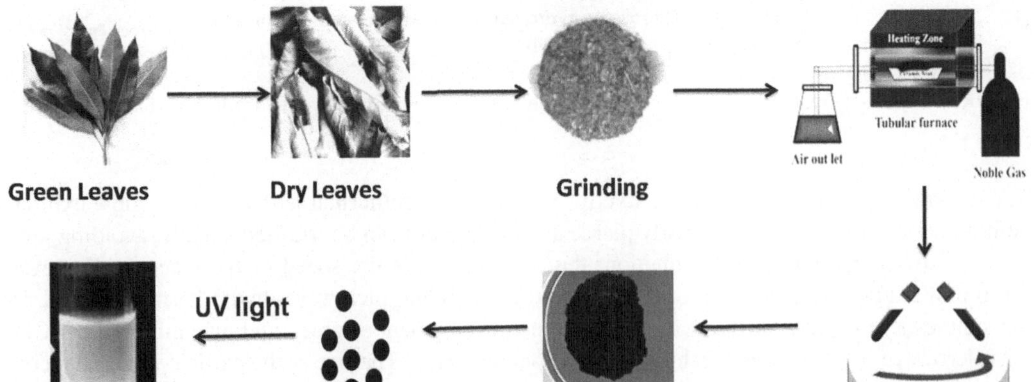

FIGURE 12.6 Schematic illustration of preparation of CQDs by pyrolysis method.

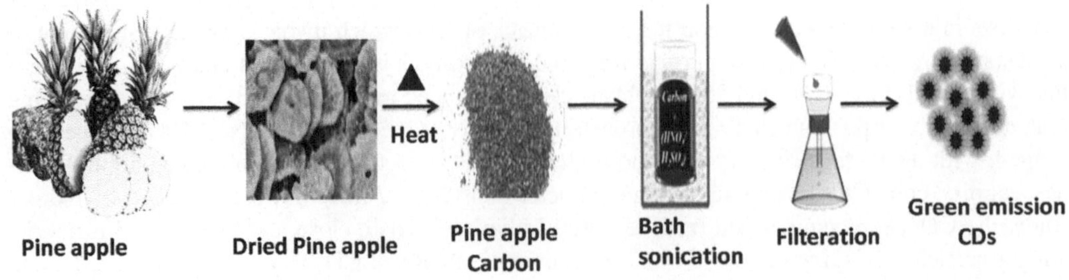

FIGURE 12.7 Schematic illustration of preparation of CQDs by chemical oxidation method.

derived from tomato using H_3PO_4 and H_2SO_4 as oxidizing agents. The blue-, green-, and yellow-color luminescent CDs exhibit high QY, are spherical in shape, range from 5 to 10 nm, and exhibit good water dispersity. Desai et al. developed multicolor blue, green, and yellow emissive carbon dots using *C. melo* fruit as carbon source via oxidation treatment.[40]

12.4 TYPES OF FLUORESCENCE

12.4.1 Blue Fluorescence

Most of the CQDs emit blue fluorescence, depending on the extent of surface functionalization and biogenic carbon source used for the synthesis. It has been reported in literature that CDs developed from lemon and onion,[27] bamboo,[29] palmyra palm leaf,[41] and peanut shell[42] show blue fluorescence characteristics. CDs synthesized from coke exhibit hydroxyl, carboxylic, and C-O-C, C-H, C=C stretching frequency vibrations due to the existence of hydroxyl moieties on the surface. In another study, CDs developed from gingko leaf suggested the -OH, -NH, C-C, C-H, C-N groups, validating the presence of -OH and $-NH_2$ groups on the surface.[43] Moreover, CDs developed from neem leaf showed the eminent O-H stretch, displaying the hydroxyl moieties presence. Xie et al. synthesized blue-luminescent CDs using ethylene diamine and barley as source of C and N containing -OH and -NH groups.[44] Zhu et al. noticed the blue luminescence of CQDs arising from amide groups. Sun et al. established the blue luminescence of CDs from hydroxyl moiety. In summary, hydroxyl groups play an important role in the origination of blue luminescence.[45]

12.4.2 Green Fluorescence

Das et al. explored the hydroxyl and carboxyl moieties' impact on green luminescence of carbon quantum dots utilizing peer juice as carbon source via one-pot solvothermal method.[46] The negative charge on the hydroxyl and oxygenated carboxyl groups acts as emission centers. Wang et al. validated the existence of C=O moiety on GQDs, and CDs are accountable for green emission.[47] In another study, Hoan et al. developed green-luminescence CDS using lemon juice via hydrothermal method.[48] The FTIR study confirm the presence of -COOH, C-O-C, and C=O groups on the CQD surface. Two major peaks correspond to -OH stretching, making CDs water-soluble, and others correspond to carbonyl and carboxyl. Green-emitting CDs were prepared from coriander leaves. The FTIR analysis showed the existence of C=C, C=O, C-N stretching, C-O bending vibrations. Overall, carboxyl and carbonyl groups dominate in the FTIR spectrum of green-luminescent CDs.

12.4.3 Red Fluorescence

Ding et al. prepared red-fluorescence CDs using pulp-free lemon juice with good QY of 28%.[49] The as-synthesized CDs exhibit an average size of 4.6 nm. FTIR spectra validate the existence of -OH, -COOH, C-O, and N-H. The presence of C=C, C-N, and C=N stretching vibration frequency employed polyaromatic structure development in carbon dots carbon core. Four types of carbon dots were prepared with changing ethanol amount and revealed that increasing ethanol content shifts the fluorescence from 518 to 616 nm. XPES data disclosed that ethanol promotes doping of nitrogen with increase in nitrogen content from 10.3% to 12.4%, giving rise to changes in the surface of CQDs. Thus, red-emissive CDs can be generated by increase in nitrogen content. Moreover, the resultant CDs acquired using lemon juice exhibit a mean size of 5.7 nm with quantum yield of 31% and nitrogen content of 20%.[50] Another study revealed that preparation of zinc-doped CDs from bougainvillea plant display red emission with QY 72%. As ultraviolet and blue light damage biological cells, red-emission luminescence materials show wide applications in the biomedical field.

12.4.4 Yellow Fluorescence

Wang et al. synthesized yellow-emitting luminescent carbon dots utilizing petroleum coke as carbon precursor using chemical oxidation method with QY 9.8%.[51] The presence of -OH, C=O, $-CH_2$, and C-H groups on CQDs was revealed by the FTIR spectrum. The origin of yellow luminescence is due to C=C aromatic bonds and defects produced by the C=O and C-OH groups. In another study, phosphorous- and nitrogen-doped yellow CDs were developed using pumpkin as a source of carbon and phosphoric acid as oxidizing agent. Surface analysis disclosed the -OH, C=O, N-H, -CH, hydrogen-bonded P=O, P-OH, and aromatic P-O-C peaks, showing the surface functionalization of CDs with phosphoric moieties.[52] Similarly, using aloe vera as the source of carbon, yellow CDs were synthesized by hydrothermal method with QY of 10.37%. The FTIR disclosed the presence of -OH, -COO, and -CH stretching frequency bands, showing the CD surface enriched with carboxylic and hydroxyl groups.[53]

12.5 APPLICATIONS OF NANODOT LUMINOGENS

12.5.1 Nanodot Luminogens for Bioimaging of Cancer Cells

Pakkath et al. demonstrate the preparation of transition metal–doped (Fe^{2+}, Ni^{2+}, Co^{2+}, and Mn^{2+}) CDs (TMCDs) utilizing lemon extract via microwave-assisted pyrolysis method within 6 min at 800 W. The lemon extract consists of citric acid (6%), ascorbic acid (64%), and trace number of flavonoid derivatives and carbohydrates (2.5%) as a source of carbon to synthesize TMCDs with good optical and physiochemical properties. The TMCDs possess good water-solubility, with 3.2 ± 0.485 nm sizes and reinforced with oxo, amino, and corresponding metal-oxide traces on the surface revealed by XPES and FT-IR spectroscopy. TMCDs display brighter fluorescence emission when excited at 460 nm, and the Mn/C, Fe/C, Co/C, and Ni/C-dots show quantum yield of 35.71, 41.72, 75.07, and 50.84% in addition to fluorescence-enhanced lifetimes (τ_{av}) of 9.4, 8.6, 9.2, and 8.9 ns, respectively (Figure 12.8). The TMCDs are biocompatible for fluorescence imaging in human colon cancer cells (SW480) and exhibit super paramagnetic and ferromagnetic nature with active T1- contrast ability. Kalita et al. designed GQDs having an average diameter of 2–6.5 nm utilizing rice grain as source of carbon. Due to quantum confinement effect, GQD size increases from 2 to 6.5 nm; a red shift from blue to cyan color has been observed in the fluorescence-emission spectra.[54] The QY of as-synthesized GQDs in aqueous medium depends on the size, which rises from 16 to 24% upon a drop in size from 6.5 to 2 nm. The as-prepared GQDs have been used as

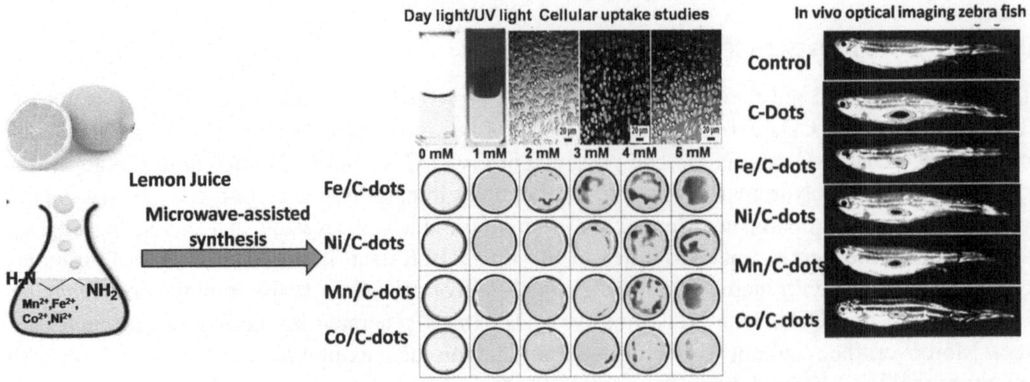

FIGURE 12.8 Schematic illustration of EDA-TMCDS as nanoprobe for fluorescent bioimaging.

Source: Reproduced with permission. Copyright 2017 American Chemical Society.

a nanoprobe for the luminescence imaging of Hella cells. The excellent biocompatibility and less cytotoxicity make GQDs an excellent candidate in biolabeling and bioimaging. Mohapatra et al. developed green-luminescent carbon dots using aqueous extract of *Tinospora cord folia* leaves with eco-friendly hydrothermal carbonization method.[28] The as-synthesized TCLCDs exhibit emission maxima at 435 nm, having a spherical shape with average size of diameter 5.47 nm. The TCLCDs display excellent biocompatibility, thermal stability, photostability, and dose-dependent bioimaging in cervical cancer (SiHa) and melanoma cell lines (B16F10). The overall outcomes confirm the potential of TLCDCs for free radical scavenging, *in vitro* bioimaging of cancer cells.

Wei et al. reported the synthesis of green, low-cost, facile fluorescent CDs using gynostemma as source of carbon without the use of toxic chemicals.[34] The as-prepared CDs show excellent dispersibility in water, good luminescence constancy in high-ionic-strength environments, average size of particle about 2.5 nm, and emit blue luminescence. Overall, luminescent carbon dots exhibit capability for disease treatment produced by oxidative damage, such as senility, cancer, aging, and bioimaging. Damera et al. presented an eco-friendly, hydrothermal green method for synthesis of multicolor light-blue, green, blue, and red emissive fluorescent carbon nanoparticles from *eucalyptus globules* twigs.[55] The eucalyptus stem contains terpenoids, tannins, flavonoids, triterpenes, and alkaloids. Moreover, light-blue and blue CNPs are biocompatible, as affirmed by cytotoxicity assay performed on MDA-MB-231 cell lines and used *in vitro* biomedical imaging due to their high photostable nature. Pal et al. reported the preparation of CDs from curcumin (*Curcuma longa*) and surface passivated with branched polyethyleneimine (bPEI), giving rise to aqueous soluble C-dots, high QY, size 4–5 nm, and good chemical stability.[56] Thus, as-prepared C-dots serve as biolabeling, contrasting agent, along with antioxidation properties. Kumawat et al. reported the synthesis of red-emitting graphene quantum dots (mGQDs) using ethanol extract of mango leaves (*Mangifera indica*) with quantum size ranging from 2 to 8 nm.[57] The mGQDs display good biocompatibility on L929 cells and demonstrated as NIR-responsive fluorescent bioimaging probe in the cell cytoplasm (Figure 12.9). Wang et al. designed water-soluble, luminescent CDs utilizing cyanobacteria as a sole source of carbon via normal hydrothermal method. The as-prepared CDs have an average diameter of 2.48 nm, with quantum yield of 9.24%, and act as a luminescent probe for imaging of PC12 cells. Sahu et al. prepared green luminescent carbon dots derived from orange juice possessing high photostability, quantum yield of 26%, and low toxicity as excellent probes on cellular imaging.[58]

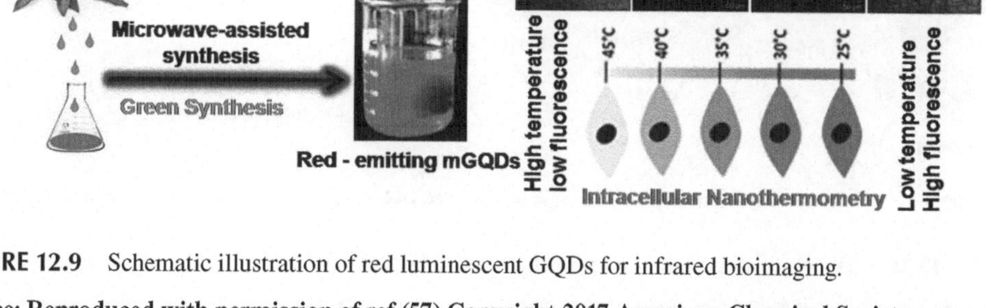

FIGURE 12.9 Schematic illustration of red luminescent GQDs for infrared bioimaging.

Source: Reproduced with permission of ref (57) Copyright 2017 American Chemical Society.

Ramanan et al. presented a rapid, green, and simple approach to prepare luminescent CDs utilizing eutrophic algal blooms as carbon source. Highly luminescent carbon dots with QY of 13% and average size of particle 8 nm exhibit low cytotoxicity, good cell permeability, high photostability, and water-solubility. Moreover, confocal microscope confirms CDs uptake by MCF-7 cells, confirming the utilization of these carbon dots as potential biomarker.[59]

12.5.2 Nanodot Luminogens for Disease Biomarker

Deb et al. illustrated the synthesis of a sandwich kind immunosensor based on luminescence emission phenomena of CDs for recognition of VEGF (vascular endothelial growth factor) disease biomarker (Figure 12.10). CDs were derived from sweet orange with good QY and conjugation, with antibody IgG–producing luminescent receptor, IgG-OCD. The biomarker VEGF was impaired on the capture antibody, and the recognition antibody connected with luminescent probe acts as a bridge to connect with VEGF biomarkers attached to capture antibody.[60] The luminescent immunosensor exhibits a low limit of detection at 5.65 pg/ml for the VEGF biomarker. Miao et al. presented a green, economical method for the preparation of CDs utilizing tomato juice as source of carbon with QY nearly 13.9%.[61] These CDs display excellent photostability and non-toxicity. Significantly, a reusable method has been presented for the recognition of carcinoembryonic antigen (CEA) on the basis of desorption and adsorption of aptamer on the surface of CDs. ssDNA gets strongly adsorbed on CD surface due to the presence of carboxyl groups and through π–π stacking interaction, which result in the quenching of luminescence by CDs–aptamer complex. The strong interaction among CEA–aptamer and CEA has been the main reason for recovery of CD fluorescence once CEA was added. Raveendran et al. derived CDs from mint leaf extract (M-CDs) via economical, green approach for application as biosensor, biomarker, or reductant. The developed CDs are highly selective and sensitive toward folic acid and applied for fluorometric recognition through inner filter effect, having detection limit of 280 nM. The C-dots are noncytotoxic, as revealed by MTT assay carried out on H8 cells, and exhibit multicolor imaging in Hella cells.[62] Kansay et al. developed an easy, cost-effective, and environment-friendly procedure for preparation of fluorescent carbon quantum dots from *Cissus quadrangularis* medicinal herb leaf extract (Figure 12.11).[63] The as-synthesized nitrogen-doped CQDs possess QY of 5%. The size of the developed CQDs was found to be 4–9 nm. The zeta potential value of CQDs was -7.8mV. CQDs have been found to be a luminescent staining agent in biolabeling of fungi, bacteria, and plant cells. Moreover, the cell survival and cytotoxicity of algae, fungus, mushrooms, leaf, and bacteria have been investigated, which is found to be 10.9 mg/ml for four days.

12.5.3 Nanodot Luminogens for Photothermal Therapy of Cancer

Mostly, carbon dots activated by blue/ultraviolet light show near-infrared/visible emission below 820 nm, reducing the *in vivo* application at low depth penetration. Li et al. designed CDs derived

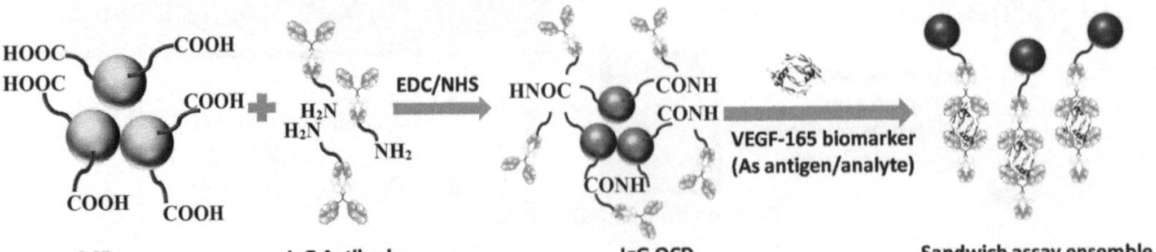

FIGURE 12.10 Schematic illustration of CQDs for disease biomarker.

Source: Reproduced with permission of ref. (60) Copyright 2023 Elsevier.

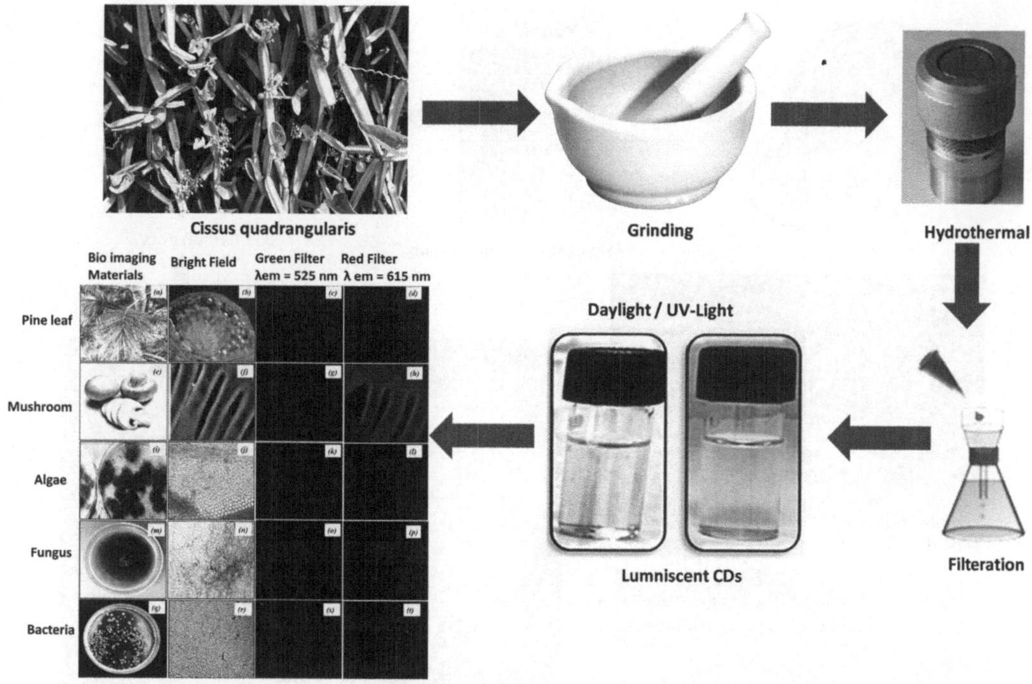

FIGURE 12.11 Schematic illustration of CQDs for multiple applications.

Source: Reproduced with permission of ref. (63) Copyright 2023 Elsevier.

from watermelon with NIR-II emission radiation as *in vivo* optical imaging agent (Figure 12.12). The developed CDs possess high QY of 0.4%, high biocompatibility, with luminescence in the 900–1,200 nm wavelength. About 65% of CDs are excreted from mouse urine within 6 h, showing quick renal clearance by CDs, the synthesized CDs show maximum photothermal efficacy of 30.6%, proving its potential as an idea material for thermal ablation of cancer.[64] Najaflu et al. developed Cu-doped CDs, a new variety of photothermal agent, via hydrothermal method utilizing $CuSO_4$ and Alcea leaf extract precursors. The Cu-dCD spheres show outstanding luminescent QY of 11.1% and photothermal conversion of 39.3%. The heat-generating Cu-dCD nanosphere exhibit excellent thermal ablation effect when prone to 808 NIR laser on 4T1 cells. Relative to GNR PTT agents, Cu-dCD-based PTT shows short time synthesis process, size less than 10 nm, single absorption peak at 800 nm, and high cellular consumption. Thus, environment-friendly preparation of copper-doped CDs and various other metals is promising in cancer therapy and imaging-guided photothermal therapy.[65]

Herein, Thakur et al. showed photodynamic along with photothermal response to near-infrared responsive (NIR) 808 nm laser on MDA-MB-231 breast cancer cells utilizing crystalline GQDs prepared from withered leaves of Indian fig tree, *Ficus racemosa*, as source of carbon.[66] The biocompatibility and cell cycle analysis studies prove the cytocompatibility of GQDs, which can be further used as labelling probe for normal and cancer cells. Upon irradiation with laser of 808 nm, production of ROS and photothermal response were noticed. In addition to this, GQDs did not lose their luminescence in spite of continual irradiation of laser on MDA-MB-231 breast cells. Therefore, it is easy to trace cell death utilizing GQD-labelled MDA-MB-231 cell after treatment utilizing photostability of GQDs. Jia et al. developed carbon dot–based phototheranostic agent using *Hypocrella bambusae* precursor, which shows good water dispersibility, red light emission, bread absorption, and low biotoxicity (Figure 12.13).[67] In addition to this, HBCDs can produce heat (27.6%) and

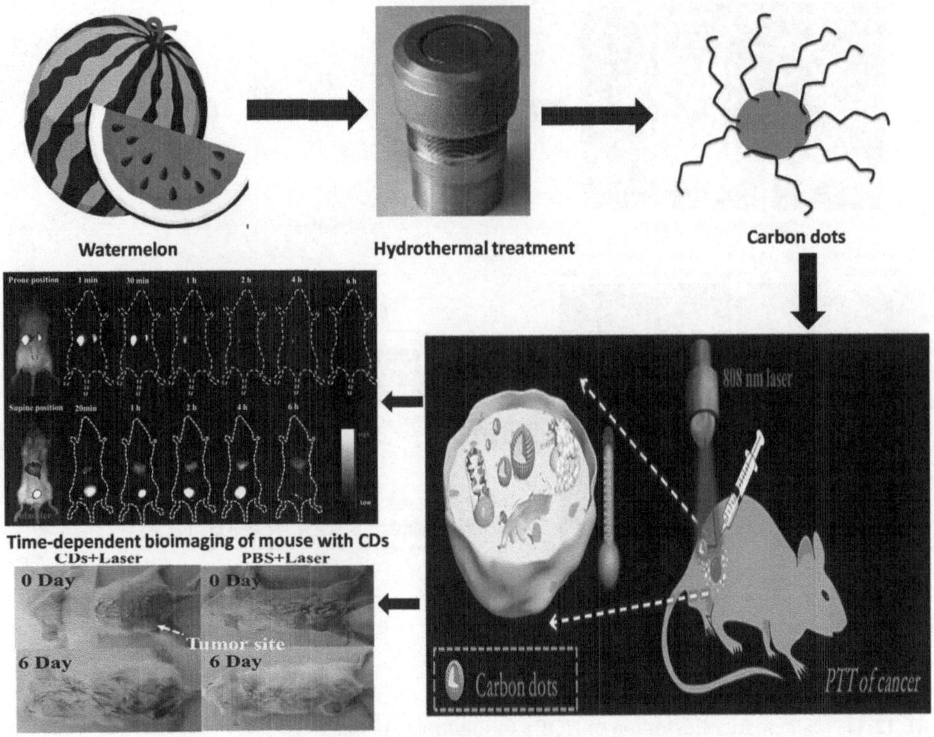

FIGURE 12.12 Schematic illustration of preparation of CQDs for photodynamic therapy.

Source: Reproduced with permission of ref. (64) Copyright 2019 American Chemical Society.

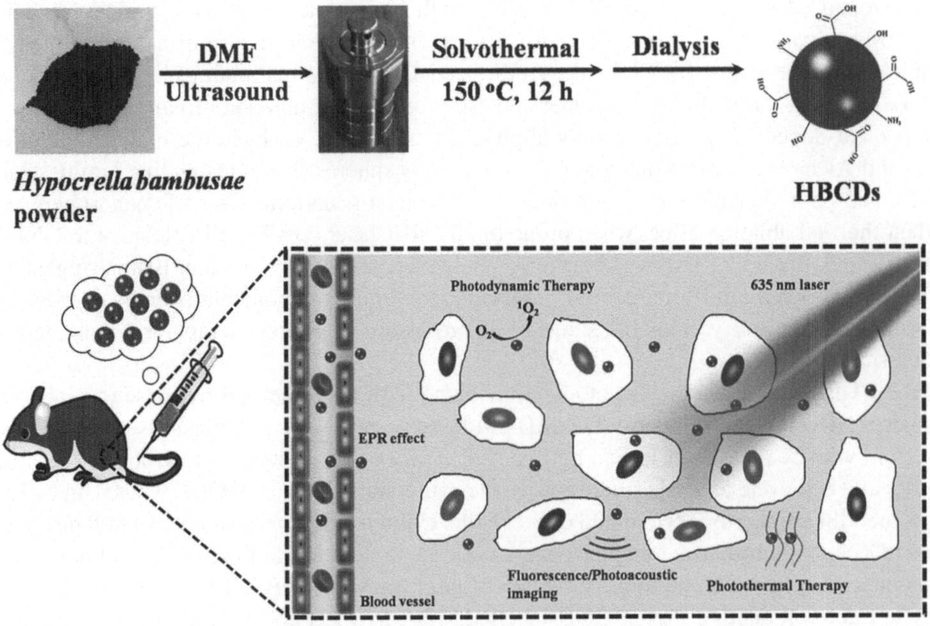

FIGURE 12.13 Schematic Illustration of HBCDs for imaging and PDT/PTT of cancer.

Source: Reproduced with permission of ref. (67) Copyright 2018 Elsevier.

$^1O_2$ (0.38) upon irradiation using 635 nm laser. Due to the excellent properties exhibited by HBCDs, it can be utilized for photoacoustic imaging/bimodal fluorescence-guided photothermal and photodynamic therapy for cancer.

12.5.4 Nanodot Luminogens for Photodynamic Therapy (PDT)

Curcumin is effectively used as a photosensitizer in PDT against a variety of bacteria. Thus, Yan et al. synthesized ST-JHCQDs using turmeric natural plant as a source of carbon, with average particle size of 10.7 nm (Figure 12.14). Antibacterial *in vitro* results show that the antibacterial activity of ST-JHCQDs upon irradiation of blue against *S. aureus* and *E. coli* purely depends upon the light, concentration, and carbonization temperature; the photodynamic antibacterial activity of ST-JHCQDs lessens with a rise of carbonation temperature. Confocal microscope analysis and biomolecule leakage displayed that ST-JHCQDs were efficient in reactive oxygen species (ROS) production upon irradiation of blue light, resulting in leakage of intracellular macromolecules and damage of bacterial cell membrane, resulting in bacterial death. Moreover, ST-JHCQDs enable luminescence imaging of *E. coli* and *S. aureus* bacteria.[68]

Elsher Binny et al. reported the use of broccoli to prepare CQDs through environment-friendly and low-cost methods. Broccoli exhibits health benefits, such as in inflammation and obesity, and has many nutrients acting as a carbon source. Moreover, *Caenorhabditis elegans*, having a short life span, has been employed as an animal model to study its photodynamic therapy effect. The broccoli-derived CQDs (BCQDs) generate singlet oxygen species ($^1O_2$) upon irradiation with wavelength of 660 nm. It has been revealed by *in vivo* examination that the photodynamic thermal efficacy of BCQDs depends on germ line induction apoptosis by cep-1/p53 pathway. Moving further, the DNA of worm damage after sufficient irradiation of light, which was checked by estimating the induction of egl-1-fold in hus-1(op244) and cep-1*(w40)* mutants, having loss of gene function associated in damage of DNA, such as and cep-1/p53 (tumor suppressor) and hus-1 (DNA checkpoint gene). Overall, SCQDs lead to PDT effect, giving rise to damage of worm DNA and promotion of PDT germ line apoptosis, depending on the cep-1/p53 route.[69] Kaur et al. developed nitrogen-doped C-dots (N@VRCD) using *Vigna radiata* sprouts serving as source of carbon via one-pot hydrothermal method and ethylene diamine using nitrogen as a dopant with QY of 58% (Figure 12.15). The fluorescent CDs have been employed for multicolor live cell imaging, intracellular and extracellular Fe^{3+} ion detection, and photothermal therapy. N@VRCD shows phototoxic behavior toward cancer cells, exhibiting its potential to be used as a theranostic agent.[70] Liu et al. prepared Bio-MnO_2@F-PDA with excellent solubility through self-polymerization on the Bio-MnO_2NPs surface. Bio-MnO_2@F-PDA can efficiently eliminate tumor by PTT treatment.[71]

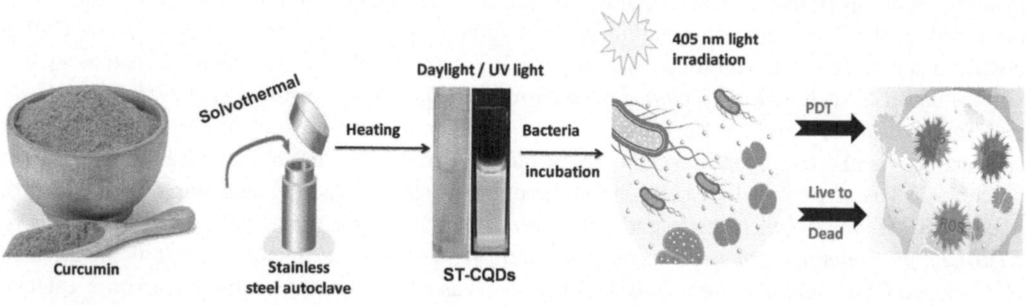

FIGURE 12.14 Schematic illustration of ST-CQDs for photodynamic antimicrobial activity against *E. coli* and *S. aureus* bacterial strains.

Source: Reproduced with permission of ref. (68) Copyright 2023 Elsevier.

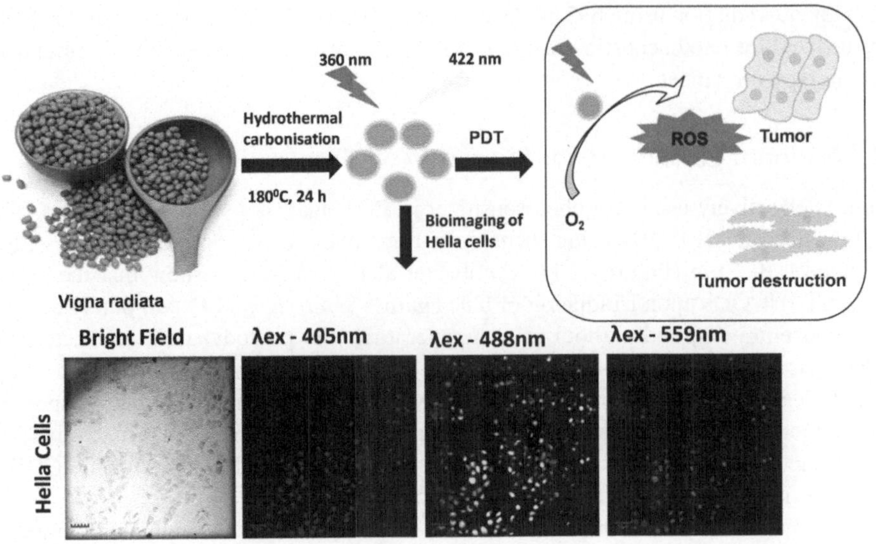

FIGURE 12.15 Schematic diagram of *Vigna radiata*–based CDs for PDT and imaging.

Source: Reproduced with permission of ref. (70) Copyright 2019 Elsevier.

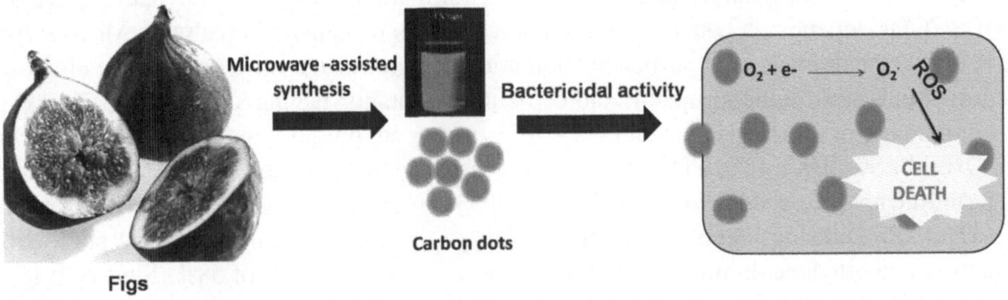

FIGURE 12.16 Schematic illustration of CQDs derived from *Ficus religiosa* plant for antibacterial activity.

12.5.5 Nanodot Luminogens for Antibacterial Activity

Roy et al. synthesized turmeric carbon dots (CDs) and sulfur-functionalized carbon dots (S-CDs) via hydrothermal method. The S-CD displays potential for antibacterial action against *E. coli* and *L. monocytogenes* as producing more reactive oxygen relative to non-functionalized CDs.[72] E. S. et al. presented synthesis of graphene quantum dots using nutmeg seeds as carbon source. These GQDS possess a spherical shape with mean size of particle 7.2 nm, displaying antimicrobial action against various bacteria, such as *Pseudomonas aeruginosa*, *Salmonella* sp., *Streptococcus mutans*, *E. coli*, and *Staphylococcus aureus*.[73] Boobalan et al. reported the preparation of luminescent blue/green carbon dots via one-step hydrothermal carbonization approach utilizing oyster mushroom as carbon precursor.[74] The as-prepared CDs exhibit numerous applications, such as anticancer action against MDA-MB-231 breast cancer cells and antibacterial action against three bacterial strains, that is, *Klebsiella pneumonia*, *Staphylococcus aureus*, and *Pseudomonas aeruginosa*. Liu et al. synthesized N, Cl-doped CDs using *Impatiens balsamina L.* stems as environment-friendly precursor in eutectic solvent.[75] The N, Cl-CDs having positive charge not only distinguish gram-positive bacteria by luminescence imaging but also kill gram-positive bacteria by production of free radicals. Tariq et al. developed red-emissive CDs from figs of the *Ficus religiosa* tree consisting of important natural products as a source of carbon (Figure 12.16). The as-prepared CDs exhibit good antibacterial action against *E.*

coli and *B. subtilis* gram-positive strain at lower concentration. The CDS derived display bactericidal activity and produce ROS, confirmed by NBT assay, which is the main cause of bacterial cell death.[76]

12.5.6 Nanodot Luminogens for Drug Delivery

Thakur et al. presented a green approach for the fabrication of water-soluble graphene quantum dots (GQDs) from cow milk as carbon precursor for bioimaging and cancer drug delivery.[77] The as-synthesized CQDs display multifluorescence and a spherical shape having the size of 5 nm. In order to link berberine hydrochloride (anticancer drug) on GQDs, cysteamine hydrochloride was used, creating GQDs@Cys-BHC complex with drug loading efficacy of 88%. The theranostic GQDs@Cys-BHC complex displays cytotoxicity effect on various cancer cell line models, such as breast cancer (MDA-MB-231), cervical cancer cell lines (Hella cells), and MTT cytotoxicity assays. Ding et al. reported the use of dried shrimps for synthesis of highly luminescent N-doped carbon dots.[78] The luminescent N-doped CDs facilitate the mapping of SH-SY5Y and MCF-7 cells, thus acting as a carrier for drug delivery to target cells. Confocal luminescence images exhibit the internalization of boldine by endosomes, where boldine is released and enter the nucleus cell, inducing MCF-7 cell death, indicating selectivity and efficacy of drug carrier system. Mehta et al. synthesized BCDs utilizing pasteurized milk as source of carbon. Then, lisinopril-loaded BCDs were prepared by lisinopril self-assembly on BCD surface.[79] The as-synthesized BCDs exhibit ability as drug vehicles with excellent biocompatibility and constant release of lisinopril, displaying that BCDs act as drug delivery system for therapeutic application. Panda et al. NIR luminescent antibacterial gel by encapsulation of papaya leaves carbon dots (PCQDs) in PVA-melamine thermosensitive hydrogel. The antibacterial gel, upon irradiation of 980 nm NIR laser, releases anticancer 5-Fu drug (Figure 12.17). The developed PCQDs

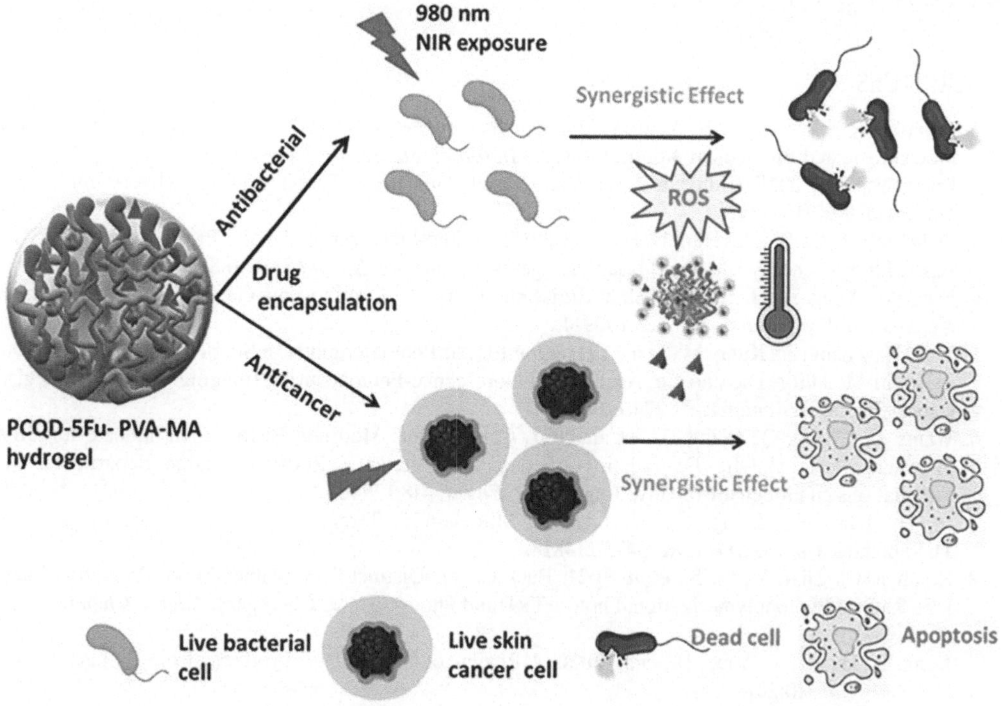

FIGURE 12.17 Schematic illustration of pH/NIR-sensitive 5-Fu released from PCQD-PVA-MA.

Source: Reproduced with permission of ref. (80) Copyright 2022 American Chemical Society.

can produce reactive oxygen species and photothermal effect under NIR, inducing antibacterial action in gram-negative and gram-positive bacteria. Thus, PCQDs act as a promising material for the release of anticancer drug 5-Fu in response to NIR, as illustrated in B16F10 melanoma cells.[80]

12.6 CONCLUSION AND FUTURE TRENDS

To conclude, recent progress in the synthetic pathway of CD-derivatized composites to biomedical applications have been summarized. Initially, the properties, synthetic routes, and applications of carbon dots were discussed briefly. Moving further, the progress of these carbon dots for biomedical applications was detailed also. Despite the advancements, preparation, applications, and characterization studies of carbon dot, composites are in early stage, exhibiting various challenges which need to be addressed. The synthetic routes for carbon dot composites require two or more steps for completion, demanding high cost for manufacturing and causing complexities in applications and characterization. In summary, biogenic materials have significant benefits over conventionally manufactured nanomaterials. Apart from the several benefits, sustainable sources suffer from various problems: (a) Tuning of the biogenic sources optical performance with chemical structure modification is a crucial task due to the complex structure of the biogenic sources, which cannot be characterized clearly. Mild biological synthesis should be applied to address this challenge, (b) Apart from the renewable and sustainable property of the biogenic sources, harsh and toxic reagents are used to convert them to fluorescent materials. More efforts should be given toward solving these issues to increase the fluorescence performance and sustainability of luminescent materials derived from biogenic sources.

ACKNOWLEDGMENTS

The authors are thankful to IIT Ropar for the research facilities. NS is thankful to the SEED project DST/SEED/SCSP/STI/2019/108(G).

REFERENCES

1. Tripathi, R. M., Chung, S. J. **2019**. Biogenic Nanomaterials: Synthesis, Characterization, Growth Mechanism, and Biomedical Applications. *J. Microbiol. Methods* 157:65–80.
2. Prabaharan, M. **2015**. Chitosan-Based Nanoparticles for Tumor-Targeted Drug Delivery. *Int. J. Biol. Macromol* 72:1313–1322.
3. Sedgwick, A. C., Wu, L., Han, H. H., et al. **2018**. Excited-State Intramolecular Proton-Transfer (ESIPT) Based Fluorescence Sensors and Imaging Agents. *Chem. Soc. Rev.* 47:8842–8880.
4. Fang, M., Yang, J., Li, Z. **2022**. Light Emission of Organic Luminogens: Generation, Mechanism and Application. *Prog. Mater. Sci.* 125:100914.
5. Wen, H., Zhang, Z., Kang, M., et al. **2021**. One-for-All Phototheranostics: Single Component AIE Dots as Multi-Modality Theranostic Agent for Fluorescence-Photoacoustic Imaging-Guided Synergistic Cancer Therapy. *Biomaterials* 274:120892.
6. Wang, Z., Cheng, Q., Wang, X., et al. **2021**. Carbon Dots Modified Bismuth Antimonate for Broad Spectrum Photocatalytic Degradation of Organic Pollutants: Boosted Charge Separation, DFT Calculations and Mechanism Unveiling. *Chem. Eng. J.* 418:129460.
7. Li, H., Kim, H., Zhang, C., et al. **2022**. Mitochondria-Targeted Smart AIEgens: Imaging and Therapeutics. *Coord. Chem. Rev.* 473:214818.
8. Kütahya, C., Zhai, Y., Li, S., et al. **2021**. Back Cover: Distinct Sustainable Carbon Nanodots Enable Free Radical Photopolymerization, Photo ATRP and Photo CuAAC Chemistry. *Angew. Chemie Int. Ed.* 60:10996–10996.
9. Kang, C., Huang, Y., Yang, H., et al. **2020**. A Review of Carbon Dots Produced from Biomass Wastes. *Nanomaterials* 10:2316.
10. Desmond, L. J., Phan, A. N., Gentile, P. **2021**. Critical Overview on the Green Synthesis of Carbon Quantum Dots and Their Application for Cancer Therapy. *Environ. Sci. Nano* 8:848–862.
11. De Oliveira, B. P., Da Silva Abreu, F. O. M. **2021**. Carbon Quantum Dots Synthesis from Waste and by-Products: Perspectives and Challenges. *Mater. Lett.* 282:128764.

12. Cui, L., Ren, X., Sun, M., et al. **2021**. Carbon Dots: Synthesis, Properties and Applications. *Nanomaterials* 11:3419.
13. Sun, Y. P., Zhou, B., Lin, Y., et al. **2006**. Quantum-Sized Carbon Dots for Bright and Colorful Photoluminescence. *J. Am. Chem. Soc.* 128:7756–7757.
14. Chu, K. W., Lee, S. L., Chang, C. J., et al. **2019**. Recent Progress of Carbon Dot Precursors and Photocatalysis Applications. *Polymers* 11:689.
15. Sk, M. A., Ananthanarayanan, A., Huang, L., et al. **2014**. Revealing the Tunable Photoluminescence Properties of Graphene Quantum Dots. *J. Mater. Chem. C* 2:6954–6960.
16. Xia, C., Zhu, S., Feng, T., et al. **2019**. Evolution and Synthesis of Carbon Dots: From Carbon Dots to Carbonized Polymer Dots. *Adv. Sci.* 6:1901316.
17. Essner, J. B., Baker, G. A. **2017**. The Emerging Roles of Carbon Dots in Solar Photovoltaics: A Critical Review. *Environ. Sci. Nano* 4:1216–1263.
18. Wang, Z., Zhang, L., Zhang, K., et al. **2022**. Application of Carbon Dots and Their Composite Materials for the Detection and Removal of Radioactive Ions: A Review. *Chemosphere* 287:132313.
19. Itoh, T. **2012**. Fluorescence and Phosphorescence from Higher Excited States of Organic Molecules. *Chem. Rev.* 112:4541–4568.
20. Jamkhande, P. G., Ghule, N. W., Bamer, A. H., et al. **2019**. Metal Nanoparticles Synthesis: An Overview on Methods of Preparation, Advantages and Disadvantages, and Applications. *J. Drug Deliv. Sci. Technol.* 53:101174.
21. Liu, W., Diao, H., Chang, H., et al. **2017**. Green Synthesis of Carbon Dots from Rose-Heart Radish and Application for Fe^{3+} Detection and Cell Imaging. *Sens. Actuator, B Chem.* 241:190–198.
22. Tyagi, A., Tripathi, K. M., Singh, N., et al. **2016**. Green Synthesis of Carbon Quantum Dots from Lemon Peel Waste: Applications in Sensing and Photocatalysis. *RSC Adv.* 6:72423–72432.
23. Ramar, V., Moothattu, S., Balasubramanian, K., et al. **2018**. Metal Free, Sunlight and White Light Based Photocatalysis Using Carbon Quantum Dots from Citrus Grandis: A Green Way to Remove Pollution. *Sol. Energy* 169:120–127.
24. Krishnaiah, P., Atchudan, R., Perumal, S., et al. **2022**. Utilization of Waste Biomass of Poa Pratensis for Green Synthesis of N-Doped Carbon Dots and Its Application in Detection of Mn^{2+} and Fe^{3+}. *Chemosphere* 286:131764.
25. Architha, N., Ragupathi, M., Shobana, C., et al. **2021**. Microwave-Assisted Green Synthesis of FLuorescent Carbon Quantum Dots from Mexican Mint Extract for Fe^{3+} Detection and Bio-Imaging Application. *Environ. Res.* 199:111263.
26. Gul, U., Kanwal, S., Tabassum, S., et al. **2020**. Microwave-Assisted Synthesis of Carbon Dots as Reductant and Stabilizer for Silver Nanoparticles with Enhanced-Peroxidase like Activity for Colorimetric Determination of Hydrogen Peroxide and Glucose. *Microchim. Acta* 187:135.
27. Monte-Filho, S. S., Andrade, S. I. E., Lima, M. B., et al. **2019**. Synthesis of Highly Fluorescent Carbon Dots from Lemon and Onion Juices for Determination of Riboflavin in Multivitamin/mineral Supplements. *J. Pharm. Anal.* 9:209–216.
28. Mohapatra, D., Pratap, R., Pandey, V., et al. **2022**. Tinospora Cordifolia Leaves Derived Carbon Dots for Cancer Cell Bioimaging, Free Radical Scavenging, and Fe^{3+} Sensing Applications. *J. Fluoresc.* 32:275–292.
29. Liang, Q., Wang, Y., Lin, F., et al. **2017**. Facile Microwave-Hydrothermal Synthesis of Fluorescent Carbon Quantum Dots from Bamboo Tar and Their Application. *Anal. Methods* 9:3675–3681.
30. Ang, W. L., Boon Mee, C. A. L., Sambudi, N. S., et al. **2020**. Microwave-Assisted Conversion of Palm Kernel Shell Biomass Waste to Photoluminescent Carbon Dots. *Sci. Rep.* 10:21199.
31. Dehvari, K., Liu, K. Y., Tseng, P. J., et al. **2019**. Sonochemical-Assisted Green Synthesis of Nitrogen-Doped Carbon Dots from Crab Shell as Targeted Nanoprobes for Cell Imaging. *J. Taiwan Inst. Chem. Eng.* 95:495–503.
32. Sanni, S. O., Moundzounga, T. H. G., Oseghe, E. O., et al. **2022**. One-Step Green Synthesis of Water-Soluble Fluorescent Carbon Dots and Its Application in the Detection of Cu^{2+} *Nanomaterials* 12: 958.
33. Rimal, V., Shishodia, S., Srivastava, P. K., et al. **2021**. Synthesis and Characterization of Indian Essential Oil Carbon Dots for Interdisciplinary Applications. *Appl. Nanosci.* 11:1225–1239.
34. Wei, X., Li, L., Liu, J., et al. **2019**. Green Synthesis of Fluorescent Carbon Dots from Gynostemma for Bioimaging and Antioxidant in Zebrafish. *ACS Appl. Mater. Interfaces* 11:9832–9840.
35 Agnol, L. D., Neves, R. M., Maraschin, M., et al. **2021**. Green Synthesis of Spirulina-Based Carbon Dots for Stimulating Agricultural Plant Growth. *Sustain. Mater. Technol.* 3: e00347.
36. Singh, J., Kaur, S., Lee, J., et al. **2020**. Highly Fluorescent Carbon Dots Derived from Mangifera Indica Leaves for Selective Detection of Metal Ions. *Sci. Total Environ.* 720:137604.

37. Boruah, A., Saikia, M., Das, T., et al. **2020**. Blue-Emitting Fluorescent Carbon Quantum Dots from Waste Biomass Sources and Their Application in Fluoride Ion Detection in Water. *J. Photochem. Photobiol. B Biol.* 209:111940.
38. Hu, Z., Jiao, X. Y., Xu, L., et al. **2020**. The N,S Co-Doped Carbon Dots with Excellent Luminescent Properties from Green Tea Leaf Residue and Its Sensing of Gefitinib. *Microchem. J.* 154:104588.
39. Gupta, D. A., Desai, M. L., Malek, N. I., et al. **2020**. Fluorescence Detection of Fe^{3+} Ion Using Ultra-Small Fluorescent Carbon Dots Derived from Pineapple (Ananas Comosus): Development of Miniaturized Analytical Method. *J. Mol. Struct.* 1216:128343.
40. Desai, M. L., Jha, S., Basu, H., et al. **2019**. Acid Oxidation of Muskmelon Fruit for the Fabrication of Carbon Dots with Specific Emission Colors for Recognition of Hg^{2+} Ions and Cell Imaging. *ACS Omega* 4:19332–19340.
41. Athinarayanan, J., Periasamy, V. S., Alatiah, K. A., et al. **2020**. Synthesis and Cytocompatibility Analysis of Carbon Nanodots Derived from Palmyra Palm Leaf for Multicolor Imaging Applications. *Sustain. Chem. Pharm.* 18:100334.
42. Xue, M., Zhan, Z., Zou, M., et al. **2016**. Green Synthesis of Stable and Biocompatible Fluorescent Carbon Dots from Peanut Shells for Multicolor Living Cell Imaging. *New J. Chem.* 40:1698–1703.
43. Zhang, Q., Zhang, X., Bao, L., et al. **2019**. The Application of Green-Synthesis-Derived Carbon Quantum Dots to Bioimaging and the Analysis of mercury(II). *J. Anal. Methods Chem.* 8183134.
44. Xie, Y., Cheng, D., Liu, X., et al. **2019**. Green Hydrothermal Synthesis of N-Doped Carbon Dots from Biomass Highland Barley for the Detection of Hg^{2+}. *Sensors* 19:3169.
45. Zhu, S., Meng, Q., Wang, L., et al. **2013**. Highly Photoluminescent Carbon Dots for Multicolor Patterning, Sensors, and Bioimaging. *Angew. Chemie – Int. Ed.* 52:3953–3957.
46. Das, G. S., Shim, J. P., Bhatnagar, A., et al. **2019**. Biomass-Derived Carbon Quantum Dots for Visible-Light-Induced Photocatalysis and Label-Free Detection of Fe(III) and Ascorbic Acid. *Sci. Rep.* 9:15084.
47. Wang, L., Zhu, S. J., Wang, H. Y., et al. **2014**. Common Origin of Green Luminescence in Carbon Nanodots and Graphene Quantum Dots. *ACS Nano* 8:2541–2547.
48. Hoan, B. T., Tam, P. D., Pham, V. H., et al. **2019**. Green Synthesis of Highly Luminescent Carbon Quantum Dots from Lemon Juice. *J. Nanotechnol.* 2852816.
49. Ding, H., Ji, Y., Wei, J. S., et al. **2017**. Facile Synthesis of Red-Emitting Carbon Dots from Pulp-Free Lemon Juice for Bioimaging. *J. Mater. Chem. B* 5:5272–5277.
50. Ding, H., Zhou, X., Qin, B., et al. **2019**. Highly Fluorescent near-Infrared Emitting Carbon Dots Derived from Lemon Juice and Its Bioimaging Application. *J. Lumin.* 21:298–304.
51. Wang, Y., Wu, W. T., Wu, M. B., et al. **2015**. Yellow-Visual Fluorescent Carbon Quantum Dots from Petroleum Coke for the Efficient Detection of Cu^{2+} Ions. *New Carbon Mater.* 30:550–559.
52. Gong, X., Lu, W., Liu, Y., et al. **2015**. Low Temperature Synthesis of Phosphorous and Nitrogen Co-Doped Yellow Fluorescent Carbon Dots for Sensing and Bioimaging. *J. Mater. Chem. B* 3:6813–6819.
53. Xu, H., Yang, X., Li, G., et al. **2015**. Green Synthesis of Fluorescent Carbon Dots for Selective Detection of Tartrazine in Food Samples. *J. Agric. Food Chem.* 63:6707–6714.
54. Mohapatra, D., Pratap, R., Pandey, V., et al. **2022**. Tinospora Cordifolia Leaves Derived Carbon Dots for Cancer Cell Bioimaging, Free Radical Scavenging, and Fe^{3+} Sensing Applications. *J. Fluoresc.* 32:275–292.
55. Damera, D. P., Manimaran, R., Krishna Venuganti, V. V., et al. **2022**. Green Synthesis of Full-Color Fluorescent Carbon Nanoparticles from Eucalyptus Twigs for Sensing the Synthetic Food Colorant and Bioimaging. *ACS Omega* 5: 19905–19918.
56. Pal, T., Mohiyuddin, S., Packirisamy, G., et al. **2018**. Facile and Green Synthesis of Multicolor Fluorescence Carbon Dots from Curcumin: *In vitro* and *in vivo* Bioimaging and Other Applications. *ACS Omega* 3:831–843.
57. Kumawat, M. K., Srivastava, R., Thakur, M., et al. **2017**. Graphene Quantum Dots from Mangifera Indica: Application in near-Infrared Bioimaging and Intracellular Nanothermometry. *ACS Sustain. Chem. Eng.* 5:1382–1391.
58. Sahu, S., Behera, B., Maiti, T. K., et al. **2012**. Simple One-Step Synthesis of Highly Luminescent Carbon Dots from Orange Juice: Application as Excellent Bio-Imaging Agents. *Chem. Commun.* 48:8835–8837.
59. Ramanan, V., Thiyagarajan, S. K., Raji, K., et al. **2016**. Outright Green Synthesis of Fluorescent Carbon Dots from Eutrophic Algal Blooms for *in vitro* Imaging. *ACS Sustain. Chem. Eng.* 4:4724–4731.
60. Deb, A., Nalkar, G. R., Chowdhury, D., et al. **2023**. Analytica Chimica Acta Biogenic Carbon Dot-Based Fluorescence-Mediated Immunosensor for the Detection of Disease Biomarker. *Anal. Chim. Acta* 1242:340808.

61. Miao, H., Wang, L., Zhuo, Y., et al. **2016**. Biosensors and Bioelectronics Label-Free Fl Uorimetric Detection of CEA Using Carbon Dots Derived from Tomato Juice. *Biosens. Bioelectron.* 86:83–89.
62. Raveendran, V., Kizhakayil, R. N., et al. **2021**. Fluorescent Carbon Dots as Biosensor, Green Reductant, and Biomarker. *ACS Omega* 6:23475–23484.
63. Kansay, V., Dutt, V., Chandan, G., et al. **2023**. Sustainable Synthesis of Nitrogen-Doped Fluorescent Carbon Quantum Dots Derived from Cissus Quadrangularis for Biomarker Applications. *Mater. Chem. Phys.* 296:127237.
64. Li, Y., Bai, G., Zeng, S., et al. **2019**. Theranostic Carbon Dots with Innovative NIR-II Emission for *in vivo* Renal-Excreted Optical Imaging and Photothermal Therapy. *ACS Appl. Mater. Interfaces* 11:4737–4744.
65. Najaflu, M., Shahgolzari, M., Bani, F., et al. **2022**. Green Synthesis of Near-Infrared Copper-Doped Carbon Dots from Alcea for Cancer Photothermal Therapy. *ACS Omega* 7:34573–34582.
66. Thakur, M., Kumawat, M. K., Srivastava, R., et al. **2017**. Multifunctional Graphene Quantum Dots for Combined Photothermal and Photodynamic Therapy Coupled with Cancer Cell Tracking. *RSC Adv.* 7:5251–5261.
67. Jia, Q., Zheng, X., Ge, J., et al. **2018**. Journal of Colloid and Interface Science Synthesis of Carbon Dots from Hypocrella Bambusae for Bimodel Fluorescence/Photoacoustic Imaging-Guided Synergistic Photodynamic/Photothermal Therapy of Cancer. *J. Colloid Interface Sci.* 526:302–311.
68. Yan, H., Li, P., Wen, F., et al. **2023**. Green Synthesis of Carbon Quantum Dots from Plant Turmeric Holds Promise as Novel Photosensitizer for *in vitro* Photodynamic Antimicrobial Activity. *J. Mater. Res. Technol.* 22:17–34.
69. Elsherbiny, S. M., Shao, C., Acheampong, A., et al. **2022**. Green Synthesis of Broccoli-Derived Carbon Quantum Dots as Effective Photosensitizers for the PDT Effect Testified in the Model of Mutant Caenorhabditis Elegans. *Biomater. Sci.* 10:2857–2864.
70. Kaur, N., Sharma, V., Tiwari, P., et al. **2019**. "Vigna Radiata" Based Green C-Dots: Photo-Triggered Theranostics, Fluorescent Sensor for Extracellular and Intracellular Iron (III) and Multicolor Live Cell Imaging Probe. *Sens. Actuator, B Chem.* 291:275–286.
71. Liu, J., Zhang, M., Wu, Y., et al. **2021**. In Situ Synthesis of Fluorescent Polydopamine on Biogenic MnO_2 Nanoparticles as Stimuli Responsive Multifunctional Theranostics. *Biomater. Sci.* 9: 5897–5906.
72. Ezati, P., Roy, S., Rhim, J. W., et al. **2022**. Pectin/gelatin-Based Bioactive Composite Films Reinforced with Sulfur Functionalized Carbon Dots. *Colloids Surf. Physicochem. Eng. Asp.* 636:128123.
73. Anooj, E. S., Praseetha, P. K. **2019**. Synthesis and Characterization of Graphene Quantum Dots from Nutmeg Seeds and Its Biomedical Application. *Int. J. Recent Technol. Eng.* 7:144–151.
74. Boobalan, T., Sethupathi, M., Sengottuvelan, N., et al. **2020**. Mushroom-Derived Carbon Dots for Toxic Metal Ion Detection and as Antibacterial and Anticancer Agents. *ACS Appl. Nano Mater.* 3:5910–5919.
75. Liu, S., Quan, T., Yang, L., et al. **2021**. N,Cl-Codoped Carbon Dots from Impatiens Balsamina L. Stems and a Deep Eutectic Solvent and Their Applications for Gram-Positive Bacteria Identification, Antibacterial Activity, Cell Imaging, and ClO^-Sensing. *ACS Omega* 6:29022–29036.
76. Tariq, M., Singh, A., Varshney, N., et al. **2022**. Biomass-Derived Carbon Dots as an Emergent Antibacterial Agent. *Mater. Today Commun.* 33:104347.
77. Thakur, M., Mewada, A., Pandey, S., et al. **2016**. Milk-Derived Multi- Fluorescent Graphene Quantum Dot-Based Cancer Theranostic System. *Mater. Sci. Eng. C* 67:468–477.
78. Stephanie, L. D., Deshmukh, B., Bhamore, J. R., et al. **2016**. Synthesis of Fluorescent Nitrogen Doped Carbon Dots from Dried Shrimps for Cell Imaging and Boldine Drug Delivery System. *RSC Adv.* 6:12169–12179.
79. Mehta, V. N., Chettiar, S. S., Bhamore, J. R., et al. **2017**. Green Synthetic Approach for Synthesis of Fluorescent Carbon Dots for Lisinopril Drug Delivery System and Their Confirmations in the Cells. *J. Fluoresc.* 27:111–124.
80. Panda, S., Chawpattnayak, B., Dash, P., et al. **2022**. Papaya-Derived Carbon-Dot-Loaded Fluorescent Hydrogel for NIR- Stimulated Photochemotherapy and Antibacterial Activity. *ACS Appl. Poly. Mater.* 4:369–380.

13 Role of Nanoferrites for the Removal of Contaminants from Environment

Güllü Kaymak
Kütahya Health Sciences University, Kütahya, Turkey

13.1 INTRODUCTION

13.1.1 Introduction to Nanoparticles

Nanotechnology has become one of the scientific and technological fields that are the focus of attention today, as it has reached wide areas of use as a result of cooperation with disciplines such as biology, physics, chemistry, medicine, pharmacy sciences, informatics, and engineering [1]. The publication of approximately 50,000 scientific articles on nanotechnology every year is another indication that this field is developing and dynamic [2]. Some examples of applications of nanotechnology that have taken place today and are targeted in the future include, but are not limited to, developing sensor and drug release systems for cancer diagnosis and treatment, purification of wastewater with drinking water quality, wearable health monitors, computers with high storage space and long-lasting batteries, sensors that can detect corrosion and stresses in bridges and send signals, and foods with a long shelf life [3].

Nanoparticles (NPs) may be defined as matter particles whose diameter lies in between 1 and 100 nanometers. If an example is given for the nanoscale to be more descriptive, it can be said that the thickness of the paper layer is approximately 105 nm [4]. The International Organization for Standardization (ISO) has defined *nanoparticles* as a material with nanoscale size or surface structure. According to the EU Commission, the term *nanoparticle* (NP) is defined as "clustered particles or material with sizes in the range of 1–100 nm" [5]. The same substance exhibits different physical, chemical, and biological properties on a large scale and in the nanoscale. As the size gets smaller, changes occur in many chemical and physical properties, such as melting point, boiling point, conductivity, tendency to enter into a chemical reaction, and color. The main reason for this is the increased surface area and surface energy as the size gets smaller [6]. These particles provide advantages with their high catalytic reactivity, chemical stability, improved mechanical strength, and antibacterial properties [7]. The NPs provide unique advantages in many areas of use due to their high surface area. Especially in surface-related physical events, such as adsorption, they allow both an increase in capacity and a shortening of the processing time. With these properties of NPs, they are also used in applications such as drug delivery systems, chemical and biological sensing, diagnostic and imaging, biotechnology, CO_2 capture, electronics, medicine, textile production, and water treatment systems [8]. The NPs have also been commercialized nowadays to improve the quality of human life in the form of bioimplants, pharmaceutical products, cosmetics, filters, biomarkers, imaging agents, biosensors, and tissue engineering scaffolds [9]. Nanoparticles represent the techno-economic sector that continues to expand in many areas of research and application.

DOI: 10.1201/9781003430087-13

13.1.2 Synthesis of Nanoparticles

Chemical methods are now widely used to create NPs in recent years, considering that this method of production produces homogeneous particles in terms of size and shape without requiring a lot of energy in the reduction process. However, in these chemical methods, risky chemicals which lead to cellular toxicity, genetic toxicity, and cancer are used. Some chemicals used in the synthesis of nanoparticles are considered toxic, irritating, flammable, and harmful to the environment [10–11]. Therefore, new environment-friendly methods are in demand for the synthesis of quite stable NPs. In line with these stated needs, the sensitive and environmentally friendly green synthesis method has emerged. Green chemistry, which connects nanotechnology and microbial biotechnology, involves the synthesis of biological NPs using living organisms such as bacteria, fungi, yeast, algae, and viruses. In the nanotechnology sector, the green synthesis method has become a rising research branch in particle production. Biogenic nanoparticle production provides an advantage in terms of being environmentally friendly, biocompatible, highly reproducible, and cost-effective due to the low use of toxic chemicals. The green synthesis method is widely used to synthesize zero-valent metal, metal oxide, and composite NPs [11–12].

13.2 CHALLENGES IN WASTEWATER TREATMENT

One of the biggest problems that we are facing today is to maintain the quality of air, water, and soil. The majority of nations deal with serious environmental problems, like the lack of access to clean drinking water, the need to treat waste and wastewater, and soil and groundwater contamination. From past to present, humans have begun to pollute, change, and disrupt the natural environment, thereby upsetting the balance that exists in nature. In our century, factors such as unregulated and uncontrolled industry, rapid population growth, and overconsumption are destroying our natural environment. The contamination of soil and groundwater with carcinogenic organic materials and/or metals is happening on a global scale. Due to pollution in the basic elements of life such as water, air, and soil, all living creatures, including humans, are harmed and adversely affected [13]. Increasing environmental problems lead to irreversible changes in ecosystems. Industrial activities negatively affect air, soil, and water ecosystems, and environmental pollution in the air not only affects the air itself but also indirectly contaminates water and soil ecosystems. The source of pollution not only affects the region where it is located but also affects other regions in the long term. The use of heavy metal ion–containing fuels for heating and energy production can also cause water pollution [14]. Traditional rehabilitation and treatment techniques typically have only modest success in lowering pollutant levels, especially in water and soil.

13.3 ROLE OF NPS IN REMEDIATION OF CONTAMINATED WATER AND SOIL

Nanotechnology offers a potential revolution in environmental remediation pathways due to the large surface area available for adsorption and degradation phenomenon. The NPs possess a higher reactivity and efficacy than their bulk counterparts. Furthermore, when compared to conventional methods, NPs have the capacity to regulate distinct surface chemistry due to their ability to be functionalized or embedded by functional groups that can target specific molecules of contaminants. Direct environmental remediation is another area where NPs have found use [15], as evidenced by the rise in publications and funding for such projects. The cost of wastewater purification can be decreased by using NPs. NPs generally offer some inherent qualities that can be used in environmental remediation processes. The surface area of NPs grows exponentially with a decrease in particle size. Additionally, due to their small size, NPs have a high degree of mobility in solution, making it possible to quickly screen the entire volume with a small number of NPs.

NPs have been extensively studied and reported in the literature for the removal of heavy metals [16–17]. The most typical use of NPs is to filter out low concentrations of particular metals from drinking water [18–19]. As a case, the As ions were removed from wastewater using the magnetic NPs of iron oxide, which get readily absorbed onto the NP's surface and collected with the help of weak magnetic field [20]. Another instance of using alumina-supported gold NPs as adsorbents for mercury ion removal from drinking water was described by Lisha et al. [21]. The main plant nutrients present in surface water to be concerned about are phosphorus and nitrogen. They can either be absorbed by NPs or undergo a change in their oxidation state.

A few of the applications of nanomaterials in environmental remediation are:

1. Photodegradation of airborne pollutants (like nitrogen oxides and volatile organic components)
2. Wastewater treatment involving desalination, hardness removal, and bioremediation (microbial treatment)
3. Soil and groundwater remediation

The use of NPs for environmental remediation can be categorized into adsorptive, reactive, in situ, or ex situ [22–23]. The nanomaterials have been explored for their potential use in each of these cases. In soil and groundwater remediation, in situ applications appear to be the most promising, as they are generally less costly. A reactive NP cloud migrating to contaminated areas or an in situ reactive zone containing immobile NPs is required. NPs can be processed on the surface of contaminated soil with the help of traditional methods of farming, for their applications in soil remediation.

Pesticides, insecticides, and other halogenated compounds are major sources of pollution in groundwater resources. Agricultural practices, regular storage leaks, and chemical accidents are the primary sources of these pollutants. Traditional remediation methods comprise chemical oxidation, thermal treatment (in situ), use of reactive barriers, soil washing (ex situ), and pump-and-treat procedures [23]. Pump-and-treat processes took roughly 18 years to operate, whereas nZVI treatment only took 1–2 years [24]. The cleaning of contaminated groundwater and soil with toxic organic compounds is an important environmental issue. Permeable reactive barriers (PRBs) using various reactive materials are used as an alternative to treatment techniques. The use of zero-valent metals as PRBs is more common, especially for nanomaterials [25]. Nanomaterials are used as PRBs for in situ purification of groundwater. Pollutants in groundwater passing through the barrier remain immobile and degrade, allowing purification to be achieved by combining with nanomaterials.

Some of the functions of NPs in environmental remediation are shown in Table 13.1.

One of the types of nanomaterials used in PRB systems is ZVI, which is similar to a combination of zero-valent iron with activated carbon, zeolite, compost, or another material, such as pumice or cotton, that purifies contaminated groundwater that passes under a natural gradient. Granular ZVI is the most widely used filler material due to its ability to decompose a wide range of organic pollutants. It reduces and immobilizes harmful metal ions. To date, ZVI has been used to degrade chlorinated solvents, particularly in the manufacture of hundreds of PRBs worldwide [49]. Iron PRBs were originally developed in the USA and Canada. Currently, several forms of PRB are commercially available worldwide [50].

ZVI-Fe0 is a reducing agent. Many studies have been conducted on the reduction reactions of organic and inorganic pollutants in the presence of ZVI in recent years due to its affordability and ease of production [51–52]. Sun et al. (2011) [53] reported that both the reduction and adsorption of As(V) under anoxic conditions were rapid when it was used alongside commercial/acid-treated ZVI. They reported that ZVI reduced arsenic ions while itself being reduced to (0) under-acidic conditions.

The most common and well-known mechanism for removing pollutants from groundwater using Fe^0 and Fe^{2+} solids involves chemical reduction, which usually requires the proximity of the contaminant to be absorbed on the iron surface (electron vacancy). For the purification of organic pollutants such as chlorinated organics and polychlorinated biphenyls, removal is usually achieved through the chemical's reductive breakdown, that is, contamination is removed physically. In contrast, purification for many heavy metals and radionuclides typically occurs through immobilization. Pollutants

TABLE 13.1
Applications of NPs for Contaminant Removal from Water and Soil

NPs	Pollutant	Environment	References
ZVI NPs	Bisphenol A	Water	[26]
	Diazepam	Water	[27]
	Amoxicillin	Water and soil	[28]
	Methylene blue	Water	[29–31]
	Dichlorvos	Water	[32]
	Lindane	Water	[33]
	Hg(II) and Cr (VI)	Water	[34]
	Nitrobenzene	Water and soil	[35]
Fe-Mn oxide NPs	As(III)	Water and soil	[36]
	Perfluorooctanoic acid	Water	[37]
Fe-Pd NPs	Trichloroethylene	Water and soil	[38]
Pd NPs	Trichloroethene	Water	[39]
GO/Pd NPs	Congo red Methylene blue Methyl orange	Water	[40]
SiO_2/Pd NPs	Oxyanion	Water	[41]
Fe-Cu NPs	Oil	Soil	[42]
Cu/β-CD/rGO	Tetracycline Oxytetracycline Doxycycline	Water	[43]
GO-MO NPs	Cd(II) and Cu(II)	Water	[44]
GAC-TiO_2 NPs	Cu	Soil	[45]
γ-Fe_2O_3	Pb(II) and Cu(II)	Water	[46]
ZVC NPs	Uranium	Water	[47]
AgNPs	Methylene blue	Water	[48]

are simply removed from the water and held in an immobilized state without physical destruction. It is important to note that when injected, nanomaterials (along with the relevant contaminant) are unrecoverable, heavy metals and radionuclides are immobilized in place, and it is important to note that the contaminants are not destroyed or eliminated from the system.

Cullen and coworkers (2011) [54] investigated the impact of nanosized and microsized zero-valent Fe particles on the activity of hydrolases and dehydrogenase enzymes and potential values in uncontaminated soil. The authors interpreted the particle reactivity and actual microbial effects interfered by the experimental parameters. The mean particle diameters of the nZVI and mZVI used are 12.5 ± 0.3 nm and 1480 ± 620 nm, respectively. In experiments, it was observed that nZVI had an inhibitory effect on the AOP values while stimulating dehydrogenase activity with little or no effect on activity of hydrolases. Overall, it was concluded that nZVI or mZVI had no adverse effects on the processes studied.

FeNPs have recently come to the fore with applications in environmental fields and are extensively researched in biomedical applications. Studies have been carried out to remove Cr(VI) from water using magnetite NPs [55]. Fe_3O_4 magnetite NPs coated with chitosan were used to remove Cu(II) ions [56]. Monodisperse Fe_3O_4 nanocrystals have been investigated for As ion removal from water by magnetic separation at very low magnetic field gradients [57]. It has been shown that a self-assembled 3D flower-shaped FeONPs have an excellent ability to remove As(V) and Cr(VI) from water [58].

According to Bokare et al. (2008) [59], Fe-Ni bimetallic NPs with a particle size of 20–40 nm (3g/L) provided complete degradation of the orange G azo dye at a contact time of 10 min. According to their optimization studies, an increase in degradation efficiency was observed as pH values decreased. Maximum efficacy was achieved within 3 min in a pH 2.0 environment. In addition, it was stated in this report that there was no change in the degradation efficiency of Fe-Ni bimetallic nanoparticles for three days, a decrease of 12% on the fifth day, and a decrease of 22% on the tenth day.

In a study published by Liu et al. (2012) [60], with the help of Ag/TiO_2 nanofiber membrane produced by depositing AgNPs on the electrospun TiO_2 nanofibrils, 99.9% of bacterial inactivation and 80.0% of dye degradation under solar radiation were successfully achieved within 30 min. The Ag/TiO_2 nanomembrane, upon comparison with the typical P25 membrane, was found to be better in terms of permeated material flow.

ZnO has been promising for the last few years due to its unique optoelectronic, catalytic, and photochemical properties [61]. As an environmentally friendly material, ZnO can be used in different areas, such as the catalyst industry [62], gas sensors [63], and solar cells [64]. ZnO as an adsorbent has mostly been applied to remove hydrogen sulfide. Recently, people have found that nanostructured ZnO can be efficient in removing heavy metals [65]. In addition to some similarity in properties with TiO_2, ZnONPs have many advantages, such as being cheap and simple to prepare and being morphologically adaptable. The quantum efficiency of ZnO is significantly greater than TiO_2 [66]. Therefore, ZnO stands out for the removal of organic pollutants [67–68].

13.4 CONCLUSION

The remediation of soil and groundwater resources with the available methods generally involves higher costs and takes much time in processing. Nanoparticles can offer unique opportunities for removing pollutants from water and soil due to their strong adsorption and advanced redox and photocatalytic properties [68]. But the use of nanoparticles for environmental remediation results in the release of nanoparticles into the environment and downstream ecosystems. Studies are needed to measure their toxicity and understand the bioavailability and potential risks of produced nanoparticles. A range of ecotoxicological effects, including on microbes, plants, invertebrates, and fish, has been reported for various produced nanoparticles. Following these data, studies using nanoparticles immobilized on a support material are emerging as a way to take advantage of the benefits of nanoparticles for remediation while preventing them from accumulating in the environment and causing secondary contamination. Thus, the use of nanomaterials for the removal of pollutants from water and soil should ensure environmental safety, lower cost, reusability, and high treatment efficiency.

REFERENCES

1. Hulla, J. E.; Sahu, S. C.; Hayes, A. W. **2015**. Nanotechnology: History and future. *Human & Experimental Toxicology* 34(12), 1318–1321.
2. Huang, H. C.; Barua, S.; Sharma, G.; Dey, S. K.; Rege, K. **2011**. Inorganic nanoparticles for cancer imaging and therapy. *Journal of Controlled Release* 155(3), 344–357.
3. López-Serrano, A.; Olivas, R. M; Landaluze, J. S.; Cámara, C. **2014**. Nanoparticles: A global vision: Characterization, separation, and quantification methods: Potential environmental and health impact. *Analytical Methods* 6(1), 38–56.
4. Iravani, S.; Korbekandi, H.; Mirmohammadi, S. V.; Zolfaghari, B. **2014**. Synthesis of silver nanoparticles: Chemical, physical and biological methods. *Research in Pharmaceutical Sciences* 9(6), 385.
5. Jeevanandam, J.; Barhoum, A.; Chan, Y. S.; Dufresne, A.; Danquah, M. K. **2018**. Review on nanoparticles and nanostructured materials: History, sources, toxicity and regulations. *Beilstein Journal of Nanotechnology* 9(1), 1050–1074.
6. Srivastava, S.; Usmani, Z.; Atanasov, A. G. et al. **2021**. Biological nanofactories: Using living forms for metal nanoparticle synthesis. *Mini Reviews in Medicinal Chemistry* 21(2), 245–265.
7. Ealia, S. A. M.; Saravanakumar, M. P. **2017**. A review on the classification, characterisation, synthesis of nanoparticles and their application. *In IOP Conference Series: Materials Science and Engineering* 263(3), 32019.

8. Chowdhury, S.; Yusof, F.; Faruck, M. O.; Sulaiman, N. **2016**. Process optimization of silver nanoparticle synthesis using response surface methodology. *Procedia Engineering* 148, 992–999.
9. Devatha, C. P.; Thalla, A. K. **2018**. Green synthesis of nanomaterials. In *Synthesis of Inorganic Nanomaterials*. Woodhead Publishing, Elsevier, Amsterdam, Netherlands, pp. 169–184.
10. Alsamhary, K. I. **2020**. Eco-friendly synthesis of silver nanoparticles by Bacillus subtilis and their anti-bacterial activity. *Saudi Journal of Biological Sciences* 27(8), 2185–2191.
11. Kalimuthu, K.; Cha, B. S.; Kim, S.; Park, K. S. **2020**. Eco-friendly synthesis and biomedical applications of gold nanoparticles: A review. *Microchemical Journal* 152, 104296.
12. Sajid, M.; Płotka-Wasylka, J. **2020**. Nanoparticles: Synthesis, characteristics, and applications in analytical and other sciences. *Microchemical Journal* 154, 104623.
13. Ok, Y. S.; Palansooriya, K. N.; Yuan, X.; Rinklebe, J. **2021**. Special issue on biochar technologies, production, and environmental applications in critical reviews in environmental science & technology during 2017–2021. *Critical Reviews in Environmental Science and Technology* 1–9.
14. Masindi, V.; Muedi, K. L. **2018**. Environmental contamination by heavy metals. *Heavy Metals* 10, 115–132.
15. Joo S.; Cheng I. **2006**. *Nanotechnology for Environmental Remediation*. Springer, New York, USA.
16. Mohmood, I.; Lopes, C. B.; Lopes, I.; Ahmad, I.; Duarte, A. C.; Pereira, E. **2013**. Nanoscale materials and their use in water contaminants removal-a review. *Environmental Science and Pollution Research* 20(3), 1239–1260.
17. Kolluru, S. S.; Agarwal, S.; Sireesha, S.; Sreedhar, I; Kale, S. R. **2021**. Heavy metal removal from wastewater using nanomaterials-process and engineering aspects. *Process Safety and Environmental Protection* 150, 323–355.
18. Fato, F. P.; Li, D. W.; Zhao, L. J.; Qiu, K.; Long, Y. T. **2019**. Simultaneous removal of multiple heavy metal ions from river water using ultrafine mesoporous magnetite nanoparticles. *ACS Omega* 4(4), 7543–7549.
19. Khoso, W. A.; Haleem, N.; Baig, M. A.; Jamal, Y. **2021**. Synthesis, characterization and heavy metal removal efficiency of nickel ferrite nanoparticles (NFN's). *Scientific Reports* 11(1), 3790.
20. Yavuz, C. T.; Mayo, J. T.; Yu, W. W. et al. **2006**. Low-field magnetic separation of monodisperse Fe_3O_4 nanocrystals. *Science* 314(5801), 964–967.
21. Lisha, K. P.; Anshup; Pradeep, T. **2009**. Enhanced visual detection of pesticides using gold nanoparticles. *Journal of Environmental Science and Health Part B* 44(7), 697–705.
22. Tratnyek, P. G.; Johnson, R. L. **2006**. Nanotechnologies for environmental cleanup. *Nano Today* 1(2), 44–48.
23. Hodson, M. E. **2010**. The need for sustainable soil remediation. *Elements* 6(6), 363–368.
24. Karn, B.; Kuiken, T.; Otto, M. **2009**. Nanotechnology and in situ remediation: A review of the benefits and potential risks. *Environmental Health Perspectives* 117(12), 1813–1831.
25. Choi, J. H.; Choi, S. J.; Kim, Y. H. **2008**. Hydrodechlorination of 2, 4, 6-trichlorophenol for a permeable reactive barrier using zero-valent iron and catalyzed iron. *Korean Journal of Chemical Engineering* 25(3), 493–500.
26. Ma, L.; He, H.; Zhu, R. et al. **2016**. Bisphenol A degradation by a new acidic nano zero-valent iron diatomite composite. *Catalysis Science & Technology* 6, 6066–6075.
27. Bautitz, I. R.; Velosa, A. C.; Pupo Nogueira, R. F. **2012**. Zero valent iron mediated degradation of the pharmaceutical diazepam. *Chemosphere* 88, 688–692.
28. Machado, S.; Pacheco, J. G.; Nouws, H.P.A. et al. **2017**. Green zero-valent iron nanoparticles for the degradation of amoxicillin. *International Journal of Environmental Science and Technology* 14, 1109–1118.
29. Parlayici, S.; Pehlivan, E. **2019**. Fast decolorization of cationic dyes by nano-scale zero valent iron immobilized in sycamore tree seed pod fibers: Kinetics and modelling study. *International Journal of Phytoremediation* 21, 1130–1144.
30. Sun, X.; Kurokawa, T.; Suzuki, M. et al. **2015**. Removal of cationic dye methylene blue by zero-valent iron: Effects of pH and dissolved oxygen on removal mechanisms. *Journal of Environmental Science and Health Part A* 50, 1057–1071.
31. Wang, X.; Wang, P.; Ma, J. et al. **2015**. Synthesis, characterization, and reactivity of cellulose modified nano zero-valent iron for dye discoloration. *Applied Surface Science* 345, 57–66.
32. Mehrotra, N.; Tripathi, R. M.; Zafar, F.; Singh, M. P. **2017**. Catalytic degradation of dichlorvos using biosynthesized zero valent iron nanoparticles. *IEEE Transactions on Nano Bioscience* 16, 280–286.
33. Elliott, D. W.; Lien, H. L.; Zhang, W. X. **2009**. Degradation of lindane by zero-valent iron nanoparticles. *Journal of Environmental Engineering* 135, 317–324.

34. Liu, T.; Wang, Z. L.; Yan, X.; Zhang, B. **2014**. Removal of mercury (II) and chromium (VI) from wastewater using a new and effective composite: Pumice-supported nanoscale zero-valent iron. *Chemical Engineering Journal* 245, 34–40.
35. Cai, Z.; Fu, J.; Du, P.; Zhao, X. **2018**. Reduction of nitrobenzene in aqueous and soil phases using carboxymethyl cellulose stabilized zero-valent iron nanoparticles. *Chemical Engineering Journal* 332: 227–236.
36. An, B.; Zhao, D. **2012**. Immobilization of As(III) in soil and groundwater using a new class of polysaccharide stabilized Fe-Mn oxide nanoparticles. *Journal of Hazardous Materials* 211(212), 332–341.
37. Liu, J.; Weinholtz, L.; Zheng, L. et al. **2017**. Removal of PFOA in groundwater by Fe_0 and MnO_2 nanoparticles under visible light. *Journal of Environmental Science and Health Part A* 52, 1048–1054.
38. He, F.; Zhao, D.; Liu, J.; Roberts, C. B. **2007**. Stabilization of Fe-Pd nanoparticles with sodium carboxymethyl cellulose for enhanced transport and dechlorination of trichloroethylene in soil and groundwater. *Industrial & Engineering Chemistry Research* 46(1), 29–34.
39. Liu, J.; He, F.; Durham, E. et al. **2008**. Polysugarstabilized Pd nanoparticles exhibiting high catalytic activities for hydrodechlorination of environmentally deleterious trichloroethylene. *Langmuir* 24(1), 328–336.
40. Omidvar, A.; Jaleh, B.; Nasrollahzadeh, M. **2017**. Preparation of the GO/Pd nanocomposite and its application for the degradation of organic dyes in water. *Journal of Colloid Interface Sciences* 496, 44–50.
41. Wang, Y.; Liu, J.; Wang, P. et al. **2014**. Palladium nanoparticles encapsulated in core–shell silica: A structured hydrogenation catalyst with enhanced activity for reduction of oxyanion water pollutants. *ACS Catalysis* 4(10), 3551–3559.
42. Vu, A. K. **2022**. *Synthesis and application of iron/copper nanoparticles and biosurfactants for remediation of oil-contaminated soil*. PhD thesis, Concordia University.
43. Yakout, A. A.; Alshitari, W.; Akhdhar, A. **2021**. Synergistic effect of Cu-nanoparticles and β-cyclodextrin functionalized reduced graphene oxide nanocomposite on the adsorptive remediation of tetracycline antibiotics. *Carbohydrate Polymers* 273, 118528.
44. Wan, S.; Ding, W.; Wang, Y. et al. **2018**. Manganese oxide nanoparticles impregnated graphene oxide aggregates for cadmium and copper remediation. *Chemical Engineering Journal* 350, 1135–1143.
45. Yin, Z.; Song, L.; Song, H. et al. **2020**. Remediation of copper contaminated sediments by granular activated carbon-supported titanium dioxide nanoparticles: Mechanism study and effect on enzyme activities. *Science of the Total Environment* 741, 139962.
46. Rajput, S.; Singh, L. P.; Pittman, C. U., Jr; Mohan, D. **2017**. Lead (Pb^{2+}) and copper (Cu^{2+}) remediation from water using superparamagnetic maghemite (γ-Fe_2O_3) nanoparticles synthesized by flame spray pyrolysis (FSP). *Journal of Colloid and Interface Science* 492, 176–190.
47. Chandra, C.; Khan, F. **2020**. Nano-scale zerovalent copper: Green synthesis, characterization and efficient removal of uranium. *Journal of Radioanalytical and Nuclear Chemistry* 324, 589–597.
48. Liao, G.; Li, Q.; Zhao, W. et al. **2018**. In-situ construction of novel silver nanoparticle decorated polymeric spheres as highly active and stable catalysts for reduction of methylene blue dye. *Applied Catalysis A: General* 549, 102–111.
49. Tosco, T.; Papini, M. P.; Viggi, C. C.; Sethi, R. **2014**. Nanoscale zerovalent iron particles for groundwater remediation: A review. *Journal of Cleaner Production* 77, 10–21.
50. Suresh, S. **2009**. Reductive remediation of pollutants using metals. *The Open Waste Management Journal* 2(1), 15.
51. Lin, K. S.; Chang, N. B.; Chuang, T. D. **2008**. Decontamination of nitrates and nitrites in wastewater by zero-valent iron nanoparticles. *Nano* 3(4), 291–295.
52. Wang, K. S.; Lin, C. L.; Wei, M. C. et al. **2010**. Effects of dissolved oxygen on dye removal by zero-valent iron. *Journal of Hazardous Materials* 182(1–3), 886–895.
53. Sun, F.; Osseo-Asare, K. A.; Chen, Y.; Dempsey, B. A. **2011**. Reduction of As (V) to As (III) by commercial ZVI or As (0) with acid-treated ZVI. *Journal of Hazardous Materials* 196, 311–317.
54. Cullen, L. G.; Tilston, E. L.; Mitchell, G. R.; Collins, C. D.; Shaw, L. J. **2011**. Assessing the impact of nano-and micro-scale zerovalent iron particles on soil microbial activities: Particle reactivity interferes with assay conditions and interpretation of genuine microbial effects. *Chemosphere* 82(11), 1675–1682.
55. Hu, J.; Chen, G.; Lo, I. M. **2005**. Removal and recovery of Cr (VI) from wastewater by maghemite nanoparticles. *Water Research* 39(18), 4528–4536.
56. Chang, Y. C.; Chen, D. H. **2005**. Preparation and adsorption properties of monodisperse chitosan-bound Fe_3O_4 magnetic nanoparticles for removal of Cu (II) ions. *Journal of Colloid and Interface Science* 283(2), 446–451.

57. Mayo, J. T.; Yavuz, C.; Yean, S. et al. **2007**. The effect of nanocrystalline magnetite size on arsenic removal. *Science and Technology of Advanced Materials* 8(1–2), 71.
58. Zhong, L. S.; Hu, J. S.; Liang, H. P.; Cao, A. M.; Song, W. G.; Wan, L. J. **2006**. Self-Assembled 3D flowerlike iron oxide nanostructures and their application in water treatment. *Advanced Materials* 18(18), 2426–2431.
59. Bokare, A. D.; Chikate, R. C.; Rode, C. V.; Paknikar, K. M. **2008**. Iron-nickel bimetallic nanoparticles for reductive degradation of azo dye Orange G in aqueous solution. *Applied Catalysis B: Environmental* 79(3), 270–278.
60. Liu, L.; Liu, Z.; Bai, H.; Sun, D. D. **2012**. Concurrent filtration and solar photocatalytic disinfection/degradation using high-performance Ag/TiO_2 nanofiber membrane. *Water Research* 46(4), 1101–1112.
61. Ayed, R. B.; Ajili, M.; Garcia, J. M.; Ziouche, A.; Kramer, J. L. C.; Turki, N. K. **2020**. First principal investigation of structural, morphological, optoelectronic and magnetic characteristics of sprayed Zn: Fe_2O_3 thin films. *Optik* 219, 165303.
62. Chen, T.; Zheng, Y.; Lin, J. M.; Chen, G. **2008**. Study on the photocatalytic degradation of methyl orange in water using Ag/ZnO as catalyst by liquid chromatography electrospray ionization ion-trap mass spectrometry. *Journal of the American Society for Mass Spectrometry* 19(7), 997–1003.
63. Jing, Z.; Zhan, J. **2008**. Fabrication and gas-sensing properties of porous ZnO nanoplates. *Advanced Materials* 20(23), 4547–4551.
64. Chou, T. P.; Zhang, Q.; Fryxell, G. E.; Cao, G. Z. **2007**. Hierarchically structured ZnO film for dye-sensitized solar cells with enhanced energy conversion efficiency. *Advanced Materials* 19(18), 2588–2592.
65. Wang, X.; Cai, W.; Lin, Y.; Wang, G.; Liang, C. **2010**. Mass production of micro/nanostructured porous ZnO plates and their strong structurally enhanced and selective adsorption performance for environmental remediation. *Journal of Materials Chemistry* 20(39), 8582–8590.
66. Adams, L. K.; Lyon, D. Y.; Alvarez, P. J. **2006**. Comparative eco-toxicity of nanoscale TiO_2, SiO_2, and ZnO water suspensions. *Water Research* 40(19), 3527–3532.
67. Habibi-Yangjeh, A.; Pirhashemi, M.; Ghosh, S. **2020**. ZnO/$ZnBi_2O_4$ nanocomposites with p-n heterojunction as durable visible-light-activated photocatalysts for efficient removal of organic pollutants. *Journal of Alloys and Compounds* 826, 154229.
68. Náfrádi, M.; Alapi, T.; Bencsik, G.; Janáky, C. **2021**. Impact of reaction parameters and water matrices on the removal of organic pollutants by TiO_2/LED and ZnO/LED heterogeneous photocatalysis using 365 and 398 nm radiation. *Nanomaterials* 12(1), 5.

14 Biogenic Nanomaterials for Chemosensor Applications

Arun Sharma, Gagandeep Singh, Narinder Singh, and Navneet Kaur

14.1 INTRODUCTION

The boomed human population disturbed the ecosystem and consequently led to reliance on petroleum products, which result in an unstainable way of living. Replacing petroleum-based chemicals/products and utilization of biomass or waste valorization are among the long pursuits of scientific research to attain a sustainable circular bioeconomy [1]. The properties and structures of green materials and systems are dynamic, making them a promising approach for sustainable solutions and advances in designing and fabricating functional materials [2]. Municipal food waste, agricultural and forestry waste, and other sea-based by-products [3] are all included in biomass, a sustainable and abundant natural resource that is widely accepted as an alternative to fossil fuels [4, 5]. Extensive research has already been done and reveals the production, properties, and varied applications of biomanufactured green nanomaterials, showcasing the breadth of the field [4, 5]. Biogenic nanomaterials [6, 7] can be derived from biomass by use of sustainable phytochemical [8, 9], algae-mediated [8], seafood waste [9], and abundant biopolymer wood [10]. Biogenic/biomass waste–derived carbonaceous nanomaterial [11] from bioderived precursors [12] holds a central place as it possess many practical applications in light-emitting materials [13] and analysis of the environment [14]. Moreover, various forms can be fabricated, namely, cellulose nanomaterials [15], graphene quantum dots [16], nanocellulose, and nanochitin [17]. Additionally, carbon nanoparticles are deeply investigated for sustainable sensing and bioimaging [18] and have optimized ways to fabrication from natural resource [19] as they have environmentally friendly precursors for optical detection [20].

The transformation of biomass resources into functional optical materials has garnered mass consideration in recent years [21]. This chapter aims to discuss the sensing capabilities of bio-based nanomaterials and their potential for understanding the recognition behavior of bioderived nanomaterials. Sensing applications are one of the promising fields that could benefit from the use of biogenic green synthesized nanomaterials. These green materials have unique physicochemical properties and can be further derived with various sensing elements to detect analytes with high sensitivity and selectivity up to attomoles [14, 15]. Biogenic nanomaterials have received extensive attention for sensor development as they possess a large surface area with tunable surface chemistry and high electron transfer ability. Thus, this chapter includes both charged and uncharged species, which are mainly of utmost environmental concern, and discussion about the various types of biomass sources that can be used to prepare luminescent materials, their structure–optical performance correlation, preparation methods, optical properties, potential applications, as well as future challenges and perspectives.

14.2 CATION SENSING

Real-time live-cell monitoring bio-cations in cellular compartments are still lacking in expertise and necessary for devising the path to study mitochondrial features. Lin and coworkers have fabricated S-P-doped carbon dots (GB-CDs) from green beans by using an eco-friendly, cost-effective, easy, and single-step hydrothermal approach that shows a "turn-off" signal in fluorescence spectroscopy

DOI: 10.1201/9781003430087-14

to detect Fe^{3+} in pure aqueous medium, and consequently, the obtained GB-CD+Fe^{3+} complex exhibits "turn-on" fluorescence signal for ATP for mitochondria cell imaging (Figure 14.1). The high selectivity toward Fe^{3+} detection using GB-CDs was revealed by its lowest sensing limit, that is, 3.6 nanomolar. Moreover, the introduction of adenosine triphosphate could reverse the diminished fluorescence of developed probe, due to which the developed nanomaterial can be utilized for mitochondrial cell imaging to detect ATP in a detection window of 50–600 micromolar, with a detection range up to 60 nanomolar [22].

In order to ensure a sustainable ecological balance, the most pressing global issue is environmental contamination, which needs immediate sustainable solution. According to Chiang et al., a method for transforming various naturally abundant bioresources into distinctive GQDs with different characteristics using microplasma technology can be demonstrated for the selective analysis of hazardous aqueous analytes, such as 4-nitrophenol, Hg^{2+}, Cu^{2+}, etc. with greater capabilities (Figure 14.2). They can provide 7.4 nM low limit of detection for the biosensor applications of harmful environmental pollutants [23].

As carbon dots are expected to have high affinity to cotton over polyester, such higher performance can be explored in multifunctional textiles for use in environmental protection. The cotton and polyester fabrics were decorated with biogenic carbon dots by Gedanken et al. They prepared carbon dots using hydrothermal treatment and sonication as unique approaches from biogenic wastes. The sonochemical approach can enhance CDs' adsorption on the fabric for improved protection against ultraviolet rays and various bacteria. The inhibition zone that surrounded the fabric

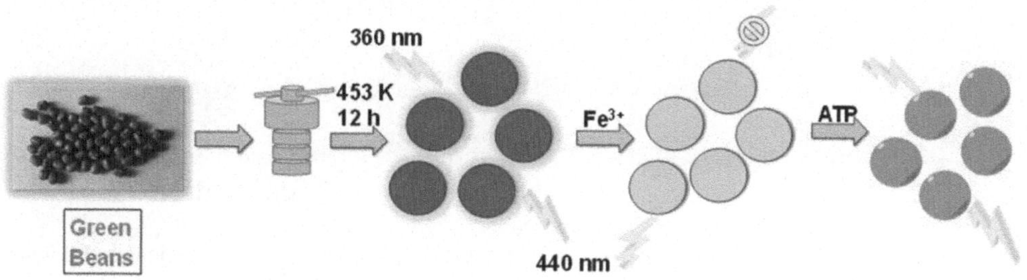

FIGURE 14.1 Synthetic procedure of GB-CDs using green beans.

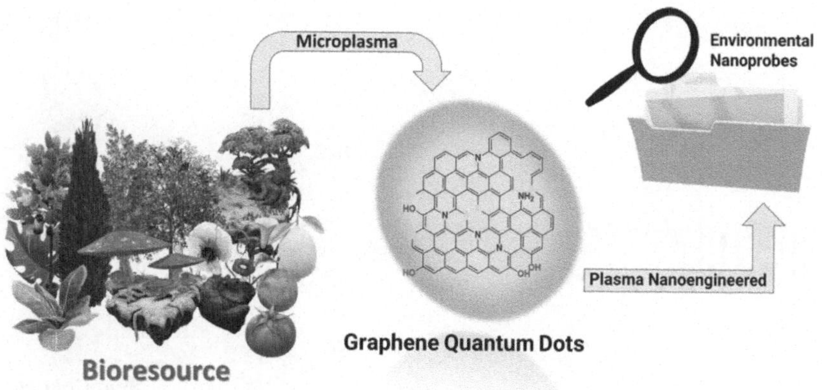

FIGURE 14.2 Pictorial representation of microplasma-enabled efficient production of heteroatom-doped GQDs produced from bioresources as nanoprobes for various environmental pollutants: metal ion sensing from bioresource GQDs.

surface demonstrated remarkable antibacterial activity of CD-coated fabrics toward *S. aureus* and *E. coli*. Moreover, the CD-coated textiles demonstrated improved UV ray–blocking properties in the 280–400 nm range. Several heavy metal cations, namely, Hg^{2+}, Fe^{3+}, and Ag^{+}, can also be detected by the developed protective clothing, which exhibits color change after interacting with these metal ions [24]. XRD patterns confirmed the purity of CDs and revealed four sharp peaks at 14.5, 16.5, 20, and 22 degrees, respectively, in CD-coated cotton/polyester fabrics.

Arun et al. devised a more environmentally friendly synthetic approach, that is, one-step hydrothermal carbonization method, for making oyster mushroom–based blue/green fluorescent carbon-dots, as shown in Figure 14.3. It was demonstrated how these C-dots can be utilized as a naked eye sensor to recognize harmful metal ions, such the heavy Pb^{2+} ions and DNA recognition. Further, it was also demonstrated that three different bacterial strains can be effectively combated by these C-dots. Ultimately, the anticancer effectiveness of C-dots is demonstrated using cells MDA-MB-231 of breast cancer [25].

Waste protein utilization is of utmost benefit due to a cocktail of biomolecules present in it [26]. Referring to Figure 14.4, recently, Ramamurthy et al. explored waste by-products from silkworm pupae (40% proteins) to develop high-end photonic biosensors. The results of their findings showed that a mixture consisting of Ag^{+} and Au^{3+} with protein derived from silkworm can be utilized in

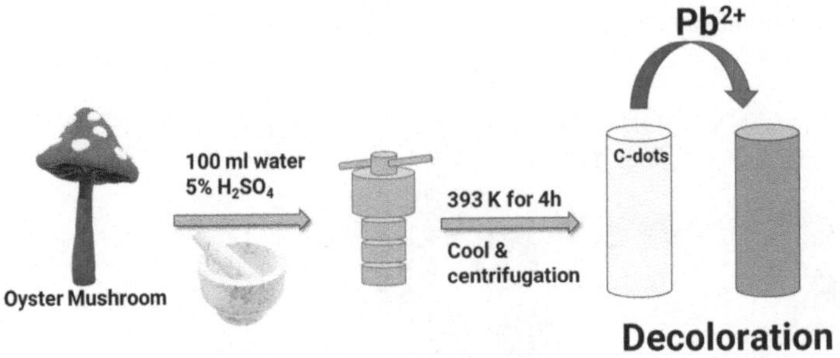

FIGURE 14.3 The fabrication of fluorescent C-dots derived from mushrooms for poisonous metal ion sensing. (Chromogenic detection of Pb^{2+} ions revealing the decoloration of light-yellow color of synthesized probe under natural light).

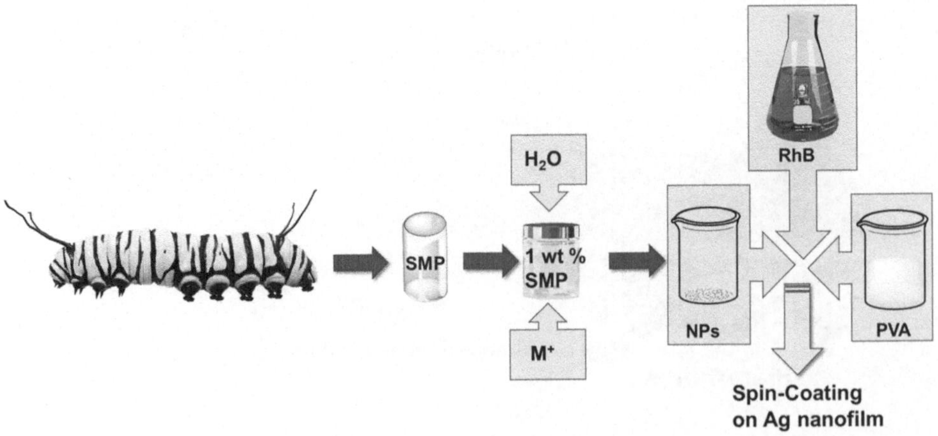

FIGURE 14.4 Outline of synthesis of NPs employing SWP: the procedures for making the various nanointerface for SPCE substrate preparation in various nanolevel interfaces. (Ag^{+} and Au^{3+} ions are referred to as M^{+} ions.)

order to develop single metallic such as, AgNPs and AuNPs, heterometallic (Ag-Au) nanohybrids; however, the ultraviolet irradiation in predetermined duration of time fabricate different nanogeometries. Further, these nanohybrids have been explored for sensing of zinc ions with detection limit up to an attomolar limit [27].

Another report by Ramamurthy's group showed the variations in the length of UV exposure led to the formation of Ag nanofractals and nanocubes, each of which has a distinctive architecture suitable for use in nanophotonic applications, pictorially represented in Figure 14.5. The SPCE technique was made possible by the materials' >900-fold SPCE improvements, which allowed for attomolar detection of dangerous Hg^{2+} ions at incredibly low concentrations utilizing a mobile phone. These materials have been studied as spacer layers, cavities, or extended cavity nanointerfaces for getting tunable plasmonic coupling effects in the technology systems comprising surface plasmon coupled emission [28]. The outline of shape evolution involving nucleation, growth, and the development of enormous aggregates is described in Figure 14.5.

The utilization of cost-effective resources with a higher degree of coalification along with numerous sp^2 carbon crystallites is required for the fabrication of an innovative carbon-based platform to add value to coal resources. In this context, anthracite fulfills all these criteria, thus, Zhang et al. revealed simple, nanosized dots (C-GQDs) from one-step cutting of anthracite coal with potent blue emission using only mechanical means. These C-GQDs exhibit two fluorescence emissions at λ429 and λ469 nm when used as-made. The 469 nm peak was blue-shifted upon reduction with sodium borohydride, which was connected to a decrease in oxygen-containing moieties and an increase in structures having sp^2-carbon domains. The graphene quantum dots fabrication was also demonstrated made from coal and could potentially function as fluorescent sensors to identify cupric ion in aqueous medium. This method promises future sensing applications and provides low cost, great selectivity, quick response, and a good window of linear response [29].

Using an inner-filter-based mechanism, Zhao et al. presented the first instance of fluorescence turn-on nanocomposite system employed for monitoring living cells live detection of cytosol Ca^{2+} level [30]. Gedanken et al. developed biocompatible N-doped carbon nanoparticles fabricated via one-pot hydrothermal method and used them to synthesize N@CDs, wherein *P. betel* leaves play an excellent nitrogen and carbon source. Additionally, the antibacterial ability of the developed N@CDs has been studied against bacteria such as *S. aureus* and *E. coli*. The antibacterial outcome shows that the developed N@CDs exhibits a superior antibacterial activity against bacteria that are resistant to antibiotics (MRSA). The N@CDs also exhibits selective, sensitive, and quick-responding fluorescent probe for Eu^{3+} detection [31].

Chen et al. presented carbon dots (CDs) that were used in a multivariate analysis technique to identify and discriminate heavy metal ions in aqueous environments. N, O, S, and Ca constituents

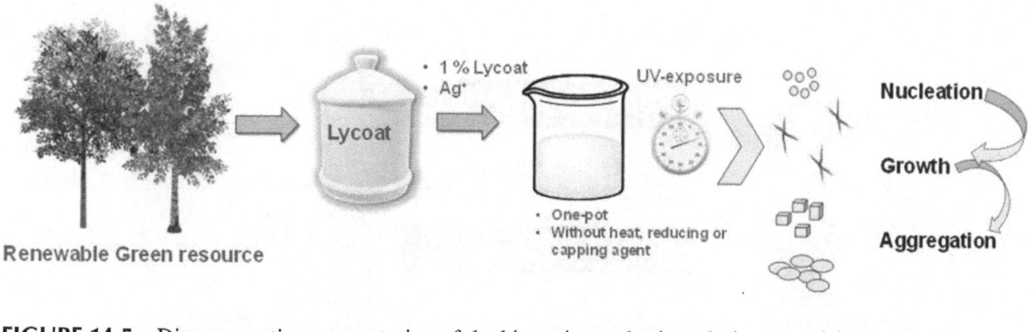

FIGURE 14.5 Diagrammatic representation of the biogenic synthesis technique requiring one pot that produces AgNPs in a ranges of geometries, also along with AgNFs and AgNCs. A mixture of 1% LYCOAT and 1 mM $AgNO_3$ can produce architecturally and morphologically distinct AgNPs when exposed to UV light at various timings.

were directly incorporated into the CDs during the hydrothermal process that created them from different origins of animal bones, pictorially represented in Figure 14.6. Pork, beef, and sheep bone CDs were employed as sensors to interact with d-block metal cations (Ag^{+}, Cu^{2+}, Hg^{2+}, Fe^{3+}, and Pb^{2+}) and revealed distinct patterns based on the variation in fluorescence intensity prior and later to the metal ion addition. The results demonstrated that Ag^{+} had the best quenching impact on biogenic CDs from sheep bone, with a detection range of 40–4000 nM, whereas Fe^{3+} had the maximum quenching behavior on the three types of biogenic carbon dots. The heavy metal ions were distinguished using analysis like hierarchical cluster and linear discriminant analysis, which both produced 100% correctness for ions as well as for both ternary and binary mixtures. The sensor array was also successful in identifying various metal ions in environmental water, offering a simple, efficient, and systematic approach to detecting multiple heavy metal ions simultaneously [32].

Thakore et al. approached a full valorization of waste, making it an environmentally friendly method for metal detection and sustainable environmental remediation. Their research findings showed a way to controlling carbon footprint by utilizing organic waste for important applications, such as detection and remediation. The researchers used a hydrothermal method to produce carbon nanodots from renewable waste sources, and the solid waste was re-employed as a biochar adsorbent. The resulting green carbon dots were successfully used to detect Cr^{6+} and Fe^{3+} metal ions with the lowest possible detection in the pM window. The biochar adsorbent was also effective in removing cationic and anionic dyes from wastewater [33].

The Li research group synthesized carbon dots using mustard seeds via a straightforward hydrothermal method and achieved a good quantum yield (17.92%) when excited at $\lambda = 468$ nm. The fabricated carbon dots were demonstrated as a label-free fluorescence-based probe to sense Fe^{3+} ions in both keratinocytes cells and in water. The lowest sensing limit for Fe^{3+} ions was reported to be 0.51 micromolar within a linear window of 0–75 micromolar. The mechanism of sensing by carbon dots toward Fe^{3+} ions was also investigated, and it was found that both mechanisms applied here, that is, a fluorescence-quenched ground state complex between the carbon dots and ferric ions is formed from the combined effects of static quenching (SQ) and aggregation-induced effect (AIE) [34].

By employing bird feathers, white and yolk of egg, and dung as precursors in the pyrolysis processes, Yan et al. presented an environmentally friendly way to synthesize four types of carbon dots doped with nitrogen and sulfur. The as-fabricated material shows better, stable fluorescence that is wavelength-dependent. They outperform other CDs made from biomass in terms of fluorescence quantum yield. In the presence of buffer solution, synthesized carbon dots were demonstrated for the sensing of Hg^{2+} and Fe^{3+} [35].

In a study, Fan and Ji et al. employed orange peel, ginkgo biloba leaves, paulownia leaves, and magnolia flower debris to produce biomass-derived carbon nanoparticles/CQDs utilizing a hydrothermal technique. The advantages of these biomass CQDs were homogenous particle size, superior

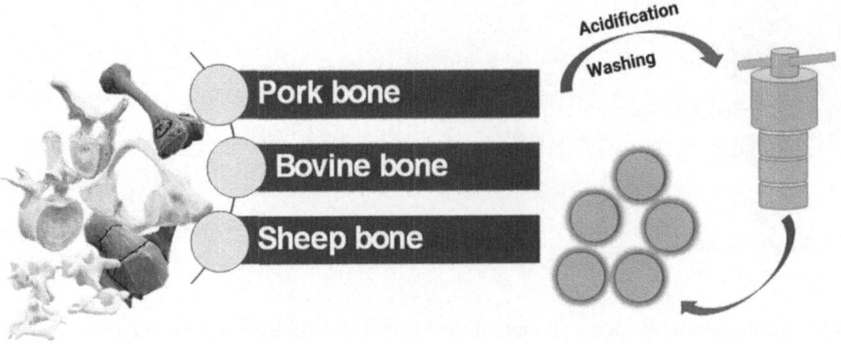

FIGURE 14.6 Scheme of formation of bone biomass carbon dots for multi-channel BCD array–based distinguishment of five metal cations.

aqueous solubility, and durability. Upon the addition of Fe^{3+} ion solution into the carbon quantum dot solution, the fluorescence was quenched. The experiment's findings revealed that LOD for the as-prepared CQDs might be as low as 0.073 μM and that they respond sensitively to ferric ions in the window of 0.2–100 μM concentration [36].

Without using any chemical reagents, a straightforward procedure for the manufacture of bright carbon dots (denoted as M-CDs) was established involving the calcination of mangosteen pulp in air by Ghao et al. The M-CDs are chemically stable, have good solubility, and could be employed as a fluorescent temperature-sensing platform. More impressively, Fe^{3+} ions can arbitrarily extinguish M-CDs fluorescence fast. The phenomenon was used to sense the Fe^{3+} using fluorescence technique with a LOD up to 52 nanomolar. The M-CDs were eventually used for cell visualization, showing its capability for a variety of applications [37].

Using *Ocimum sanctum* leaves, Sahu et al. presented luminous carbon dots fabrication with brilliant fluorescence (quantum yield = 9.3%) and good aqueous medium stability. As Pb^{2+} cations turn off fluorescence signal, so they have demonstrated carbon dots for sensing lead ion with a LOD of 0.59 nanomolar having good selectivity and sensitivity toward Pb^{2+} ions. Using triple-negative cancer cells, the usefulness of manufactured carbon dots for Pb^{2+} ion practical detection is established [38].

Karimi and Samimi used green marine crude extract for an eco-synthetic approach of Ag@AgCl nanomaterial development without any reducing/capping agents or external halide sources and organic solvent. On the opposite side, while several approaches for analyzing harmful Hg^{2+} in drinking water have been documented, chromogenic sensor tests of such green produced Ag@AgCl-NPs revealed an intriguing property for detecting dangerous Hg^{2+} in water. The detection of mercury ion presence is relied on the deformation of synthesized Ag@AgCl-NPs in a concentration-dependent manner. The detection limit of this low-cost test is 4.19 nM, which is lower than both the Chinese agency's value and, more importantly, lower than the US Environmental Protection Agency's standard for drinkable water [39].

Coi et al. did an environment-friendly fabrication of lignin-assisted AgNPs as a selective sensor for mercury ion sensing [40]. A viable catalytic reduction method for harmful nitro-based pollutant elimination using L-AgNPs is reported. The capacity of the synthesized L-AgNPs to detect H_2O_2 and Hg^{2+} in an aqueous system was investigated. Additionally, the quick formation of amino derivative from hazardous 4-nitrophenol and nitrobenzene indicated improved catalytic efficacy of L-AgNP. These studies supported previous research by demonstrating that L-AgNPs are a sensor that also can be used as a catalyst in ecosystem cleanup as well as innovative biomedical applications.

Arumugam and Kim presented a pathway for the fabrication of water-soluble carbon quantum dots (CQDs) from edible plant broccoli via an eco-friendly and economic means. The chemical structure, physical appearance, and resistance to changes in pH of the produced CQDs were analyzed. It was found that Ag^{+} turns off the photoluminescence of these carbon quantum dots better than other heavy metals and possesses LOD up to 0.5 μM, which indicates their potential use as an Ag^{+}-sensing tool [41].

Green and inexpensive CDs were easily obtained by Yao et al. using a single hydrothermal step using *Momordica charantia* as the only precursor. Due to a likely static quenching effect, Pd^{2+} and Fe^{3+} could efficiently reduce fluorescence from the CDs. The cutting-edge nanoprobe displayed exceptional selectivity and a wide linear range. The acquired probe might be able to find Pd^{2+} and Fe^{3+} in aqueous environmental conditions [42].

Kailsa et al. synthesized gold nanoclusters utilizing root extract from *Curcuma longa* (CE) and 11-mercapto undecanoic acid, the fluorescence sensing platform created for the simultaneous analysis of three dipositive ions, namely, Cd^{2+}, Zn^{2+}, and Cu^{2+} ions, in a watery environment. Several masking agents were utilized when analyzing the target cation in the coexisting metal ions presence. By cadmium and zinc ions, a fluorescence "enhancement" process was seen; however, a fluorescence "turn-off" was observed in the case of Cu^{2+} ions [43].

In one study, Parmar and group introduced a new method for detecting heavy metals in drinking water, which is especially important in developing countries, where these metals can pose significant health risks. Pictorially represented in Figure 14.7, the researchers used a hydrothermal method for banana leaf extract to create chlorophyll-functionalized carbon quantum dots (ChlCQDs) that act as nanoprobe sensors for detecting As^{3+} and Hg^{+} ions in the aqueous medium. Through the variation in temperature conditions, that is, ranging from 120°C to 230°C, they were able to produce CQDs with various optical properties. It was discovered that CQDs created at 120°C and 160°C had effective chlorophyll groups, whereas those made at 200°C and 230°C lacked them. The fluorescence band of chlorophyll-functionalized CQDs was observed in the red region, which was either suppressed or intensified by the incorporation of metal cation. They found that ChlCQDs produced at 160°C functioned as both turn-off and on sensors, exhibiting low limits of detection for Hg^{+} and As^{3+} ions. XPS measurements conferred that binding affinity of As^{3+} toward carbonyl group of chlorophyll is more, whereas for Hg^{+} is less. Further, they used DFT calculations to get an understanding of the atomic-level interactions between Hg^{+} and As^{3+} with ChlCQD. The calculations indicated that As^{3+} displayed an absorption at around λ660 nm, which was a result of the extensive interaction between As^{3+} and the quantum dot surface. In contrast, this peak was absent for Hg^{+} because of the negligible interaction between mercury ion and surface of Chl–carbon dots. The findings revealed that Hg^{+} and As^{3+} displayed comparable interactions with different sites on the CQDs in a high-temperature model structure, consistent with their observed experimental absorption spectra [44].

Saleh et al. presented a new method for synthesizing highly luminous gold nanoclusters (AuNCs) using a reducing and stabilizing agent from extract of coffee Arabica seeds (CASE), outline shown in Figure 14.8. They utilized an eco-friendly method to improve the specificity of AuNCs for sensing cupric and mercuric cation. The CASE-AuNCs detect Cu^{2+} and Hg^{2+} ions by fluorescence quenching

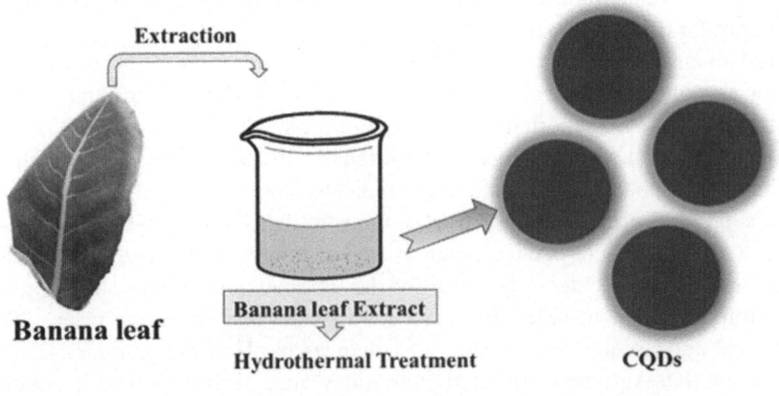

FIGURE 14.7 Outline explaining biofabrication of nanodots for dual sensing.

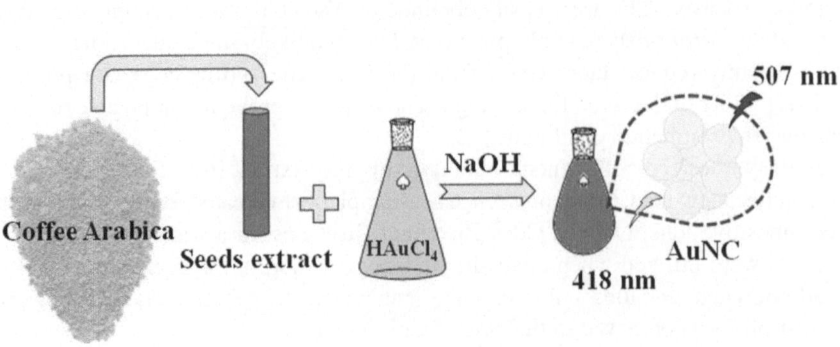

FIGURE 14.8 Pictorial representation of biofabrication of nanocluster for dual sensing.

of gold nanoclusters. To enhance the specificity, EDTA and $NaBH_4$ were employed as capping agents. EDTA was observed to form a metal complex by binding with Cu^{2+} ions and shielding the quenching effect on gold nanocluster fluorescence via transfer of electron or energy. Similarly, sodium borohydride was able to reduce Hg^{2+} to mercury Hg^0, which prevented the Hg^{2+}-CASE-AuNCs system quenching effect. They achieved significant specificity of AuNCs toward Cu^{2+} and Hg^{2+} ions, enabling the determination of these ions even in the presence of one another in the environment. The nanomolar range was the detection limit for Cu^{2+} and Hg^{2+} using the as-prepared CASE-AuNCs. Further, they detected Cu^{2+} and Hg^{2+} in real samples utilizing gold nanoclusters [45].

14.3 ANION SENSING

Biogenic nanomaterials have the unexplored potential for anion-sensing applications. A plant biomass–derived nanosensor was reported by Viswanathan et al. by utilizing the pseudo-stem of the banana plant as the green precursor to create carbon nanodots with a quantum efficiency of 48%. Further prepared nanosensor demonstrated selective Fe^{3+} detection without the interference of competing analytes viz. metal ions. A "turn-off" sensing mode with great specificity with respect to cellular imaging of Fe^{3+} CD also possesses a "turn-on" property toward $S_2O_3^{2-}$ anion with limit of detection 6.4 nM [46].

Ghao et al. synthesized carbon dots by using *Impatiens balsamina L.* stems as a biogenic precursor in a deep eutectic solvent (DES) (as depicted in Figure 14.9). The N,Cl-CDs were utilized for the sensing of ClO^- via fluorescence approach with a lowest possible detection up to 30 nM. The DES-doped N,Cl-CDs were selective toward gram-positive bacteria, as nitrogen and chlorine elements are present. The potential and morphological studies revealed that positively charged N,Cl-CDs were capable of differentiating between gram-positive and gram-negative bacteria during selective fluorescence microscopy and demonstrated antibiotic properties against gram-positive bacteria [47].

14.4 NEUTRAL MOLECULES

Because of the high sensitivity provided by the surface plasmon-coupled emission (SPCE) framework, fabrication of bionanomaterials has evolved as a viable technology due to its simplicity and adaptability toward many applications, such as biosensing platforms. The development of bioinspired heterometallic nanohybrids is critical to achieve high sensitivity in fluorescence-based sensing methods. Recently, Ramamurthy et al. utilized a single-pot, eco- friendly method in which UV irradiation is used to enhance the photo-induced electron transfer pathway, depicted in Figure 14.10. This method yields nanoscale-level materials of amendable optoelectronic functionalities, such as homometallic and heterometric nanohybrids, with better localized SPR. Eventually, these nanoantennae are employed for real-time applications with lower detection limits (attomolar). The research

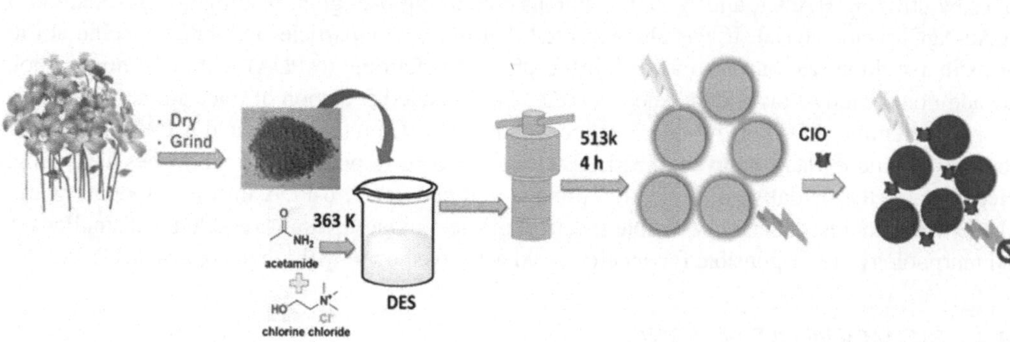

FIGURE 14.9 Pictorial representation of fabrication of N,Cl-CDs using *Impatiens balsamina L.* stems as green precursor.

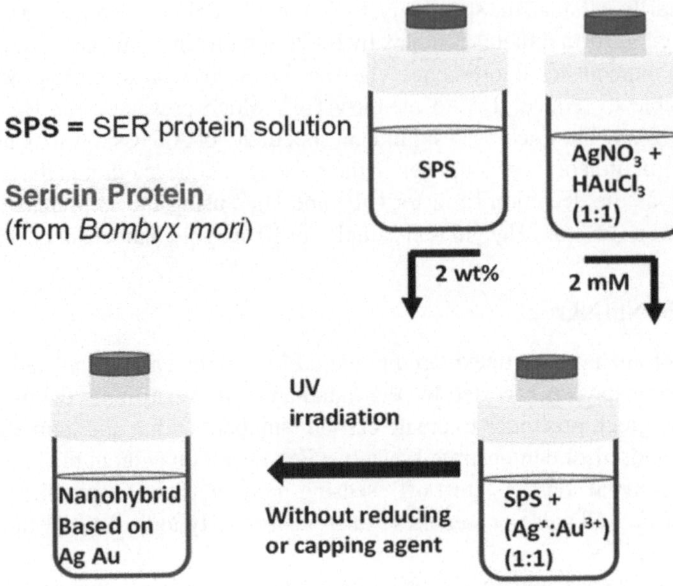

FIGURE 14.10 Pictorial illustration of Ag-Au nanohybrid fabrication employing SER protein solution (SPS) via a one-step pathway.

proposed a reasonable design for in situ nanoparticle production employing biocompatible, non-harmful sericin protein as the capping and reducing agent. This approach produces nanosystems with adjustable optoelectronic functions that exhibit strong light-restricting properties when interfaced with the SPCE platform [48].

According to Sen et al., an antibacterial and anticancer agent can be generated from the bio-derived AgNPs (from *Linum usitatissimum* extract), and that could be employed as well as a catalyst for commercial wastewater remediation. AgNPs were synthesized using green synthesis strategy with *Linum usitatissimum* as reducing agent. AgNPs displayed strong peaks in the 435 nm range when examined with UV-visible spectroscopy. Silver NPs demonstrated DNA cleavage capability and demonstrated great anticancer characteristics against MDA-MB-231 cells. Furthermore, biogenic AgNPs were found to have strong electrocatalytic ability for hydrogen peroxide (H_2O_2) detection. Linear range varies from 1 μM to 5 μM, as reported, with a limit of detection up to 1.323 μM for H_2O_2 [49].

Dhamodharan et al. developed a novel, simple, single one-pot synthesis process of Au-AgCl cubes by utilizing $HAuCl_4$ and $AgNO_3$ solutions containing bael gum, resulting in the creation of an Au-AgCl nanomaterial. It was also reported that these nanoparticles exhibited specific ability for oxidative dimerization reaction with ortho-phenylenediamine (o-PDA) within 30 min without any additional catalyst or oxidant, such as H_2O_2. This enabled detection of trace amounts of ortho-phenylenediamine (PDA), using fluorescence technique LOD up to 2 ppm, and if UV-visible spectroscopy technique was employed, then LOD came out to be 15 ppm. These nanocubes have a high selectivity for the oxidative dimerization process and can detect o-PDA in trace amounts within 30 min using fluorescence or UV-visible spectroscopy techniques. Results revealed that smaller size and morphology are responsible for specific rapid activity shown by these nanoparticles [50].

14.5 BIOMOLECULE SENSING

Wen and his colleagues devised a single-step biogenic fabrication of biochar (BC) from kudzu vine that was doped with molybdenum selenide ($MoSe_2$), having applications similar to oxidase as a smart nanozyme detection system for sensing of hesperetin (HP), a flavanone, in waste covering of citrous fruit, that is, orange. These studies employed hydrothermal fabrication procedure for the value conversion of waste kudzu vine. The structure as well as the characteristics of $MoSe_2$-BC were studied, and it was discovered that biochar enhanced stability of the electrochemical cycle, area of electrochemical activity, and electrocatalytic characteristics of $MoSe_2$. The behavior similar to oxidase enzyme of $MoSe_2$-BC was demonstrated by the formation of blue products from the oxidation of the colorless substrate, that is, 3,3′,5,5′-tetramethylbenzidine (TMB) and the alteration in absorption coefficient of UV-Visible absorption spectra. Further, by using differential pulse voltammetric technique, the $MoSe_2$-BC demonstrated good electrochemical sensing capabilities for hesperetin (HP) in linear range of 10 nanomolar to 9.5 micromolar, with a sensing range up to 2 nanomolar. This research would give direction to the manufacture and deployment of biochar-adorned graphene-like nanoscale materials having oxidase-like activity, as well as the construction of a smart nanozyme sensing platform [51].

Gautam et al. concluded that natural extract-assisted synthesis can be used to manipulate nanostructures and create novel properties. This work discussed three distinct polyaniline samples: PANI-HCl (inorganic acid synthesis), PANI–citric acid (organic acid synthesis), and PANI-lemon (synthesized using crude lemon juice). These polyaniline samples successfully detect catechol molecules in an electrocatalytic way, leading to the generation of new redox centers on carbon paste electrodes, exhibiting high sensitivity, specificity, etc. Compared to its other counterparts, PANI-lemon is a better electrical conductor because of high surface area, desired interfacial resistance, porous design, and high electrocatalytic function [52]. Furthermore, this paper only focused on polyaniline and does not consider other types of polymers or nanoparticles. Another limitation is that while lemon juice was used to manipulate nanostructures, there may be other natural extracts which could produce similar effects but have yet to be explored in detail.

Zhang et al. developed a new process for producing recyclable, ecologically friendly carbon quantum dots having good fluorescence quantum efficiency, that is, 19.5%. They employed eggshell membrane as an economic and feasible precursor to microwave-using pathway to create C-dots, taking advantage of its high protein content as a common waste material. When activated at a wavelength of 235 nm, the C-dots emitted at a maximum of 375 nm. Using the fluorescence turn-off-on technique, they detected Hg^{2+} and glutathione with developed probe. Developed probe, after interacting with Hg^{2+}, quenched the fluorescence of the C-dots, whereas after interacting with glutathione, fluorescence quenching did not occur due to the peculiar structure of GSH [53].

14.6 ENVIRONMENTAL POLLUTANTS

Environmental pollutants such as toxins, dyes, drugs, etc. enter the ecosystem because of human activities and cause harmful effects to human health. In this regard, their detection is necessary and has been explored in the following reports.

14.6.1 Toxins

Singh et al. established a limit of detection, recorded lowest to date, that is, up to 3 ppb concentration of chloroform, via simple thermal processing from biomass waste like hairs with no use of solvents (Figure 14.11), the best sensitivity of chloroform in water analysis utilizing bright fluorescence showing CDs (quantum yield 38%). They discovered that the unique OCDs pyridinic N oxide confers quick and selective detecting feature set. The fluorescence-based detecting feature is governed by both the factors, that is, binding capacity of the analyte to the carbon dot and also by the electronic structures offered by binding elements. This was discovered through comparative analysis on autoclave-fabricated CDs (OCDs) and their microwave-fabricated counterpart. This investigation provided a thorough analysis of the many characteristics of biowastes and how to provide CDs special functions for creating innovative composite material. The identical hair sample

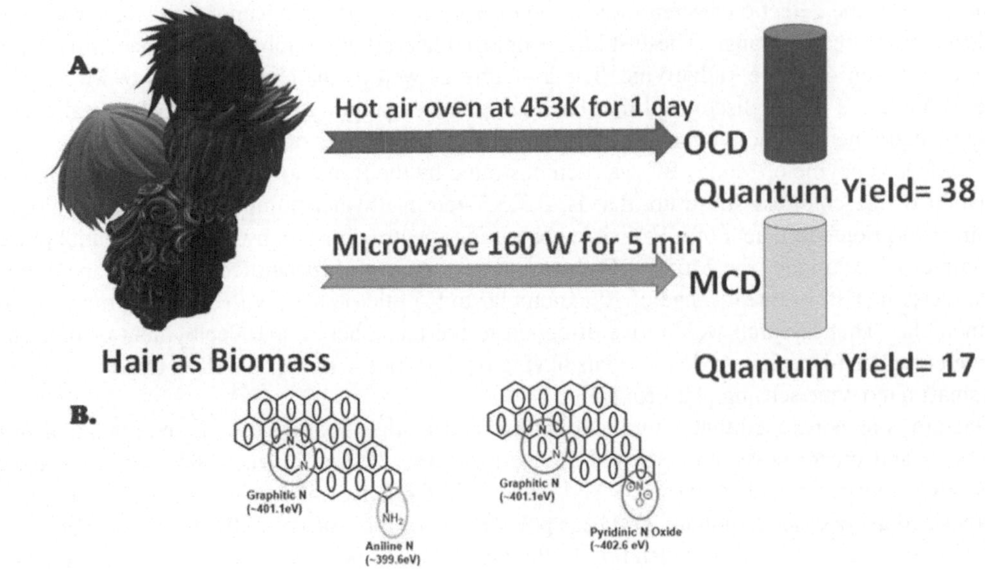

FIGURE 14.11 (a) Fabrication outline of different surface functionalities, imparted by the synthesis methods. (b) Representation of altered ways of nitrogen bonding in MCD and OCD, respectively (inferred from high-resolution N1s XPS spectrums).

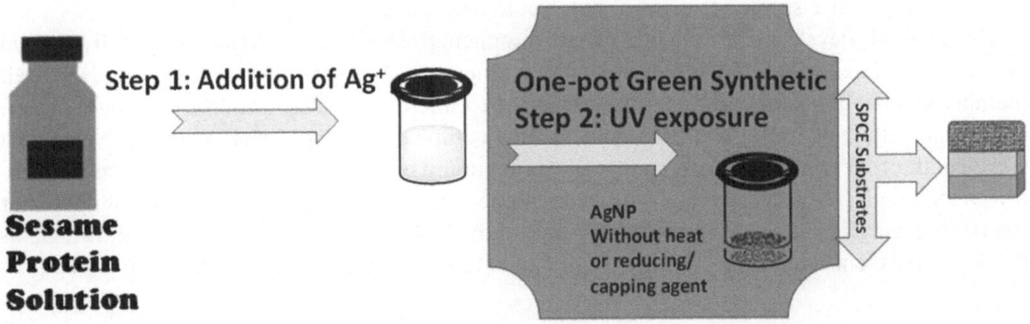

FIGURE 14.12 Single-pot eco-friendly fabrication of AgNPs–SPCE optical system.

was conventionally autoclaved and microwave fast-processed, yielding two types of CDs with distinctly different physicochemical features. It should be noted that the synthesis technique can be adjusted to change the N-containing functional groups containing species along with different functional moieties, and this synthetic difference produces different functionalities, leading to altered sensing. Particularly, the OCDs' pyridinic N oxide has provided exceptional chloroform-detecting performance [54].

14.6.2 Dyes

Ramamurthy and Reddy, using a bioinspired approach with sesame oil seed cake proteins, referring to Figure 14.12, discussed the synthesis and evaluation of anisotropic sharp-edged silver nanoparticles (AgNPs). These AgNPs showed augmented emission when used in surface plasmon-coupled emission (SPCE) platform, for cellphone-based femtomolar detection applications with rhodamine B as reporter molecule [55]. Further research can be done to optimize the synthesis process as well as

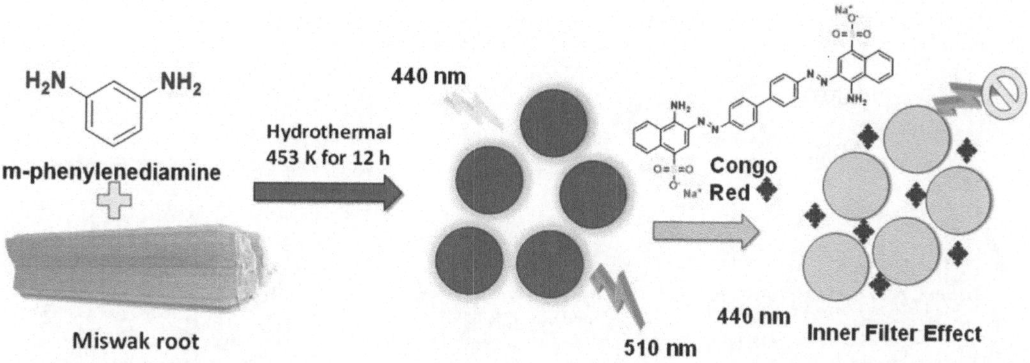

FIGURE 14.13 Schematic illustration of fabrication of Ca-, N-, S-doped CQDs for universal cell imaging and intracellular CR sensing.

to explore more eco-friendly approaches for synthesizing these kinds of SPCE optical nanointerface platform and using them in other functional sensing applications with existing reporter molecules.

Since there is no in situ intracellular assay for Congo red (CR) detection, Lin and his team are the first to use carbon dots for probing of intracellular dyes using fluorescence method. The carbon dots (abbreviated as Mis-mPD-CDs), which emit a green light when negatively charged, were made from plants and m-phenylenediamine (mPD) via a simple hydrothermal carbonization process (Figure 14.13). High sensitivity and selectivity were demonstrated by the green emissive, which was Ca-, N-, and S-doped, with good quantum yield (19%) and limit of detection up to 58 nM. aMis-mPD-CDs have also demonstrated the capacity to enter and label a range of cells as well as fungi and bacteria, with appreciable biocompatibility and long-term stability. Mis-mPD-CDs are a great option for the detection and imaging of Congo red in living cells as well as in zebra fish due to their distinct properties. Mis-mPD-CDs have also been used successfully to quantify CR in actual samples, such fish tissues and commercial effluent [56].

Yang et al. concluded in their findings that by the means of one-step hydrothermal treatment, employing notorious tobacco as provider of carbon for facile fabrication of highly illuminating carbon dots (CDs) can be used as a form of sensor to specifically differentiate three tetracyclines, as they exhibit varied fluorescence emissions (mild blue, navy blue, and chartreuse) with different antibiotics. As illustrated in Figure 14.14, CD fluorescence was quenched by tetracycline without any shift in its emission via an inner filter effect. The addition of chlortetracycline resulted into a blue shift, while oxytetracycline led to a red shift in the fluorescence emissions from these CDs [57]. Additionally, further research needs to be done to enhance the sensitivity and accuracy for detecting these compounds at lower concentrations.

A novel, simple, inexpensive, and one-pot hydrothermal synthesis was obtained to produce bright fluorescent carbon dots (CDs), as demonstrated by Rezaei and coworkers. The *Lawsonia inermis* (henna) plant was utilized in this research to create CDs for the first time without the use of chemical reagents, and this was distinguished by various procedures. The as-fabricated CDs have great solubility in solvents of hydrophilic character, as well as high stability under a variety of circumstances. Also, the carbon dots were employed as a biosafe sensing element for methotrexate (denoted as MTX) determination in the concentration window ranging from 0.02 μmol/L to 18 μmol/L, with the lowest possible spotting of 7 nmol/L. FRET mechanism was utilized for determination of MTX in a very sensitive and selective manner [58].

Kailasa and colleagues reported the synthesis of fluorescent platinum nanoclusters (CEW-Pt NCs) employing chicken egg white as both a template and stabilizing material. Under a 365 nm UV light, the CEW-Pt NCs displayed fluorescence (blue), and when triggered at 330 nm, it emitted at 406 nm. Further, with the help of developed probe, they detect carbidopa, followed by fluorescence

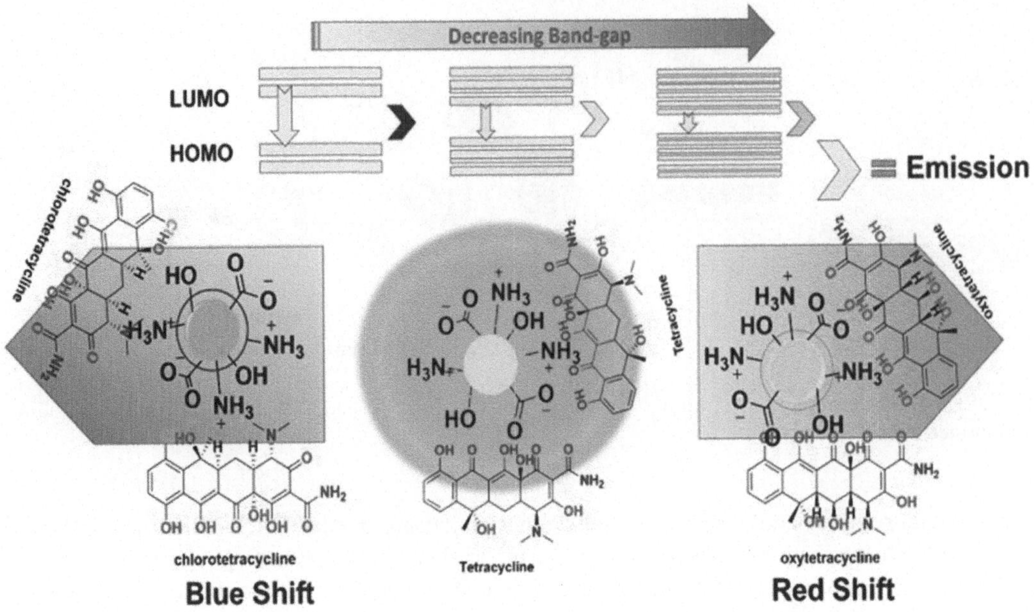

FIGURE 14.14 Illustration of modulation of bandgap of the CDs by interaction with chlortetracycline, tetracycline, and oxytetracycline, respectively.

turn-off approach. The linear response range for carbidopa was found to be 5.0–35 micromolar, with a lowest feasible analysis up to 1.71 micromolar. The developed probe was found to be selective toward carbidopa as compared to similar kinds of drugs, chemicals, and biomolecules and further used for real samples, such as pharmaceutical samples and biofluids [59].

Xie et al. fabricated N-doped CD from pork liver through a single hydrothermal step. The prepared N-CDs showed appreciable quantum efficiency of 11.74%. The synthesized CDs demonstrated 6-thioguanine (6-TG) recognition, and the results revealed that the excitation band of the synthesized probe overlapped significantly with the absorption band of target analyte, which led to the formation of an inner filter effect (IFE). The analyte worked as an absorber in the IFE, leading to the quenching of N-CD fluorescence. The quenching intensity of the system's fluorescence was linearly varying with the amount of 6-TG within the window of 1–200 micromolar, with a low limit of sensing up to 0.75 μM, and synthesized probe was utilized for real sample analysis, such as human serum.

14.6.3 Pesticides

Park et al. concentrated on the single-pot straightforward and practical approach for carbon dots (CDs) fabrication utilizing leaves from *Calotropis procera*. Having a good quantum yield of 71.95% and impressive excitation-dependent emission spectra, the synthesized CDs do not need surface modulations to increase their fluorescence. At 416 nm, the CDs exhibit a strong emission peak. According to the results obtained, adding isoprothiolane dramatically reduced the emission peak of CDs. Furthermore, under various pH circumstances, the probe's fluorescence intensity was not significantly hindered. The isoprothiolane detection threshold was 3.36 ppb, and the developed sensor was utilized to analyze target analyte in real samples, such as fruit and rice [60].

Heidari and group, for the sensing of diazinon, developed a straightforward, quick, and affordable nanoprobe utilizing biogenic carbon dots (CDs). When stimulated at 393 nm, the resultant CDs displayed their emission at 485 nm due to the fluorescence. When diazinon was added, the emission was subsequently diminished, with dependence on concentration [61].

14.6.4 Both Cation and Biomolecules

Doped carbon dots (N/S-CDs) with nitrogen along with sulfur were synthesized by Zhuo et al. employing gardenia fruit as the starting material. The synthesized probe has exceptional fluorescence durability over a broad pH range, high NaCl concentrations, and UV light irradiation. No surface modification was done; still, the absolute quantum yield of the CDs reached 10.7%. The fluorescence signal of synthesized probe can be turned off by Hg^{2+} and recovered on the introduction of amino acid, that is, cysteine. The detection values for Hg^{2+} and cysteine are 2–20 μM and 0.1–2.0 μM, respectively [62].

Ni et al. described an analysis method for mercury ions (Hg^{2+}) and cysteine (Cys) using as fluorescent graphene quantum dots. When Cys is added, it forms strong metal thiol bonds with Hg^{2+}, which results in an enhancement of the GQD's fluorescence; Hg^{2+} cations were absorbed on their negatively polarized surface due to electrostatic attraction, which quenched fluorescence by 92%. When Cys was added, it formed a strong metal thiol bond with Hg^{2+}, displacing weak electrostatic interaction, resulting in an enhancement of 510% at 450 nm for 8 μmol L^{-1} solution [63].

14.6.5 Miscellaneous

In addition to being an increasingly common type of fisheries waste, marine shells also provide a special class of hybrid nanomaterials that are produced extensively and sustainably. Li et al. presented a novel approach to synthesize carbon nanomeshes from marine shells, which can be used as an alternative to graphene nanomeshes (Figure 14.15). The synthesis process is green and low-cost, with high production rate due to the availability of raw materials in nature. Carbon membranes produced through this method have layered structures that enable them for applications such as filtration, absorption, electrochemical sensing, and strain sensing with good sensitivity. The resulting nanomeshes are 1 nm thick, with hierarchical porosity and heteroatom doping and large surface area (378 $m^2\ g^{-1}$) [64].

Early detection of cancer is now achievable because of the analysis of cancer-related biomarkers, which paves a path to the development of strategies to prevent it from advancing to more severe stages. Manickam et al. synthesized a nanocomposite based on a component from reduced graphene oxide-silver (RGO-Ag), using a simple, eco-friendly one-step synthesis method and an extract from the health-promoting *Ganoderma lucidum* mushroom to create a sensitive technique for the sensing of H_2O_2. Additionally, the synthesized nanocomposite was used for the enzymeless sensing of H_2O_2, showing also excellent electrocatalytic activity for H_2O_2 reduction, with the lowest possible detecting limit up to 136 nM [65].

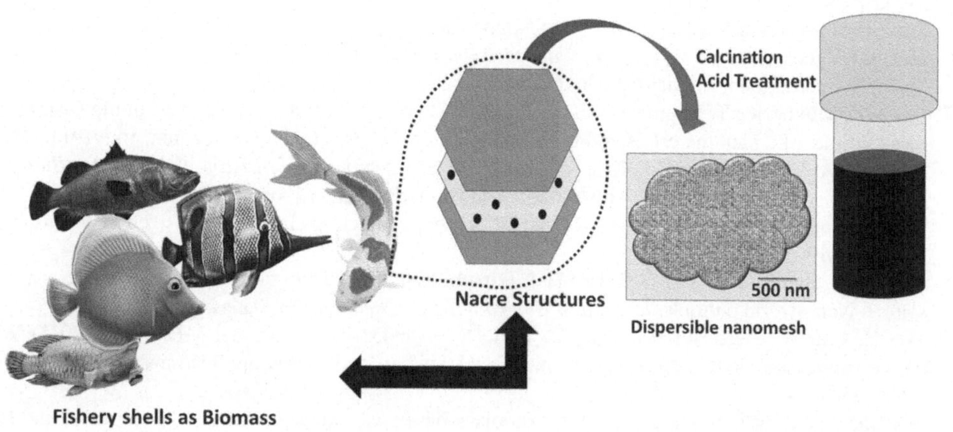

FIGURE 14.15 Synthesis pathway of carbon nanomesh from nacre carbon nanomesh for sensing membrane.

14.7 CONCLUSION

For chemosensor applications, biogenic nanomaterials have significant benefits over conventionally manufactured nanomaterials. Researchers are continuously investigating strategies to control the morphology of biogenic nanostructure manufacturing as well as new biogenic nanomaterials and their use for diverse sensing applications. Further research and thorough evaluations are needed to create new classes of nanoscale materials with high photostability, safety, and considerable biocompatibility that can be exploited in biomedical applications. Upcoming investigation in the area of green-synthesized nanocatalysts and nanomaterials may focus on wastewater remediation. Exploring novel ways for synthesizing these compounds that are more efficient, cost-effective, and ecologically benign is one promising avenue. Another possibility is to look at how different types of pollutants interact with these materials at the molecular level in order to better understand their methods of action. Furthermore, there may be potential to improve existing technologies' efficacy by fine-tuning factors such as pH levels or reaction times. Finally, more research into the bio-safety risks related with the use of biogenic nanomaterials in water treatment applications would be beneficial.

ACKNOWLEDGMENTS

The authors thank DST/NM/TAP-AGRI/01/2019(G) grant from the Department of Science and Technology, government of India.

REFERENCES

1. Geng Y, Sarkis J, Bleischwitz R (2019) How to globalize the circular economy. *Nature* 565:153–155.
2. Sarma H, Joshi SJ, Prasad R, Jampilek J (eds) (2021) Biobased nanotechnology for green applications. https://doi.org/10.1007/978-3-030-61985-5
3. Aswathi VP, Meera S, Maria CGA, Nidhin M (2022) Green synthesis of nanoparticles from biodegradable waste extracts and their applications: A critical review. *Nanotechnology for Environmental Engineering* 8:377–397. https://doi.org/10.1007/s41204-022-00276-8
4. Kumar N, Balamurugan A, Balakrishnan P, Vishwakarma K, Shanmugam K (2020) Biogenic nanomaterials: Synthesis and its applications for sustainable development. In: Ghorbanpour M, Bhargava P, Varma A, Choudhary DK (eds) *Biogenic Nano-Particles and Their Use in Agro-Ecosystems*. Springer, Singapore, pp. 99–132.
5. Saratale RG, Saratale GD, Shin HS, Jacob JM, Pugazhendhi A, Bhaisare M, Kumar G (2018) New insights on the green synthesis of metallic nanoparticles using plant and waste biomaterials: Current knowledge, their agricultural and environmental applications. *Environmental Science and Pollution Research* 25:10164–10183.
6. Srivastava SK, Ogino C, Kondo A (2015) Nanoparticle synthesis by biogenic approach. In: Basiuk VA, Basiuk EV (eds) *Green Processes for Nanotechnology: From Inorganic to Bioinspired Nanomaterials*. Springer International Publishing, Cham, pp. 237–257.
7. Gul MZ, Mukherjee T, Rupula K, Rao BS (2023) Chapter 1 – Strategies involved in bio-inspired synthesis of metallic nanomaterials and their applications – a comprehensive account. In: Ozturk M, Roy A, Bhat RA, Vardar-Sukan F, Policarpo Tonelli FM (eds) *Synthesis of Bionanomaterials for Biomedical Applications*. Elsevier Science, Amsterdam: The Netherlands, pp. 1–36.
8. Thangadurai D, Sangeetha J, Prasad R (eds) (2021) Bioprospecting algae for nanosized materials. https://doi.org/10.1007/978-3-030-81557-8
9. Sharma A, Sharma RK, Kim Y-K, Lee H-J, Tripathi KM (2021) Upgrading of seafood waste as a carbon source: Nano-world outlook. *Journal of Environmental Chemical Engineering* 9:106656.
10. Zhu H, Luo W, Ciesielski PN, Fang Z, Zhu JY, Henriksson G, Himmel ME, Hu L (2016) Wood-derived materials for green electronics, biological devices, and energy applications. *Chemical Reviews* 116:9305–9374.
11. Wareing TC, Gentile P, Phan AN (2021) Biomass-based carbon dots: Current development and future perspectives. *ACS Nano* 15:15471–15501.

12. Shaik SA, Sengupta S, Varma RS, Gawande MB, Goswami A (2021) Syntheses of N-doped carbon quantum dots (NCQDs) from bioderived precursors: A timely update. *ACS Sustainable Chemistry & Engineering* 9:3–49.
13. Li W, Chen Z, Yu H, Li J, Liu S (2021) Wood-derived carbon materials and light-emitting materials. *Advanced Materials* 33:2000596.
14. Chawla S, Rai P, Garain T, Uday S, Hussain CM (2022) Green carbon materials for the analysis of environmental pollutants. *Trends in Environmental Analytical Chemistry* 33:e00156.
15. Peng Z, Lin Q, Tai Y-AA, Wang Y (2020) Applications of cellulose nanomaterials in stimuli-responsive optics. *Journal of Agricultural and Food Chemistry* 68:12940–12955.
16. Abbas A, Mariana LT, Phan AN (2018) Biomass-waste derived graphene quantum dots and their applications. *Carbon* 140:77–99.
17. Bai L, Liu L, Esquivel M, Tardy BL, Huan S, Niu X, Liu S, Yang G, Fan Y, Rojas OJ (2022) Nanochitin: Chemistry structure, assembly, and applications. *Chemical Reviews* 122:11604–11674.
18. Sharma V, Tiwari P, Mobin SM (2017) Sustainable carbon-dots: Recent advances in green carbon dots for sensing and bioimaging. *Journal of Materials Chemistry B* 5:8904–8924.
19. Das R, Bandyopadhyay R, Pramanik P (2018) Carbon quantum dots from natural resource: A review. *Materials Today Chemistry* 8:96–109.
20. Ullal N, Muthamma K, Sunil D (2022) Carbon dots from eco-friendly precursors for optical sensing application: An up-to-date review. *Chemical Papers* 76:6097–6127.
21. Ge M, Liu S, Li J, Li M, Li S, James TD, Chen Z (2023) Luminescent materials derived from biomass resources. *Coordination Chemistry Reviews* 477:214951.
22. Durrani S, Zhang J, Mukramin, Wang H, Wang Z, Khan LU, Zhang F, Durrani F, Wu F-G, Lin F (2023) Biomass-based carbon dots for Fe^{3+} and adenosine triphosphate detection in mitochondria. *ACS Applied Nano Materials* 6:76–85.
23. Kurniawan D, Sharma N, Rahardja MR, Cheng Y-Y, Chen Y-T, Wu G-X, Yeh Y-Y, Yeh P-C, Ostrikov KK, Chiang W-H (2022) Plasma nanoengineering of bioresource-derived graphene quantum dots as ultrasensitive environmental nanoprobes. *ACS Applied Materials & Interfaces* 14:52289–52300.
24. Durairaj A, Maruthapandi M, Luong JHT, Perelshtein I, Gedanken A (2022) Enhanced UV protection, heavy metal detection, and antibacterial properties of biomass-derived carbon dots coated on protective fabrics. *ACS Applied Bio Materials* 5:5790–5799.
25. Boobalan T, Sethupathi M, Sengottuvelan N, Kumar P, Balaji P, Gulyás B, Padmanabhan P, Selvan ST, Arun A (2020) Mushroom-derived carbon dots for toxic metal ion detection and as antibacterial and anticancer agents. *ACS Applied Nano Materials* 3:5910–5919.
26. Peydayesh M, Bagnani M, Soon WL, Mezzenga R (2022) Turning food protein waste into sustainable technologies. *Chemical Reviews* 7:2112–2154. acs.chemrev.2c00236
27. Rai A, Bhaskar S, Reddy N, Ramamurthy SS (2021) Cellphone-aided attomolar zinc ion detection using silkworm protein-based nanointerface engineering in a plasmon-coupled dequenched emission platform. *ACS Sustainable Chemistry & Engineering* 9:14959–14974. https://pubs.acs.org/doi/10.1021/acssuschemeng.1c05437. Accessed 3 March 2023.
28. Rathnakumar S, Bhaskar S, Rai A, Saikumar DVV, Kambhampati NSV, Sivaramakrishnan V, Ramamurthy SS (2021) Plasmon-coupled silver nanoparticles for mobile phone-based attomolar sensing of mercury ions. *ACS Applied Nano Materials* 4:8066–8080.
29. Zhang Y, Li K, Ren S, Dang Y, Liu G, Zhang R, Zhang K, Long X, Jia K (2019) Coal-derived graphene quantum dots produced by ultrasonic physical tailoring and their capacity for Cu(II) detection. *ACS Sustainable Chemistry & Engineering* 7:9793–9799.
30. Chen D, Zhao J, Zhang L, Liu R, Huang Y, Lan C, Zhao S (2018) Capsicum-derived biomass quantum dots coupled with alizarin red S as an inner-filter-mediated illuminant nanosystem for imaging of intracellular calcium ions. *Analytical Chemistry* 90:13059–13064.
31. Maruthapandi M, Das P, Saravanan A, Natan M, Banin E, Kannan S, Michaeli S, Luong JHT, Gedanken A (2021) Biocompatible N-doped carbon dots for the eradication of methicillin-resistant S. aureus (MRSA) and sensitive analysis for europium (III). *Nano-Structures & Nano-Objects* 26:100724.
32. Fu L, Liu T, Yang F, Wu M, Yin C, Chen L, Niu N (2022) A multi-channel array for metal ions discrimination with animal bones derived biomass carbon dots as sensing units. *Journal of Photochemistry and Photobiology A: Chemistry* 424:113638.
33. Das M, Thakkar H, Patel D, Thakore S (2021) Repurposing the domestic organic waste into green emissive carbon dots and carbonized adsorbent: A sustainable zero waste process for metal sensing and dye sequestration. *Journal of Environmental Chemical Engineering* 9:106312

34. Zhang Q, He S, Zheng K, Zhang L, Lin L, Chen F, Du X, Li B (2022) Green synthesis of mustard seeds carbon dots and study on fluorescence quenching mechanism of Fe^{3+} ions. *Inorganic Chemistry Communications* 146:110034.
35. Ye Q, Yan F, Luo Y, Wang Y, Zhou X, Chen L (2017) Formation of N, S-codoped fluorescent carbon dots from biomass and their application for the selective detection of mercury and iron ion. *Spectrochimica Acta Part A: Molecular and Biomolecular Spectroscopy* 173:854–862.
36. Wang C, Shi H, Yang M, Yan Y, Liu E, Ji Z, Fan J (2020) Facile synthesis of novel carbon quantum dots from biomass waste for highly sensitive detection of iron ions. *Materials Research Bulletin* 124:110730.
37. Yang R, Guo X, Jia L, Zhang Y, Zhao Z, Lonshakov F (2017) Green preparation of carbon dots with mangosteen pulp for the selective detection of Fe^{3+} ions and cell imaging. *Applied Surface Science* 423:426–432.
38. Kumar A, Chowdhuri AR, Laha D, Mahto TK, Karmakar P, Sahu SK (2017) Green synthesis of carbon dots from Ocimum sanctum for effective fluorescent sensing of Pb^{2+} ions and live cell imaging. *Sensors and Actuators B: Chemical* 242:679–686.
39. Karimi S, Samimi T (2019) Green and simple synthesis route of Ag@AgCl nanomaterial using green marine crude extract and its application for sensitive and selective determination of mercury. *Spectrochimica Acta Part A: Molecular and Biomolecular Spectroscopy* 222:117216.
40. Lee SJ, Begildayeva T, Yeon S, Naik SS, Ryu H, Kim TH, Choi MY (2021) Eco-friendly synthesis of lignin mediated silver nanoparticles as a selective sensor and their catalytic removal of aromatic toxic nitro compounds. *Environmental Pollution* 269:116174.
41. Arumugam N, Kim J (2018) Synthesis of carbon quantum dots from Broccoli and their ability to detect silver ions. *Materials Letters* 219:37–40.
42. Dong Y, Zhang Y, Zhi S, Yang X, Yao C (2021) Green synthesized fluorescent carbon dots from Momordica charantia for selective and sensitive detection of Pd^{2+} and Fe^{3+}. *ChemistrySelect* 6:123–130.
43. Desai ML, Basu H, Saha S, Singhal RK, Kailasa SK (2020) One pot synthesis of fluorescent gold nanoclusters from Curcuma longa extract for independent detection of Cd^{2+}, Zn^{2+} and Cu^{2+} ions with high sensitivity. *Journal of Molecular Liquids* 304:112697.
44. Bayazeed Alam M, Hassan N, Sahoo K, Kumar M, Sharma M, Lahiri J, Singh Parmar A (2022) Deciphering interaction between chlorophyll functionalized carbon quantum dots with arsenic and mercury toxic metals in water as highly sensitive dual-probe sensor. *Journal of Photochemistry and Photobiology A: Chemistry* 431:114059.
45. Saleh SM, Almotiri MK, Ali R (2022) Green synthesis of highly luminescent gold nanoclusters and their application in sensing Cu(II) and Hg(II). *Journal of Photochemistry and Photobiology A: Chemistry* 426:113719.
46. Vandarkuzhali SAA, Jeyalakshmi V, Sivaraman G, Singaravadivel S, Krishnamurthy KR, Viswanathan B (2017) Highly fluorescent carbon dots from Pseudo-stem of banana plant: Applications as nanosensor and bio-imaging agents. *Sensors and Actuators B: Chemical* 252:894–900.
47. Liu S, Quan T, Yang L, Deng L, Kang X, Gao M, Xia Z, Li X, Gao D (2021) N,Cl-codoped carbon dots from impatiens Balsamina L. stems and a deep eutectic solvent and their applications for gram-positive bacteria identification, antibacterial activity, cell imaging, and Clo– sensing. *ACS Omega* 6:29022–29036.
48. Bhaskar S, Rai A, Ganesh KM, Reddy R, Reddy N, Ramamurthy SS (2022) Sericin-based bio-inspired nano-engineering of heterometallic AgAu nanocubes for attomolar mefenamic acid sensing in the mobile phone-based surface plasmon-coupled interface. *Langmuir* 38:12035–12049.
49. Karimi F, Tiri RNE, Aygun A, Gulbagca F, Özdemir S, Gonca S, Gur T, Sen F (2023) One-step synthesized biogenic nanoparticles using Linum usitatissimum: Application of sun-light photocatalytic, biological activity and electrochemical H_2O_2 sensor. *Environmental Research* 218:114757.
50. Balasubramanian S, Bezawada SR, Dhamodharan R (2016) Facile aqueous phase synthesis of (200) faceted Au-AgCl cubes using Bael gum and its activity toward oxidation and detection of o-PDA. *ACS Sustainable Chemistry & Engineering* 4:2960–2968.
51. Rao L, Lu X, Xu L, Zhu Y, Xue T, Ge Y, Duan Z, Duan X, Wen Y, Xu J (2022) Green synthesis of kudzu vine biochar decorated graphene-like MoSe2 with the oxidase-like activity as intelligent nanozyme sensing platform for Hesperetin. *Chemosphere* 289:133116.
52. Gautam V, Singh KP, Yadav VL (2018) Multicomponent template effects – preparation of highly porous polyaniline nanorods using crude lemon juice and its application for selective detection of catechol. *ACS Sustainable Chemistry & Engineering* 6:2256–2268.

53. Zhang H, Wu S, Xing Z, Wang H-B (2021) Turning waste into treasure: Chicken eggshell membrane derived fluorescent carbon nanodots for the rapid and sensitive detection of Hg^{2+} and glutathione. *Analyst* 146:7250–7256.
54. Singh A, Eftekhari E, Scott J, et al. (2020) Carbon dots derived from human hair for ppb level chloroform sensing in water. *Sustainable Materials and Technologies* 25:e00159.
55. Rai A, Bhaskar S, Battampara P, Reddy N, Sathish Ramamurthy S (2022) Integrated photo-plasmonic coupling of bioinspired sharp-edged silver nano-particles with nano-films in extended cavity functional interface for cellphone-aided femtomolar sensing. *Materials Letters* 316:132025.
56. Durrani S, Zhang J, Yang Z, Pang A-P, Zeng J, Sayed SM, Khan A, Zhang Y, Wu F-G, Lin F (2022) Plant-derived Ca, N, S-doped carbon dots for fast universal cell imaging and intracellular congo red detection. *Analytica Chimica Acta* 1202:339672.
57. Miao H, Wang Y, Yang X (2018) Carbon dots derived from tobacco for visually distinguishing and detecting three kinds of tetracyclines. *Nanoscale* 10:8139–8145.
58. Shahshahanipour M, Rezaei B, Ensafi AA, Etemadifar Z (2019) An ancient plant for the synthesis of a novel carbon dot and its applications as an antibacterial agent and probe for sensing of an anti-cancer drug. *Materials Science and Engineering: C* 98:826–833.
59. Borse S, Murthy ZVP, Kailasa SK (2020) Chicken egg white mediated synthesis of platinum nanoclusters for the selective detection of carbidopa. *Optical Materials* 107:110085.
60. Ghosh S, Gul AR, Park CY, Xu P, Baek SH, Bhamore JR, Kim MW, Lee M, Kailasa SK, Park TJ (2021) Green synthesis of carbon dots from Calotropis procera leaves for trace level identification of isoprothiolane. *Microchemical Journal* 167:106272.
61. Khaledian S, Noroozi-Aghideh A, Kahrizi D, Moradi S, Abdoli M, Ghasemalian AH, Heidari MF (2021) Rapid detection of diazinon as an organophosphorus poison in real samples using fluorescence carbon dots. *Inorganic Chemistry Communications* 130:108676.
62. Sun D, Liu T, Wang C, Yang L, Yang S, Zhuo K (2020) Hydrothermal synthesis of fluorescent carbon dots from gardenia fruit for sensitive on-off-on detection of Hg^{2+} and cysteine. *Spectrochimica Acta Part A: Molecular and Biomolecular Spectroscopy* 240:118598.
63. Li Z, Wang Y, Ni Y, Kokot S (2015) A rapid and label-free dual detection of Hg (II) and cysteine with the use of fluorescence switching of graphene quantum dots. *Sensors and Actuators B: Chemical* 207:490–497.
64. Kong M, Li M, Shang R, Wu J, Yan P, Xu D, Li C (2018) Nacre-templated synthesis of highly dispersible carbon Nanomeshes for layered membranes with high-flux filtration and sensing properties. *ACS Applied Materials & Interfaces* 10:2850–2858.
65. Bai RG, Muthoosamy K, Shipton FN, Pandikumar A, Rameshkumar P, Huang NM, Manickam S (2016) The biogenic synthesis of a reduced graphene oxide–silver (RGO–Ag) nanocomposite and its dual applications as an antibacterial agent and cancer biomarker sensor. *RSC Advances* 6:36576–36587.

Index